Lecture Notes on Data Engineering and Communications Technologies

Volume 148

Series Editor

Fatos Xhafa, Technical University of Catalonia, Barcelona, Spain

The aim of the book series is to present cutting edge engineering approaches to data technologies and communications. It will publish latest advances on the engineering task of building and deploying distributed, scalable and reliable data infrastructures and communication systems.

The series will have a prominent applied focus on data technologies and communications with aim to promote the bridging from fundamental research on data science and networking to data engineering and communications that lead to industry products, business knowledge and standardisation.

Indexed by SCOPUS, INSPEC, EI Compendex.

All books published in the series are submitted for consideration in Web of Science.

More information about this series at https://link.springer.com/bookseries/15362

Ngoc-Thanh Nguyen · Nhu-Ngoc Dao ·
Quang-Dung Pham · Hong Anh Le
Editors

Intelligence of Things: Technologies and Applications

The First International Conference
on Intelligence of Things (ICIT 2022), Hanoi,
Vietnam, August 17–19, 2022, Proceedings

Springer

Editors
Ngoc-Thanh Nguyen 🄳
Wroclaw University of Science
and Technology
Wrocław, Poland

Quang-Dung Pham
Vietnam National University of Agriculture
Hanoi, Vietnam

Nhu-Ngoc Dao
Sejong University
Seoul, Korea (Republic of)

Hong Anh Le
Hanoi University of Mining and Geology
Hanoi, Vietnam

ISSN 2367-4512 ISSN 2367-4520 (electronic)
Lecture Notes on Data Engineering and Communications Technologies
ISBN 978-3-031-15062-3 ISBN 978-3-031-15063-0 (eBook)
https://doi.org/10.1007/978-3-031-15063-0

This Springer imprint is published by the registered company Springer Nature Switzerland AG
The registered company address is: Gewerbestrasse 11, 6330 Cham, Switzerland

Preface

This volume contains the proceedings of the First International Conference on Intelligence of Things (ICIT 2022), held in Hanoi, Vietnam, during August 17–19, 2022. The conference was co-hosted by the Hanoi University of Mining and Geology (HUMG) and Vietnam National University of Agriculture (VNUA), Vietnam, in cooperation with the Ho Chi Minh City University of Technology (HCMUT), Ho Chi Minh City Open University (HCMOU), and the University of Danang, Vietnam–Korea University of Information and Communication Technology (VKU), Vietnam. Due to the COVID-19 pandemic, the conference was organized in a hybrid mode to allow both on-site and online paper presentations. This event also marks the 20th anniversary of the Faculty of Information Technology, HUMG. Since its establishment in 2002, the Faculty of Information Technology has set the goal of becoming one of the prestige research and training centers, especially in applications of information technology for Earth sciences, mining, environment, and energy. Currently, the faculty has 64 staff members working in seven departments and one laboratory. It will continue to develop high-quality human resources in IT and make a positive contribution to society in Vietnam.

In recent years, we have witnessed important changes and innovations that the Internet of things (IoT) enables for emerging digital transformations in human life. Continuing impressive successes of the IoT paradigms, things now require an intelligent ability while connecting to the Internet. To this end, the integration of artificial intelligence (AI) technologies into the IoT infrastructure has been considered a promising solution, which defines the next generation of the IoT, i.e., the intelligence of things (AIoT). The AIoT is expected to achieve more efficient IoT operations in manifolds such as flexible adaptation to environmental changes, optimal trade-off decisions among various resources and constraints, and friendly human–machine interactions. In this regard, the ICIT 2022 was held to gather scholars who address the current state of technology and the outcome of ongoing research in the area of AIoT.

The organizing committee received over 100 submissions from 12 countries. Each paper was reviewed by at least two members of the program committee (PC) and external reviewers. Finally, we selected 40 best papers for oral presentation and publication.

We would like to express our thanks to the keynote speakers: Schahram Dustdar from Vienna University of Technology, Austria, Gottfried Vossen from the University of Muenster, Germany, and Jiming Chen from Zhejiang University/Zhejiang University of Technology, China, for their world-class plenary speeches.

Many people contributed toward the success of the conference. First, we would like to recognize the work of the PC co-chairs for taking good care of the organization of the reviewing process, an essential stage in ensuring the high quality of the accepted papers. In addition, we would like to thank the PC members for performing their reviewing work with diligence. We thank the local organizing committee chairs, publicity chair, multimedia chair, and technical support chair for their fantastic work before and during the conference. Finally, we cordially thank all the authors, presenters, and delegates for their valuable contribution to this successful event. The conference would not have been possible without their support.

Our special thanks are also due to Springer for publishing the proceedings and to all the other sponsors for their kind support.

Finally, we hope that ICIT 2022 contributed significantly to the academic excellence of the field and will lead to the even greater success of ICIT events in the future.

August 2022   Ngoc Thanh Nguyen
                                                                    Nhu-Ngoc Dao
                                                                  Quang-Dung Pham
                                                                     Hong Le Anh

Organization

Organizing Committee

Honorary Chairs

Thanh Hai Tran	Hanoi University of Mining and Geology, Vietnam
Lan Nguyen T.	Vietnam National University of Agriculture, Vietnam
Thanh Thuy Nguyen	Vietnam National University, Vietnam
Huynh Cong Phap	The University of Danang, Vietnam–Korea University of Information and Communication Technology, Vietnam

General Chairs

Nhu-Ngoc Dao	Sejong University, South Korea
Quang-Dung Pham	Vietnam National University of Agriculture, Vietnam
Hong Anh Le	Hanoi University of Mining and Geology, Vietnam
Tran Vu Pham	Ho Chi Minh City University of Technology, Vietnam

Program Chairs

Takayuki Okatani	Tohoku University, Japan
Vo Nguyen Quoc Bao	Posts and Telecommunications Institute of Technology, Vietnam
Duy T. Ngo	The University of Newcastle, Australia
Shin Nakajima	National Institute of Informatics, Japan
Diep N. Nguyen	University of Technology Sydney, Australia

Steering Committee

Ngoc Thanh Nguyen (Chair)	Wroclaw University of Science and Technology, Poland
Hoang Pham	Rutgers University, USA
Sungrae Cho	Chung-Ang University, South Korea
Hyeonjoon Moon	Sejong University, South Korea
Jiming Chen	Zhejiang University, China
Dosam Hwang	Yeungnam University, South Korea
Gottfried Vossen	Muenster University, Germany
Manuel Nunez	Universidad Complutense de Madrid, Spain
Torsten Braun	University of Bern, Switzerland

Track Chairs

Dinh-Thuan Do	The University of Texas at Austin, USA
Ngoc Thanh Dinh	Soongsil University, South Korea
Khac-Hoang Ngo	Chalmers University of Technology, Sweden
Kim Phuc Tran	University of Lille, ENSAIT, GEMTEX, France
Trung V. Phan	Chemnitz University of Technology, Germany
Quoc-Viet Pham	Pusan National University, South Korea
Thuy Nguyen	Vietnam National University of Agriculture, Vietnam
Mai Dung Nguyen	Hanoi University of Mining and Geology, Vietnam
Ngoc Thinh Tran	Ho Chi Minh City University of Technology, Vietnam
Thanh Tho Quan	Ho Chi Minh City University of Technology, Vietnam
Thanh-Binh Nguyen	The University of Danang, Vietnam–Korea University of Information and Communication Technology, Vietnam
The-Son Tran	The University of Danang, Vietnam–Korea University of Information and Communication Technology, Vietnam
Quang-Vu Nguyen	The University of Danang, Vietnam–Korea University of Information and Communication Technology, Vietnam

Liaison Chairs

Tien-Hien Nguyen	Vietnam National University of Agriculture, Vietnam
Duy Huy Nguyen	Hanoi University of Mining and Geology, Vietnam

Finance Chair

Tuyet-Lan Nguyen T.   Vietnam National University of Agriculture,
                                      Vietnam

Local Arrangement Chairs

Trong Kuong Nguyen   Vietnam National University of Agriculture,
                                      Vietnam
Bich-Thuy Nguyen T.   Vietnam National University of Agriculture,
                                      Vietnam
Tuan-Anh Ngo   Vietnam National University of Agriculture,
                                      Vietnam
Hoang Long Nguyen   Hanoi University of Mining and Geology,
                                      Vietnam
Hoai Nga Nguyen   Hanoi University of Mining and Geology,
                                      Vietnam

Publication Chairs

Laihyuk Park   Seoul National University of Science
                                      and Technology, South Korea
Woongsoo Na   Kongju National University, South Korea
Luong Vuong Nguyen   Chung-Ang University, South Korea
Trong-Hop Do   Vietnam National University, Vietnam
Ngoc-Lam Dao   The University of Danang, Vietnam–Korea
                                      University of Information and Communication
                                      Technology, Vietnam

Registration Chairs

Xuan-Thao Nguyen   Vietnam National University of Agriculture,
                                      Vietnam
Hien Thanh T. Duong   Hanoi University of Mining and Geology,
                                      Vietnam
Cong Hoang Diem   Hanoi University of Mining and Geology,
                                      Vietnam

Multimedia Chairs

Vu-Ha Tran   Vietnam National University of Agriculture,
                                      Vietnam
Cong-Thang Ngo   Vietnam National University of Agriculture,
                                      Vietnam
Thu Hang Nguyen   Hanoi University of Mining and Geology,
                                      Vietnam

The Loc Nguyen   Hanoi University of Mining and Geology,
                                      Vietnam
Minh-Nhut Pham-Nguyen   The University of Danang, Vietnam–Korea
                                      University of Information and Communication
                                      Technology, Vietnam

Program Committee

Daisuke Ishii
Takako Nakatani
Kozo Okano
Trong-Minh Hoang
Ngoc Phi Nguyen
Mai The Vu
Bui Nam Hoai Khac
Khac-Hoang Ngo
Laihyuk Park
Trong-Hop Do
Tieu Long Mai
Nam D. Vo
The-Vinh Nguyen
Woongsoo Na
Trung V. Phan
Xuan Dieu Tran
Ngoc Thanh Dinh
Thanh Phung Truong
Thuy Nguyen
Hai Chien Pham
Ngoc-Son Dang
Thu-Hong Phan T.
Duc-Kien Thai
Nguyen Xuan Mung
Hoang-Huy Nguyen
Luong Vuong Nguyen
Tri-Hai Nguyen
Huu-Du Nguyen
Anh-Tien Tran
Tran Van Tien Si
Demeke Shumeye Lakew
Arooj Masood

Van Hung Le
Dieu Tien Bui
The Vinh Bui
Khanh Hoang Doan
Ninh-Thuan Truong
Duong Quang Khanh Ngoc
Van-Quan Pham
Shariq Bashir
Daphne Teck Ching Lai
Thongchai Surinwarangkoon
Kittikhun Meethongjan
Kietikul Jearanaitanakij
Satya Ranjan Dash
Phuong T. Nguyen
Thanh-Lanh Le
Nguyen Thi Lieu
Quoc Cuong Pham
Vinh Truong Hoang
Dat Nguyen Tien
Xuan Le
Huu-Thanh Duong
Viet-Tuan Le
Thanh Binh Nguyen
The-Son Tran
Quang-Vu Nguyen
Ngoc-Tho Huynh
Duc-Hien Nguyen
Anh-Quang Nguyen-Vu
Cuu Long Le-Phuoc
Van-Phi Ho
Sy-Thin Nguyen
Dai-Tho Dang

Contents

Intelligence Services and Applications

Theoretical Intelligence Analyses

Structural Health Monitoring and IoT: Opportunities and Challenges

Marzuki Kamal$^{(\boxtimes)}$ and Atif Mansoor

The University of Western Australia, Perth, Australia
{marzuki.kamal,atif.mansoor}@uwa.edu.au

Abstract. As structures like sky scrappers get taller and bridges are getting longer, there is a need to closely monitor the health of the structures, particularly under varying environmental effects. The traditional wire-based structural health monitoring systems that require laying down cables are costly and time-consuming. New and miniaturised sensors coupled with Internet of Things (IoT) and powerful cloud computing platforms lead to a new cost-effective approach to SHM. This paper introduces Structural Health Monitoring (SHM), its conventional approaches of Visual, Destructive and Non-Destructive evaluations. After discussing the limitations of conventional SHM approaches, Internet of Things and its components are introduced. SHM with IoT, its strengths and challenges are reviewed in light of published literature. This is evident that SHM will benefit enormously from IoT, provided technical challenges like energy consumption, scalability, data security and reliability are handled.

Keywords: Structural Health Monitoring (SHM) · Internet of Things (IoT) · Non-destructive evaluation · Safety · Data security

1 Introduction

Since ancient times, humans have been building structures for living. With the advancement of technology, the structures improved from wood, stone, mud to iron, steel and new construction technologies. Additionally, structures are also getting taller, longer and bigger. The tallest structure in the world is currently the Burj Khalifa in Dubai, United Arab Emirates with a height of 828 m [1] and the longest bridge is the Danuang-Kunshan Grand Bridge, located between Shanghai and Nanjing, China [2]. Proper measures and planning are made to ensure the safety and integrity of the structures. However, with age or environmental effects, the structure deteriorates and needs to be maintained otherwise accidents may happen. For instance, between 2020 and 2021, four bridges collapsed including the 111-year-old bridge in northern Italy which injured 2 persons [3]. Similarly, 98 people were killed in a collapse of Champlain Towers South apartment building in Florida, USA on 24 June 2021 [4]. There are many aging structures in the world and among them, many are used on daily basis. This demonstrates that regular maintenance and inspection need to be done to maintain its structural integrity.

© The Author(s), under exclusive license to Springer Nature Switzerland AG 2022
N.-T. Nguyen et al. (Eds.): ICIT 2022, LNDECT 148, pp. 3–15, 2022.
https://doi.org/10.1007/978-3-031-15063-0_1

Traditionally, the structures are visually inspected by experienced profession-als for their health [5]. The visual inspections require a trained human resource that needs travelling and accessibility to the structure. A regular and timely visual inspection of the scattered structures in a vast area will require huge costs and human effort. This subjective evaluation at times may overlook important indications. Further with bigger structures, visual inspection of the whole struc-ture becomes very difficult. The health of the structure is also observed through specialized evaluations. There are two types of such evaluations; destructive eval-uation (DE) and non-destructive evaluation (NDE). In destructive evaluation, the evaluation is performed on a sample of structures to gauge the health of the whole structure. In non-destructive evaluation, testing such as Ultrasonic and Acoustic Emission are performed to evaluate the health of the structure [6].

Both destructive and non-destructive evaluations, like visual inspections, are limited by the availability of technical human resources and equipment. This led to Structural Health Monitoring (SHM), which is a non-destructive evaluation technique [7] that performs a continuous observation of the structure [8]. The data gathered from SHM system over time give an insight that was not earlier available and thus aids in proper maintenance action to be taken. Although SHM provides benefits over NDE, it has limitations of high initial cost and later continuous maintenance due to installation and cables [9]. This is where IoT can be integrated into SHM with its miniaturised sensors and wireless communication technologies. With a low initial cost of IoT, SHM can be deployed and maintained easily. Combine with intelligent sensors, ubiquitous connectivity and a powerful cloud computing platform, IoT based SHM allows seamless and effective data collection, storage and processing.

This paper aims to review the literature to provide an overview of the use of Structural Health Monitoring in combination with IoT. The rest of the paper is structured as follows: Sect. 2 introduces Internet of Things (IoT), Sect. 3 gives details of Structural Health Monitoring while Sect. 4 explains the combination SHM and IoT. Section 5 discusses the Challenges of IoT in SHM and the paper is concluded in Sect. 6.

2   Internet of Things (IoT)

There is no agreed upon definition of Internet of Things. IBM describes IoT as things that connect to the Internet, and other devices [10], whereas Oracle describes it as an ordinary object which has sensors and software and connects to the Internet and exchange information with other devices [11].

Internet of Things devices typically consists of 4 components; sensors, com-munication, data processing and user interface. The sensor collects information from its environment. Then, the information is stored and processed by the data processing system. Data storage and processing can be done either onsite or offsite. The data is transmitted via a communication protocol. Various commu-nication protocols can be used to transmit the data depending on the require-ments such as data rate, range, cost and power consumption. Table 1 compares different IoT communication protocols on the basis of these parameters.

Table 1. IoT communication protocols [12–14]

Technology	Frequency	Data rate	Range	Power usage	Cost
2G/3G	Cellular Bands	10 Mbps	35 to 150 km	High	High
Bluetooth/BLE	2.4 GHz	1,2,3 Mbps	50–150 m	Low	Low
LoRa	sub-GHz	<50 kbps	2.5–15 km	Low	Medium
LTE-M	Cellular Bands	1–10 Mbps	35 to 150 km	Medium	High
NB-IoT	Cellular Bands	0.1–1 Mbps	35 to 150 km	Medium	High
SigFox	sub-GHz	<1 kbps	3–50 km	Low	Medium
WiFi	2.4 GHz, 5GHz	0.1–54 Mbps	50–100 m	Medium	Low
Zigbee	2.4 GHz	250 kbps	10–100 m	Low	Medium
Z-Wave	sub-GHz	40 kbps	∼30 m	Low	Medium

The data transmitted over the communication channel is then processed in a server or a cloud computing platform to give meaningful insights through data visualisation and alerts.

3 Structural Health Monitoring (SHM)

The objective of SHM is to enhance structural safety by continuous monitoring of the structure. In SHM system, multiple sensors are placed at the structure, which normally sends parameters related to structure health to a data processing system via a communication channel. Once the data is obtained, the condition of the structure can be evaluated by an expert or a machine learning algorithm that analysed the data.

SHM can deliver real-time information of the condition of the structures or even predict future events [15,16]. Malekloo et al. [17] discussed the use of machine learning and SHM by employing machine learning algorithms such as k-nearest neighbour (kNN), support vector machine (SVM), k-means, random forest and neural network for damage assessment. Zhang et al. [18] proposed the use of Acoustic Emission (AE) sensors to collect data and predict the state of blades on a gas turbine engine.

Typically, the current approach of structure monitoring and maintenance is scheduled on time. With SHM, the maintenance schedule can be based on the condition of the structure, facilitating preemptive maintenance [19]. Cusati et al. [20] have demonstrated SHM to monitor the structure of aircraft through a condition-based maintenance approach resulting in cost lowering.

Kim et al. [21] have designed and deployed 64 nodes accelerometer sensors on the Golden Gate Bridge, San Francisco, USA. The sensors network placed on the bridge has provided an accurate, high-frequency sampling with low jitter data for analysis of bridge health. Diamanti and Soutis [22] discussed the use of ultrasonic transducers permanently attached to aircraft for its condition monitoring. Hodge et al. [23] reviewed numerous use of SHM in the railway industry which can be divided into movable monitoring (i.e. train) and fixed monitoring (i.e. rail tracks). Schubel et al. [24] compared multiple structure health monitoring methods for wind turbine blades. One interesting application of SHM is monitoring the structure of historical buildings. Vestroni et al. [25] installed 12 accelerometers on the Colosseum building in Rome to monitor the vibration induced by the environment and relate it to structural health. Pierdicca et al. [26] demonstrated the use of vibration-based sensors to monitor the structure of the historical building "Palazzo Comunale di Castelfidardo" in Italy. These examples show that SHM is applied in diverse fields.

3.1   Wired and Wireless SHMs

According to Aygün and Gungor [27], SHM systems were originally designed to have an array of sensors wired to the system. But, the wired SHM systems were not gaining popularity due to the high cost to deploy and subsequent maintenance. Celebi et al. [28] in their paper informed the total cost of wired SHM systems for the Bill Emerson Memorial Bridge in Cape Girardeau, Missouri, consisting of 86 accelerometer sensors to be about US$1.3 million. This makes the cost of each sensor approximately US$15,000. Cao and Liu [9] analyzed the high cost of a wired SHM system is largely due to hardware and installation.

Table 2 shows the comparison between wired and wireless SHM. According to Noel et al. [29] and Muttillo et al. [30], the benefits of a wired SHM system is that it provides a high data rate and high bandwidth. But due to the high cost, the deployment of wired sensors is limited. As an example, Cao and Liu [9] mentioned that the Tsing Ma Bridge in Hong Kong has 39 accelerometers for the 2 km suspension bridge, thus limiting the accuracy.

Table 2. Wired & wireless SHM comparison [29]

Metric	Wired	Wireless
Cost	Very high	Low
Deployment time	Very long	Short
Lifespan	Long	Short
Number of sensors	Low	High
Connection bandwidth	High	Limited
Data rate	High	Low

With recent advancements in wireless technology and sensors, wireless SHM has become more feasible. They cost less and have comparatively shortened deployment time. According to Cao and Liu [9], the cost of a wireless SHM sensor node is typically less than US$500, while the wired-based system can cost more than US$10,000. Additionally, the deployment is also significantly reduced. The wireless SHM systems have low bandwidth and reduced data rate compared to the wired monitoring systems, but this shortcoming can be overcome through intelligent designs. Additionally, the technology is systematically bringing improvements.

4   IoT Components in SHM

As discussed in Sect. 2, IoT architecture comprises four components; sensors, communication, data storage and processing, and user interface. This section will discuss the sensors, data processing hardware and communication protocols used for SHM and IoT.

4.1   Sensors Used in SHM

In most structures, the parameters to monitor are vibration, strain and internal structural integrity. Different sensors are used to monitor these parameters such as accelerometer, strain sensor, acoustic emission sensor and optical fibre-based sensor.

Accelerometer sensor measures the acceleration levels of the system. With the acceleration measurement, various parameters can be obtained, such as vibration, frequency and motion [31,32]. Komarizadehasl et al. [33] mentioned different types of accelerometer used in SHM as Piezoresistive, Piezoelectric and Capacitive. These sensors could be now miniaturized as Micro-Electro-Mechanical Systems (MEMS) sensors. Villacorta et al. [34] stated that the Piezoelectric accelerometer is the most commonly used due to its high accuracy and sensitivity.

Initially, MEMS accelerometers suffered from limited bandwidth, noise and measurement range restrictions [35], but have improved with time. One of the benefits of MEMS-based accelerometer compared to piezoelectric is lower cost and lower energy consumption. Table 3 shows the comparison between the piezoelectric accelerometer and MEMS accelerometer [35]. Sabato et al. [36] analysed and surveyed the use of MEMS-based accelerometers in structural health monitoring systems. The authors concluded that the MEMS accelerometers demonstrate the same performance as the piezoelectric accelerometer. On the other hand, Bassoli et al. [37] stated that the MEMS accelerometer sensors placed on an ancient masonry bell tower showed better accuracy compared to the piezoelectric accelerometer.

Table 3. Piezoelectric accelerometer and MEMS accelerometer comparison [35]

Sensor	Cost	Potential battery life
Piezoelectric accelerometer	$25 to $500+	Short to medium
MEMS accelerometer	$10 to $30	Medium to long

Muttillo et al. [30] used digital accelerometer ADXL355 in their bridge health monitoring system in combination with a Stochastic Subspace Identification method. Pierleoni et al. [19] also proposed a structural health monitoring system using the accelerometer ADXL355 by evaluating its dynamic response. Other type of accelerometers used in different SHM projects are ADXL335 [38], LIS3L02AS4 [39], KXR94-2050 [40], SD1221L [41] and ADXL345 [42].

Strain sensors are also used in SHM. The sensor measures the structure's strain that can lead to a deformation like a crack. There are two types of strain sensors; electric strain sensors and optical fibre-based sensors. Similar to the piezoelectric accelerometer sensor, electric strain gauges are used for a considerable time in SHM. Electric strain sensors are typically arranged in a long, thin conductive strip. It works by measuring the resistance change between two terminals when strain is applied. Chanv et al. [42] and Naraharisetty et al. [43] proposed the use of strain sensors together with an accelerometer for SHM applications. Chanv et al. [42] used strain sensors BF350-3AA in their research. Although electric strain sensors have been used for many years in SHM, these have the limitation of durability and accuracy in case of continuous usage [44].

Fibre Bragg grating (FBG) sensors work differently than an electrical strain sensor by fixed index modulation. Tennyson et al. [45] have shown the use of fibre optic strain sensors for bridge health monitoring on 16 bridges. In the 6 years operational period, the sensors showed good performance compared to traditional strain gauges in terms of durability and performance.

Some SHM systems are based on acoustic emission (AE) measurements. Piezoelectric acoustic transducers are used to evaluate the internal structural damage. Dai and He [46] stated that using an array of the piezoelectric transducer and ultrasonic guided wave, the position of internal damage in the structure can be located.

4.2   Data Storage and Processing Hardware

The sensors need a platform to store, process and send the information for further analysis. Traditionally, sensors in SHM systems were connected to a wired data logger to record the sensors' data. The data loggers used in SHM are typically expensive. In IoT-based systems, a cost-effective microcontroller with some memory is normally used to process the sensors' data. Microcontroller platforms such as Arduino can connect multiple sensors and transmit the data to the users by attaching a communication module with it.

Many research works in SHM used Arduino-based microcontrollers for interfacing with sensors [42,47,48]. Chanv et al. [42] developed an SHM system that connected accelerometer, strain sensors and moisture sensor with Arduino UNO coupled with WiFi module. Paul et al. [47] used Arduino 101 to connect flex sensors, measuring the amount of deflection. STM32 Nucleo board were used by Pierleoni et al. [19] and Di Nuzzo et al. [49] for its low cost, performance and low power consumption. Patil and Patil [38] used PIC16F877 microcontroller, while Muttillo et al. [30] used SAM3X8E for their SHM systems.

There are also commercial boards to interface with the sensors. Balestrieri and Picariello [50] proposed an SHM system that used a Waspmote platform to interface with WS-3000 weather station, temperature and humidity sensors. The Waspmote itself has an embedded accelerometer. Rice and Spencer [39] used an Imote2 platform developed by Intel for their SHM system, having a variable processing speed to optimise power consumption. The authors used accelerometer sensors to interface with the Imote2 in their SHM system.

With the improvement of technology, computers are getting smaller, more powerful and energy-efficient. With a wide range of connectivity, i.e. Bluetooth and WiFi, tiny computers like Raspberry Pi are commonly used in SHM and IoT. Mahmud et al. [7] and Abdelgawad & Yelamarthi [51] proposed the use of Raspberry Pi with the help of analog to digital converter (ADC) to interface with a piezoelectric sensor. Raspberry Pi requires more energy compared to a microcontroller and therefore proper energy consideration is required when using Raspberry Pi.

4.3   Communication Protocols

In an IoT based SHM, the sensors' data need to be transmitted to a central powerful computing entity for further processing. The transmission is usually done wirelessly as there might be multiple sensor nodes in the structure [29]. As given in Table 1, there are different communication protocols to transfer the sensor's information, from short-range to long-range. Another important thing when deciding the communication besides the range is power consumption.

WiFi is also used as a communication protocol particularly if the structure is a building in a city. Chanv et al. [42] used WiFi module ESP8266 to transmit the Arduino UNO connected sensors' readings. Balestrieri and Picariello [50] used a WiFi module that connected the Waspmote board to the distant gateway. The benefit of using WiFi is high bandwidth and low latency, but it suffers from range and power consumption. WiFi also requires a router or a gateway to interface with the internet. It is suitable for IoT devices that are connected to power.

Another commonly used communication protocol is Zigbee. The Zigbee works at 433 MHz ISM (Industrial, scientific and medical) radio bands. It is a low-power and low-cost communication protocol [52]. Zigbee also requires a gateway to receive the communication from different sensors. One benefit of using Zigbee compared to WiFi is low energy consumption. Patil and Patil [38] have used Zigbee protocol in bridge health monitoring as a communication protocol between sensors. Cho et al. [53] used the Zigbee protocol to connect 70 sensor nodes, which covered the 484-meters-long Jindo Bridge in South Korea. Seventy sensors were divided into 2 sub-networks, where each was controlled by a separate base station.

Cellular communication is also used in SHM as demonstrated in [48,50]. Niranjan and Rakesh [48] used a GSM SIM800A cellular module that interfaced with Arduino UNO and flex sensor. The data received from the sensor is transmitted to a ThinkSpeak server via cellular communication. Balestrieri et al. [50] proposed an SHM system that sent the sensors' data to the gateway via WiFi and transmitted the information from the gateway to the user over the internet via a cellular network. Cellular communication has a longer range but higher energy consumption. Communication protocols such as Narrowband-IoT (NB-IoT) handles the higher energy consumption issue in cellular communication. It is designed to be a low-power and long-range communication protocol built for IoT using a cellular network. Di Nuzzo et al. [49] proposed an SHM system that utilised the NB-IoT communication protocol.

5   Challenges of SHM and IoT

According to Lamonaca et al. [54], SHM and IoT bring many benefits such as cost efficiency and improved safety compared to the traditional SHM. However, the benefits also come with their own challenges. In this section, we discuss a few of the challenges and related literature.

5.1   Energy Consumption

Noel et al. [29] and Wang et al. [55] discussed in their paper that one of the challenges of SHM and IoT is energy consumption. Most devices do not have the access to power sockets. Also, being wirelessly connected means that each device has limited energy through the batteries. Further, the data transmission consumes considerable energy. Additionally, replacing depleted batteries for each node is not feasible since some sensors are difficult to access. Noel et al. [29] discussed that there are techniques to manage energy consumption such as energy harvesting, radio optimisation, data reduction and transmission duty cycle. Ghosh et al. [56] have proposed an event-based wake-up to optimise energy consumption for a railway bridge health monitoring system. The use of an event-based wake-up conserves the energy as the system sleeps if there is no train crossing the bridge. Aygün and Gungor [27] have discussed energy harvesting like solar panels to maintain continuous energy supply.

5.2   Scalability

IoT devices can be deployed easily due to their wireless connectivity. As the number of devices increases the amount of data increases too. This large amount of data then needs to be stored and processed. According to Noel et al. [29], traditional data processing systems are inefficient and expensive to handle a large amount of data. One solution to handle the large data is to use the cloud for storage and processing. Cloud computing platforms such as Amazon AWS, Google Cloud and Microsoft Azure etc. have now specific services to cater for IoT traffic. Further cloud services can scale much easier, more reliable and secure compared to on-site data processing systems.

5.3 Security

As with any other IoT systems, security is one of the challenges that need to be addressed. Alonso et al. [57] emphasize the proper security measures for the protection of sensitive SHM data. In IoT, devices' processing power, memory and energy are limited. Attackers can make use of these constraints. Mahmoud et al. [58] stated that each component of an IoT system is vulnerable to security attacks such as Denial of Service attacks (DoS), Replay attacks, Timing attacks, Node Capture attacks and Man-in-the-Middle attacks. The security in IoT is an open challenge and considerable research is being undertaken in this area.

5.4 Reliability

One of the benefits of the traditional wired approach is reliability. As everything is connected through a cable, data transmission is reliable and fast with a high data rate and throughput. With wireless connectivity, measures are needed to cater transmission to data loss or corruption. A communication protocol with retransmission will consume the energy of the devices, so a lightweight reliable communication protocol for IoT is needed. Kim et al. [21] designed and implemented the lightweight communication protocol STRAW (Scalable Thin and Rapid Amassment Without loss) to monitor structural health at Golden Gate Bridge.

Another challenge that needs to be addressed is data accuracy and noise from sensors. With the advancement in sensors technology, the sensors are smaller and cheaper. Sensors come differently with operational ranges and accuracy. Some sensors might be susceptible to noise and changes within the environment more than others. Mahmud et al. [7] in their paper mentioned several signal processing techniques to handle the noise in sensors with techniques like Wavelet denoising, Fast Fourier Transform (FFT), Wavelet transform and Cross-Correlation (CC).

6   Conclusion

We reviewed the published literature related to Structural Health Monitoring (SHM) using Internet of Things (IoT). The review demonstrated that IoT can bring many benefits to SHM, removing the limitations of the conventional time-based SHM approach. This will lead to cost reduction, wider and more sophisticated structural monitoring and real-time alerts. The powerful analysis tools will give insights that are presently not available. Thus, aiding in preemptive maintenance and consequently improved structural and human safety. The combination of SHM and IoT comes with its own challenges of data security, reliability, data denoising, reliable lightweight communication protocols and limitations attached with constrained devices. These challenges are in fact opportunities and the research community is continuously working on them. It is evident that the Structural Health Monitoring and Internet of Things will certainly lead to enhanced safety of structures and human lives.

Acknowledgment. This research is funded by the Planning and Transport Research Centre (PATREC) and the iMOVE CRC and supported by the Cooperative Research Centres program, an Australian Government initiative.

References

1. Burj khalifa - facts & figures (2021). www.burjkhalifa.ae/en/the-tower/facts-figures/
2. Britannica: Bridge (2021). www.britannica.com/technology/bridge-engineering/U-S-designs#ref1261381
3. Italy bridge collapse: Two drivers survive (2020). www.bbc.com/news/world-europe-52213898
4. Miami building collapse (2021). https://edition.cnn.com/2021/07/15/us/miami-dade-building-collapse-thursday/index.html
5. Humar, J., Bagchi, A., Xu, H.: Performance of vibration-based techniques for the identification of structural damage. Struct. Health Monit. **5**(3), 215–241 (2006)
6. Attar: What is destructive and non-destructive testing? (2020). www.attar.com.au/what-is-destructive-and-non-destructive-testing/
7. Mahmud, M.A., Bates, K., Wood, T., Abdelgawad, A., Yelamarthi, K.: A complete Internet of Things (IoT) platform for structural health monitoring (SHM). In: 2018 IEEE 4th World Forum on Internet of Things (WF-IoT), pp. 275–279 (2018)
8. Edwards, G.R., Tuncbilek, K., Walker, B.: Development of a multi-purpose wireless network for the structural health monitoring of a suspension bridge. In: IET Conference on Wireless Sensor Systems (WSS 2012), pp. 4B1–4B1 (2012)
9. Cao, J., Liu, X.: Structural health monitoring using wireless sensor networks. In: Mobile and Pervasive Computing in Construction, pp. 210–236 (2012)
10. Clark, J.: What is the internet of things, and how does it work? (2021). www.ibm.com/blogs/internet-of-things/what-is-the-iot/
11. What is the internet of things (IoT)? (2021). www.oracle.com/au/internet-of-things/what-is-iot/
12. 802.15.4 wireless for internet of things developers (2021). https://blog.helium.com/802-15-4-wireless-for-internet-of-things-developers-1948fc313b2e
13. Sonee, S.: Top IoT communication protocols updated 2021 (2020). https://hashstudioz.com/blog/top-iot-communication-protocols-2020/
14. Mobile base stations (2021). https://mobilenetworkguide.com.au/mobile_base_stations.html
15. Tudosa, I., Picariello, F., Balestrieri, E., Carnì, D.L., Lamonaca, F.: A flexible DAQ hardware architecture using SoCs for IoT based structural health monitoring systems. In: 2019 II Workshop on Metrology for Industry 4.0 and IoT (MetroInd4.0&IoT), pp. 291–295 (2019)
16. Lynch, J.P.: A summary review of wireless sensors and sensor networks for structural health monitoring. Shock Vibr. Digest **38**(2), 91–128 (2006)
17. Malekloo, A., Ozer, E., AlHamaydeh, M., Girolami, M.: Machine learning and structural health monitoring overview with emerging technology and high-dimensional data source highlights. Struct. Health Monit. **21**, 1906–1955 (2021)
18. Zhang, Z., Yang, G., Hu, K.: Prediction of fatigue crack growth in gas turbine engine blades using acoustic emission. Sensors (Basel, Switzerland) **18**(5), 1321 (2018)

19. Pierleoni, P., et al.: IoT solution based on MQTT protocol for real-time building monitoring. In: 2019 IEEE 23rd International Symposium on Consumer Technologies (ISCT), pp. 57–62 (2019)
20. Cusati, V., Corcione, S., Memmolo, V.: Impact of structural health monitoring on aircraft operating costs by multidisciplinary analysis. Sensors (Basel, Switzerland) **21**(20), 6938 (2021)
21. Kim, S., et al.: Health monitoring of civil infrastructures using wireless sensor networks. In: 2007 6th International Symposium on Information Processing in Sensor Networks, pp. 254–263 (2007)
22. Diamanti, K., Soutis, C.: Structural health monitoring techniques for aircraft composite structures. Progr. Aeros. Sci. **46**(8), 342–352 (2010)
23. Hodge, V.J., O'Keefe, S., Weeks, M., Moulds, A.: Wireless sensor networks for condition monitoring in the railway industry: a survey. IEEE Trans. Intell. Transp. Syst. **16**(3), 1088–1106 (2015)
24. Schubel, P., Crossley, R., Boateng, E., Hutchinson, J.: Review of structural health and cure monitoring techniques for large wind turbine blades. Renew. Energy **51**(C), 113–123 (2013)
25. Vestroni, F., De Sortis, A., Pau, A.: Measurements of the colosseum response to environmental actions. In: XI International Conference on Structural Dynamics (2020)
26. Pierdicca, A., Clementi, F., Isidori, D., Concettoni, E., Cristalli, C., Lenci, S.: Numerical model upgrading of a historical masonry palace monitored with a wireless sensor network. Int. J. Masonry Res. Innov. **1**(1), 74 (2016)
27. Aygün, B., Gungor, V.C.: Wireless sensor networks for structure health monitoring: recent advances and future research directions. Sensor Rev. **31**(3), 261–276 (2011)
28. Celebi, M., et al.: Seismic instrumentation of the bill emerson memorial mississippi river bridge at cape girardeau (mo): a cooperative effort. In: Proceedings of the 4th International Seismic Highway Conference (2004)
29. Noel, A.B., Abdaoui, A., Elfouly, T., Ahmed, M.H., Badawy, A., Shehata, M.S.: Structural health monitoring using wireless sensor networks: a comprehensive survey. IEEE Commun. Surv. Tutor. **19**(3), 1403–1423 (2017)
30. Muttillo, M., et al.: Structural health monitoring: an IoT sensor system for structural damage indicator evaluation. Sensors **20**(17), 4908 (2020)
31. Accelerometer sensors (2021). www.rohm.com/electronics-basics/sensor/accelerometer-sensor
32. Hanly, S.: Accelerometers: Taking the guesswork out of accelerometer selection (2021). https://blog.endaq.com/accelerometer-selection
33. Komarizadehasl, S., Mobaraki, B., Ma, H., Lozano-Galant, J.A., Turmo, J.: Development of a low-cost system for the accurate measurement of structural vibrations. Sensors **21**(18), 6191 (2021)
34. Villacorta, J.J., et al.: Design and validation of a scalable, reconfigurable and low-cost structural health monitoring system. Sensors **21**(2), 648 (2021)
35. Why mems accelerometers are becoming the designer's best choice for cbm applications (2021). www.analog.com/en/technical-articles/why-memes-acceler-are-best-choice-for-cbm-apps.html
36. Sabato, A., Niezrecki, C., Fortino, G.: Wireless MEMS-based accelerometer sensor boards for structural vibration monitoring: a review. IEEE Sens. J. **17**(2), 226–235 (2016)
37. Bassoli, E., Vincenzi, L., Bovo, M., Mazzotti, C.: Dynamic identification of an ancient masonry bell tower using a MEMS-based acquisition system. In: 2015

IEEE Workshop on Environmental, Energy, and Structural Monitoring Systems (EESMS) Proceedings, pp. 226–231 (2015)

38. Patil, P.K., Patil, S.R.: Structural health monitoring system using WSN for bridges. In: 2017 International Conference on Intelligent Computing and Control Systems (ICICCS), pp. 371–375 (2017)

39. Rice, J.A., Spencer, J.B.F.: Structural health monitoring sensor development for the Imote2 platform. In: Sensors and Smart Structures Technologies for Civil, Mechanical, and Aerospace Systems 2008, pp. 693234–693234-12 (2008)

40. Bedon, C., Bergamo, E., Izzi, M., Noè, S.: Prototyping and validation of MEMS accelerometers for structural health monitoring-the case study of the pietratagliata cable-stayed bridge. J. Sens. Actuator Netw. **7**(3), 30 (2018)

41. Hu, X., Wang, B., Ji, H.: A wireless sensor network-based structural health monitoring system for highway bridges. Comput.-Aided Civil Infrastruct. Eng. **28**(3), 193–209 (2013)

42. Chanv, B., Bakhru, S., Mehta, V.: Structural health monitoring system using IoT and wireless technologies. In: 2017 International Conference on Intelligent Communication and Computational Techniques (ICCT), pp. 151–157 (2017)

43. Naraharisetty, V., Talari, V.S., Neridu, S., Kalapatapu, P., Pasupuleti, V.D.K.: Cloud architecture for IoT based bridge monitoring applications. In: 2021 International Conference on Emerging Techniques in Computational Intelligence (ICETCI), pp. 39–42 (2021)

44. White paper: Optical fiber sensors vs. conventional electrical strain gauges for infrastructure monitoring applications (2021). www.hbm.com/en/6482/white-paper-optical-fiber-sensors-vs-conventional-electrical-strain-gauges/

45. Tennyson, R.C., Mufti, A.A., Rizkalla, S., Tadros, G., Benmokrane, B.: Structural health monitoring of innovative bridges in Canada with fiber optic sensors. Smart Mater. Struct. **10**(3), 560 (2001)

46. Dai, D., He, Q.: Structure damage localization with ultrasonic guided waves based on a time-frequency method. Signal Process. **96**, 21–28 (2014)

47. Paul, P., et al.: An internet of things (IoT) based system to analyze real-time collapsing probability of structures. In: 2018 IEEE 9th Annual Information Technology, Electronics and Mobile Communication Conference (IEMCON), pp. 1070–1075 (2018)

48. Niranjan, D., Rakesh, N.: Early detection of building collapse using IoT. In: 2020 Second International Conference on Inventive Research in Computing Applications (ICIRCA), pp. 842–847 (2020)

49. Di Nuzzo, F., Brunelli, D., Polonelli, T., Benini, L.: Structural health monitoring system with narrowband IoT and mems sensors. IEEE Sens. J. **21**(14), 16371–16380 (2021)

50. Balestrieri, E., Vito, L.D., Picariello, F., Tudosa, I.: IoT system for remote monitoring of bridges: measurements for structural health and vehicular traffic load. In: 2019 II Workshop on Metrology for Industry 4.0 and IoT (MetroInd4.0&IoT), pp. 279–284 (2019)

51. Abdelgawad, A., Yelamarthi, K.: Internet of Things (IoT) platform for structure health monitoring. Wirel. Commun. Mob. Comput. **2017**, 1–10 (2017)

52. Why zigbee? (2021). https://zigbeealliance.org/why-zigbee

53. Cho, S., et al.: Structural health monitoring of a cable-stayed bridge using wireless smart sensor technology: data analyses. Smart Struct. Syst. **6**(5–6), 461–480 (2010)

54. Lamonaca, F., Sciammarella, P., Scuro, C., Carnì, D., Olivito, R.: Internet of Things for structural health monitoring. In: 2018 Workshop on Metrology for Industry 4.0 and IoT, pp. 95–100 (2018)

55. Wang, P., Yan, Y., Tian, G.Y., Bouzid, O., Ding, Z.: Investigation of wireless sensor networks for structural health monitoring. J. Sens. **2012**, 1–7 (2012)
56. Ghosh, S.K., Suman, M., Datta, R., Biswas, P.K.: Power efficient event detection scheme in wireless sensor networks for railway bridge health monitoring system. In: 2014 IEEE International Conference on Advanced Networks and Telecommuncations Systems (ANTS), pp. 1–6 (2014)
57. Alonso, L., Barbarán, J., Chen, J., Díaz, M., Llopis, L., Rubio, B.: Middleware and communication technologies for structural health monitoring of critical infrastructures: a survey. Comput. Stand. Interfaces **56**, 83–100 (2018)
58. Mahmoud, R., Yousuf, T., Aloul, F., Zualkernan, I.: Internet of things (IoT) security: current status, challenges and prospective measures. In: 2015 10th International Conference for Internet Technology and Secured Transactions (ICITST), pp. 336–341 (2015)

Semantic Interoperability Issues and Challenges in IoT: A Brief Review

Devamekalai Nagasundaram[1]([✉]) [iD], Selvakumar Manickam[2] [iD],
and Shankar Karuppayah[2] [iD]

[1] National Advanced IPv6 Centre, University Sains Malaysia (USM), Penang, Malaysia
mekalai.deva@gmail.com
[2] National Advanced IPv6 Centre, University Sains Malaysia (USM), Penang, Malaysia
{selva,kshankar}@usm.my

Abstract. Semantic interoperability is one of the enormous challenges that need to be addressed to achieve the Internet of Things (IoT) vision. Accomplishing semantic interoperability in a heterogeneous IoT environment will allow a billion devices to exchange meaningful data in a form understandable by multiple devices. The paper presents the challenges faced by IoT due to the lack of semantic interoperability and the available method of achieving semantic interoperability in IoT. It also discusses the limitation of the current solution and recommendations on the future work needed.

Keywords: Internet of Things · Interoperability · Semantics · Semantic interoperability · Ontology · Semantic network

1 Introduction

IoT is a system of interconnected devices through the Internet to collect, share and communicate data with each other [1]. IoT technologies provide valuable benefits and significant inventions for society, and it is one of the pillars of the fourth industrial revolution [2]. Cisco has visioned that by 2030, around 500 billion smart devices are expected to be connected to the Internet globally [3]. IoT devices are expected to generate a large amount of data due to the exponential growth of IoT and increased adaptation [4]. According to IDC, the data collected by IoT will reach 73.1 ZB by 2025, which equals 422% of the 2019 output, when 17.3 ZB of data was produced [4].

IoT devices are highly heterogeneous, especially in data format; therefore, data management has always been an important topic [5]. Furthermore, no proper standard defined and availability of various data formats, cause data interoperability issue, also called semantic interoperability remains a substantial problem and unresolved for many years [6]. For example, a body temperature sensor records and transfers data in Degree Celsius in an IoT healthcare system, while the IoT healthcare controller system was developed to accept temperature in Fahrenheit. Therefore, the issue arises when the data from the temperature sensor need to be shared with the healthcare controller system. The

N.-T. Nguyen et al. (Eds.): ICIT 2022, LNDECT 148, pp. 16–31, 2022.
https://doi.org/10.1007/978-3-031-15063-0_2

data from the temperature sensor is not understandable by the healthcare controller system due to differences in the data formats. Semantic interoperability means the presents of a common understanding of the information exchanged [7].

IoT ecosystem facing challenges caused by the disintegration of existing platforms, protocols, data formats, and standards [8]. Furthermore, Covid-19 highlighted the vital need for interoperability as the end-users around the globe seek to integrate many of the technologies required to continue business as usual. For example, technologies such as video analytics to assist with occupancy management, enforcement of social distancing, and compliance with facial covering requirements increasing in demand results in tremendous growth in data and turning up requests for interoperability as the organisations looking to deploy the best solutions [9]. Various research work has been done to address methods and resolve semantic interoperability issues in IoT [10–12]. Still, there is a lack in the implementation of semantic interoperability due to some reason that will be discussed in this paper.

Research contribution

This paper will briefly explain the interoperability technology and issues faced in IoT due to lack of interoperability. We reviewed the past research work related to enabling semantic interoperability in IoT to help readers understand the current status and future trends of semantic interoperability. The main contributions of this paper are:

a. To expose the definition and models of IoT Interoperability, Semantic Interoperability and semantic technologies
b. To survey the semantic interoperability handling approach in IoT by past research work
c. To highlight the challenges faced in IoT due to lack of semantic interoperability
d. To discuss the open challenge and future work in the context of semantic interoperability in IoT

The rest of this paper is organised as follows. Section 2 explains the interoperability dimension in IoT, followed by semantic interoperability technology in Sect. 3 and semantic technologies in Sect. 4. Section 5 comprehensively reviews the existing works on semantic interoperability in IoT. In Sect. 6, challenges due to the lack of semantic interoperability in IoT are highlighted. Finally, we provide an overview of the open challenges and potential future research direction in IoT semantic interoperability.

Research Methodology

The study design selected for this research was a semi-systematic review. This type of review is also known as a traditional literature review. The semi-systematic review identifies and explains all research patterns that are theoretically important to the subject under consideration [13]. In addition, a semi-systematic review was selected for this study to provide a clear and crucial factual overview of existing knowledge on this subject [14]. We followed a general steps in the literature review process in conducting the review, including research objective creation, screening, literature review, study assessment, and data analysis.

The first step was recognising the research objective to guide the review process. The research objectives steered the type of information, keywords, and search terms pursued in the literature. The second step elaborates a comprehensive search of the existing literature on google scholar and various electronic databases such as scopus, science direct, and directory of open access journals. for relevant information based on the research objective. Several keywords and terms related to the semantic interoperability concepts in IoT were appropriately combined in the search.

The following steps are the studies and collecting the related literature on objectives, taxonomies and application of the interoperability concepts in IoT to identify potential studies eligible for review. After the screening process, all potential papers were evaluated. As the fourth step, the assessment and analysis of studies collected in step 3 were performed. The articles were read and reviewed to identify the current trends, challenges, and research work on semantic interoperability in IoT.

As a final step, studies extracted from the included studies were assembled, summarised and interpreted. The findings contribute to the body of knowledge and are presented in the next section (Fig. 1).

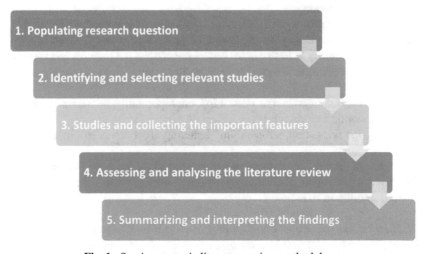

Fig. 1. Semi-systematic literature review methodology

2   Interoperability in IoT

Interoperability in IoT is defined as the ability of two or multiple devices or systems to exchange knowledge and use the information shared [7]. A basic illustration of interoperability, divided into four dimensions, is in Fig. 2. Technical Interoperability, syntactical Interoperability, semantic interoperability, and organisational interoperability are the four mentioned dimensions of interoperability in IoT.

Technical interoperability is the ability of the IoT devices in terms of hardware and software components and platforms to communicate with each other. It is also known as device interoperability, primarily focuses on communication protocols and infrastructure needed to enable machine-to-machine communications successfully [7]. Syntactical interoperability is related to the protocols for representing data such as HTML and XML

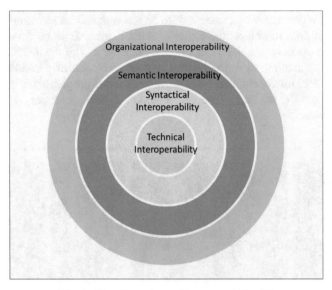

Fig. 2. The dimensions of Interoperability [7]

[15]. This ensures that the message transferred by communication protocol has proper syntax and encoding. Semantic interoperability is the ability of an IoT application to share and exchange data between two or multiple devices [7]. The exchanged information must be in an understandable form and share a common meaning of the content across the diverse application. Organisational interoperability covers technical, syntactical, and semantic interoperability [16]. It enables IoT organisations or applications with various infrastructure, communication protocols, and domains to effectively communicate and exchange meaningful information in an understandable form [17].

3 Semantic Interoperability

Semantic interoperability is more than just data transfer format or explicit definition of information models between IoT devices [18]. It is about a mechanism that automates this without requiring any specific configuration, programming, and reviewing of thick manuals to understand the meaning of particular data. Semantic interoperability is the ability of multiple devices, services, or applications to exchange information, data, or knowledge in a meaningful way [19]. The data is not enough to be only exchanged between two or more systems but also needs to be understood by each system. The data generated by sensors or devices have self-defined data formats such as JSON, XML, and others. However, the data and schemas used by different devices are different and not always compatible. Moreover, the data is represented in different units of measurement. The semantic mismatch between data representations leads to IoT systems causing interoperability issues due to varying forms of data description [5].

The current trend moving towards Industry 4.0 makes interoperability significantly important [20]. For example, the manufacturing industry needs interoperability to implement flexible production and seamless integration of new production machines at a reduced cost. When a new machine needs to be incorporated into a manufacturing area containing multiple machines, the foremost thing that needs to be done is to understand the data the new machine produces and integrate it with the manufacturing area management application to support better process flow such as decision making and failure prediction. Interoperability helps prevents repeated programming efforts for new machines, where the semantic layer can provide a specific information model of the particular device.

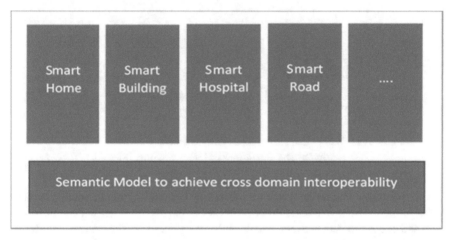

Fig. 3. Optimised smart city model with interoperability

Cross-platform or cross-domain interoperability plays an essential role in IoT growth [21]. Various domains such as hospitality, transportation, building management work together to provide impactful IoT solutions such as the smart city. Therefore, these systems have to be integrated and connected horizontally to exchange data with cross-domain applications. To achieve cross-domain semantic interoperability, semantic models and annotations implemented in each application help IoT integrate information from a different source and consistently understand them [22]. These improve the scalability, sustainability, efficiency and provide an optimised application for the smart city [23]. Figure 3 illustrates the smart city framework with a semantic model implemented. Challenges and gaps in data analysis and reasoning, data integration, annotation of data, data management, data discovery, and data visualisation are highlighted as the main contributor to the lack of semantic interoperability.

4   Ontology and Semantic Network

Semantic interoperability of IoT devices in the presence of heterogeneous data structures can be addressed using ontologies or semantic networks [24].

Ontology is described as a common machine-readable data or representation of knowledge. Ontology is a formal, explicit specification of a shared conceptualisation [25]. Ontology describes explicit relationships between things and contains semantic information intended for automatic processing by machines [26]. Ontology can be represented in different languages such as RDF, RDF Schema, and OWL. It represents composite relations and information about things or sets of things.

RDF is a short form for Resource Description Framework. It is a simple ontology language used for standardised data interchange that the World Wide Web Consortium developed during the evolution of semantic technologies [27]. RDF offers the groundwork for connecting these data as an established, broadly used, and tested technology for modelling data [25]. RDF was only previously part of the semantic web stack, but it is currently used to represent high-quality connected data [28]. Information represented by RDF is easy to be identified, disambiguate, and interconnected by the software application and various systems to read, analyse and act upon [28].

RDF as a graph-based format represents named links between resources in triples form, Subject, Object, and Predicate, where the predicate denotes the relationship between subject, which denotes an entity, and object denotes an entity or value [29]. The triple explains that subject has a relation to an object or the subject has an attribute whose value is the object [19]. The relationship is always directional from subject to object, as shown in Fig. 4. Figure 5 presents an example RDF of wind sensor ws01 and its observation represented in turtle format, such as "wind sensor ws01 is a sensor observes wind speed" and "wind sensor ws01 measured the wind speed to be 30 km/h at 2020–12-17T08:05:36" [30].

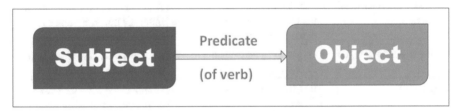

Fig. 4. RDF representation

1.	:windsensor_ws01 a	sosa: Sensor	
2.		sosa: observes	:windspeed
3.		sosa: madeObservation	:windspeedObs01
4.	:winspeedObs01	sosa: observedProperty	:windspeedRate
5.		sosa: hasSimpkeResult	"30km/h";
6.		sosa: resultTime	"2020-12-17T08:05:36"

Fig. 5. Sample RDF statement in turtle format [30]

The Web Ontology Language (OWL) is the best-known language for expressing ontologies for their use on the web [27]. OWL is available in several variants with improvised expressiveness. Only the most expressive variant, OWL Full, has proper layering over RDF Schema that enables meta-modelling and verification features. In contrast, the OWL DL variant is focused on the formalism of description logic (DL) [27].

A semantic network is a graph structure representing knowledge in patterns of inter-connected nodes [31]. The implementation of the semantic network in a computing environment was first used for artificial intelligence and machine translation, but it has also been used in philosophy, psychology, and linguistics [32]. All semantic networks are used as a declarative graphical representation that is used to represent knowledge or to support an automated application for knowledge inference [32]. Ontology and semantic networks do serve a similar purpose. Still, there is vast differentiation on characteristics and how these both work. Table 1 will briefly compare the characteristic of both ontology and semantic networks.

Table 1. Comparison between semantic network and ontology [32]

	Semantic network	Ontology
Specify property characteristic	Not able to specify the properties characteristic	Able to identify the properties characteristic
Unique name assumption	Two objects with different names assumed to be different	There is no assumption made unless explicitly stated about the relationship
Open vs. closed network assumption	Nothing can be entered into the network unless there is a place for it in the network	Anything can be added into the ontology unless it violates any of its constraints
Ability to add new nodes	Only can be added if the object or property is related	Can be added new object and properties can be described
Visualisation	Easy to visualise the network content	Complicated network which is a challenge to be interpreted by human

5   Related Work on Semantic Interoperability in IoT

Achieving semantic interoperability is an essential aspect of the IoT framework [33]. Several works and research ideas are ongoing in IoT domains on reusing, combining, and abstracting existing solutions to achieve semantic interoperability [34]. Most available solutions use ontologies to help implement semantic interoperability [35]. Some researchers have used fog computing to help to achieve semantic interoperability.

Dr Joy Long Zong Chen has proposed IoT semantic interoperability with the assistance of fog computing [36]. This framework is expected to improve energy usage, latency, network utilisation and reduce processing costs [36]. He planned a two-layer hierarchical fog network to execute the semantic process, and the data mapped are presented to the cloud for processing. As per Fig. 6, all the processing includes filtration, aggregation, composing, modelling, mapping data in fog layers and utilising single-owl file ontology [36]. Even with the proposed fog-based semantic framework whole level of interoperability could not be achieved with growing data in the current IoT environment. The lightweight framework proposed does not accept new IoT sensors as single-owl file ontology does not support dynamic data mapping. With such a situation full level of semantic interoperability could not be achieved.

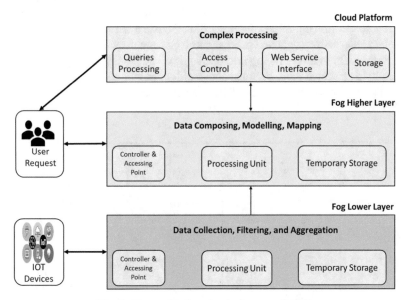

Fig. 6. Fog-assisted semantic framework [36].

Hafizur Rahman proposed a lightweight ontology to achieve semantic interoperability in IoT. This paper primarily focuses on lightweight Ontology (LiO-IoT) regarding Sensors, Actuators, and RFID as IoT concepts. This solution concentrates only on normalcy used and available ontologies to provide semantic interoperability. The ontology used was designed as simple as possible to achieve lightweight [35]. But the dynamic semantic ontology could not be achieved, and this is considered full heterogeneity not achieved in this paper. The author also had proposed a lightweight dynamic ontology for the IoT using the machine learning framework [35].

The proposed model can detect and add new concepts to the ontology automatically. This dynamic ontology comprises a mechanism for dynamic interaction and real-time functions between multiple IoT devices [35]. Machine learning exploits the volume of the information within an ontology to detect novelties within ontologies and discover semantic knowledge patterns. The technique of clustering used accelerates the search and discovery of new concepts in the IoT infrastructure. Dynamic ontology poses a clustering delay which causes additional time consumption and higher response time.

Lightweight dynamic ontology is more effective for large-scale networks than SSN ontology and uses less memory [37]. The average response time of the overall system with dynamic semantics is slightly higher than without dynamic semantics due to the overhead of clustering and novel node discovery and integration. Additionally, delays were observed to add new clusters.

Hafizur Rahman has proposed a Fog computing-based semantic mechanism to improve interoperability in IoT. The proposed model offloads some commonly used cloud semantic processes to the network edge. An effective offloading technique developed between fog-to-fog and fog-to-cloud devices reduces task execution time and energy consumption of fog nodes [10]. Fog nodes were organised hierarchically based on the fog node's functionalities. Figure 7 illustrates that the level two Fog nodes (L2-Fog) collect raw sensory data and process some tasks such as filtering and aggregation, then forward them to level one fog nodes(L1-Fog). The L1-Fog nodes then receive this high-quality aggregated data, perform composition, modeling, and linking operations, and take appropriate actions. The modelled data's semantic annotation is made with the ontology [33].

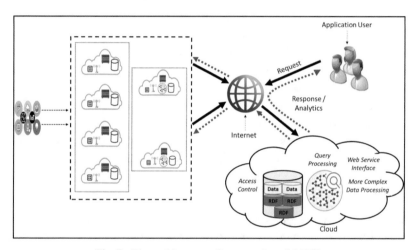

Fig. 7. The architecture of proposed model [35].

The ontology available in the L1 Fog devices is a lightweight single owl file that delivers a semantic annotation of the IoT data [10]. The lightweight middleware performs the ontology implementation and integration in the proposed architecture. The ontology used is static and cannot add any new component to the system. In the traditional semantic model, computations are performed only in the cloud, while Fog devices pre-process the data packets in the Fog computing architecture. This avoids significant delays and energy loss, network usage, and computational cost. However, it is observed that as the number of fog nodes increases, service delays also increase. The energy consumption is less than other solutions as the semantic operations are performed in the nearby fog nodes compared to the cloud. The proposed model reduces the service delay, energy consumption, and overall cost and improves the network bandwidth utilisation of the system [10].

Amelie Gyrard has proposed a unified semantic engine for the IoT and smart cities. Her work integrated the semantic web of things into smart cities to achieve interoperability in the Smart cities environment [11]. The web semantic proposed in Fig. 8 is expected to provide unifying IoT data, unifying modelling, and unifying reasoning to interpret IoT data. With this approach, the author can provide the unified semantic engine that helps achieve semantic interoperability, but this system will compose high latency due to transmission through web technology. This will also possess high bandwidth.

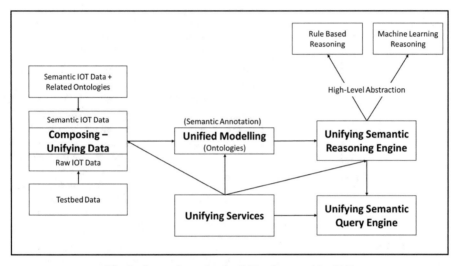

Fig. 8. Semantic engine for IoT and smart cities [11]

5.1 Finding and Discussion on Related Work

Semantic interoperability can be divided into dynamic and static. Static semantic interoperability is meant for IoT applications that do not have new IoT nodes or devices to be added to the framework. Therefore, the ontology used for static semantic interoperability can be defined and trained before being implemented into the IoT framework. Dynamic semantic interoperability requires new IoT devices or nodes constantly added to the IoT framework where the ontology used in this environment needs to be updated from time to time to meet the application's needs.

Ontology has been used widely by researchers to achieve semantic interoperability in IoT. Finding and discussion on the related work have been done briefly in Table 2. There are lightweight and heavyweight ontology models available. Most of the researchers tend to use lightweight ontology models due to the nature of IoT devices like low power and resource constraints. Lightweight ontology models are static and could not support the processing capability for dynamic ontology models [12]. Therefore, there is a research gap on implementing dynamic ontology in IoT framework to achieve dynamic semantic interoperability.

Table 2. Comparison of related work

Solution	Ontology	Pros	Cons
Interoperability improvements In IoT using fog assisted semantic framework	Single owl	• Better latency, energy usage • Lightweight model • Multi-level fog enables high-level processing • Processing took place in the fog layer; therefore, less processing cost and network usage	• A new addition could not be added • A multi-layer model might need an additional device which might increase the costing • The single owl model could not support dynamic data mapping
LiO-IoT	NA	• Lightweight • Focus on sensors, actuators, and RFID • High latency • Low processing cost	• Not able to achieve full level semantic interoperability as new domain could not be added into the available system • Not suitable for larger IoT framework • Processing needs to take place in the cloud, so high bandwidth and network usage
Dynamic LiO-IoT	NA	• Lightweight • Automatically detect new concepts and add them into the ontology • Support real-time model • Machine learning technique used accelerates the searching and discovering for new concepts • Perform better for large scale network	• Possess clustering delay • High response time • Additional delay for new node detection and inclusion compared to non-dynamic
Semantic-fog	Single owl	• Using lightweight ontology • Reduced execution time and energy consumption • Edge processing at the fog nodes help reduce net-work usage and response time	• Static ontology and could not add a new component into the system

(*continued*)

Table 2. (*continued*)

Solution	Ontology	Pros	Cons
OpenIoT	SSN	• Provide unified modelling • Provide integrated environment • Handling mobile sensors • Is an open-source project, able to be accessed easily	• High latency due to transmission through web technology • A heavyweight ontology that in-curs high processing power • Not suitable for small scale

6 Challenges in IoT Due to Lack of Semantic Interoperability

There are multiple challenges identified in IoT due to the lack of semantic interoperability in a system or framework.

1. Vendor lock-in

   Vendor lock-in happens when customers are forced to use IoT devices, applications, or systems from a single supplier. Some IoT devices are only compatible with a particular system, which causes a situation where the framework does not function properly with different models of IoT devices [8]. Therefore, the consumer has to use a single supplier to upgrade their system. This situation leads to a monopoly in the IoT market and hinders switching or upgrading the system to a more cost-effective solution in the future.

2. Technology instability

   Technology instability is another issue that arises from a lack of interoperability in the IoT ecosystem. There is a risk that vendors fail to deliver agreed functionality and quality of service [38]. Due to a lack of interoperability, consumers cannot switch or use other vendor's services and end up using an unstable technology solution.

3. Increased cost

   Lack of Interoperability also causes an increase in IoT ecosystem implementation costs. IoT devices and applications are kept on evolving, and the massive technology advancement causes the cost of IoT systems to become inexpensive [8]. But not all consumers can move to cheaper and advanced IoT solutions as they could not upgrade their application due to lack of interoperability. They have been forced to change to a new IoT system that will incur a higher cost than just updating some of the IoT devices in their application.

4. Hampered scalability

   These closed IoT ecosystems hamper the effective integration of new IoT devices and systems to resolve a broader range of IoT issues. Due to the IoT applications are not interoperable, we cannot scale the IoT ecosystem with advanced or additional new IoT devices or applications as it will cause compatibility issues [38]. Therefore, the scalability and productivity of IoT systems are hampered.

5. Lower accessibility

Lower accessibility is another severe issue IoT systems face due to a lack of interoperability. Closed IoT ecosystems, better known as silos, hinder the accessibility availability of data and information of an IoT application to other systems.

6. Storage wastage

IoT applications work independently, and data collected is not shared across the ecosystem. An Interoperable IoT ecosystem can provide data sharing within IoT applications and lead to less storage. With the current IoT trend facing an increase in data, the larger storage capacity required and sharing data within the IoT ecosystem can reduce storage waste.

7   Open Challenges and Future Work

We have summarised the challenges faced by IoT due to the lack of semantic interoperability and discussed several methods proposed by researchers to tackle semantic interoperability in IoT. There are six significant challenges or issues identified in the IoT framework caused by the absence of semantic interoperability. The identified issues can cause the overall IoT application to perform ineffectively. Due to this, technology advances in IoT that are currently booming, such as Industrial revolution 4.0, cannot be implemented entirely by many sectors. Therefore, complete IoT technology advancement cannot be scaled and cause the IoT ecosystem to be left behind. In order to address scalability issues, an important consideration has to be given to achieve semantic interoperability in IoT architecture.

We have discussed various approaches proposed by researchers to achieve semantic interoperability in IoT and discussed the main limitations addressed by all researchers. The primary concern placed by most of the researchers is on the need for the lightweight ontology to achieve semantic interoperability in IoT. The need for a lightweight model emphasised by researchers is because IoT devices are tiny with resources and have limited storage, such as actuators, RFID, and sensor. But from another point of view, the lightweight ontology model cannot provide complete semantic interoperability support as it has to be in its simplest form. Research is needed to achieve comprehensive semantic interoperability with a lightweight model or implement a complete heavyweight ontology model in IoT devices.

Furthermore, most semantic interoperability solutions are designed to be implemented in a specific domain of experts such as healthcare, smart city, and others. A single model that can cater to all the domains has to be developed to provide a standardised mechanism to achieve semantic interoperability in IoT, which is another potential research gap to consider. Nevertheless, the need for static semantic interoperability and dynamic semantic interoperability in the heterogeneous IoT environment is crucial. Most researchers have focused on achieving static semantic interoperability, while less research work is performed for dynamic semantic interoperability.

Dynamic semantic interoperability in IoT is crucial as IoT poses real-time applications, and scalability is one of the main aspects of IoT. Therefore, more research is needed to provide dynamic semantic models to improve and optimise the IoT framework. Currently, there are quite a few different academia and standardisation bodies working on solving interoperable semantic issues by defining some standards that can be universally accepted. But standards involving semantics are still a challenge as there is multiple domains for the dataset which require diverse attributes and classification.

In conclusion, semantic interoperability plays a vital role in achieving an effective and successful IoT framework. Challenges and research gaps in tackling the challenges caused by semantic interoperability have been briefly discussed in this paper.

References

1. Ahmadh, R.K., Kariapper, R.: Awareness of internet of thing among students of south eatern university of Sri Lanka. Journal of Critical Reviews **7**(15), 4673–4678 (2020)
2. Nizetic, S., Solic, P., Lopez-de-Ipina Gonzalez-de-Artaza, D., Patrono, L.: Internet of Things (IoT): Opportunities, issues and challenges towards a smart and sustainable future. Journal of Cleaner Production **274**, 122877 (Nov 2020). https://doi.org/10.1016/j.jclepro.2020.122877
3. Cisco: Internet of Things (2016). [Online]. Available: https://www.cisco.com/c/en/us/products/collateral/se/internet-of-things/at-a-glance-c45-731471.pdf
4. Jovanovic, B.: Internet of Things statistics for 2021 – Taking Things Apart. DataProt (2021) https://dataprot.net/statistics/iot-statistics/. Accessed 18 Jun 2021
5. Noura, M., Atiquzzaman, M., Gaedke, M.: Interoperability in internet of things: taxonomies and open challenges. Mobile Networks and Applications **24**(3), 796–809 (2019). https://doi.org/10.1007/s11036-018-1089-9
6. Hazra, A., Adhikari, M., Amgoth, T., Srirama, S.N.: A Comprehensive Survey on Interoperability for IIoT: Taxonomy, Standards, and Future Directions. ACM Computing Surveys **55**(1), 1–35 (2023). https://doi.org/10.1145/3485130
7. Agarwal, R., et al.: Unified IoT ontology to enable interoperability and federation of testbeds. In: 2016 IEEE 3rd World Forum on Internet of Things, WF-IoT 2016, pp. 70–75 (2017). https://doi.org/10.1109/WF-IoT.2016.7845470
8. Laura, D.: Semantic Technologies to address interoperability challenges of IoT (2021). https://2021-eu.semantics.cc/semantic-technologies-address-interoperability-challenges-iot accessed 15 Jun 2021
9. Bjorkdahl, P.: AI and IoT continue to increase interoperability demands in 2021 (Feb 2021). [Online] Available: https://www.securitymagazine.com/articles/94659-ai-and-iot-continue-to-increase-interoperability-demands-in-2021
10. Rahman, H., Hussain, M.I.: Fog-based semantic model for supporting interoperability in IoT. IET Communications **13**(11), 1651–1661 (2019). https://doi.org/10.1049/iet-com.2018.6200
11. Gyrard, A., Serrano, M.: A unified semantic engine for internet of things and smart cities: from sensor data to end-users applications. In: Proceedings - 2015 IEEE International Conference on Data Science and Data Intensive Systems; 8th IEEE International Conference Cyber, Physical and Social Computing; 11th IEEE International Conference on Green Computing and Communications and 8th IEEE International Conference on Internet of Things, DSDIS/CPSCom/GreenCom/iThings 2015, no. August 2017, pp. 718–725 (2015). https://doi.org/10.1109/DSDIS.2015.59
12. Rahman, H., Hussain, I.: A light-weight dynamic ontology for internet of things using machine a light-weight dynamic ontology for internet of things using machine learning technique. In: ICT Express, no. December (2020). https://doi.org/10.1016/j.icte.2020.12.002
13. Snyder, H.: Literature review as a research methodology: an overview and guidelines. J. Bus. Res. **104**(July), 333–339 (2019). https://doi.org/10.1016/j.jbusres.2019.07.039
14. Johnson, A.T., Hennessy, E.A.: Systematic reviews and meta-analyses in the health sciences: Best practice methods for research syntheses. Social Science and Medicine **233**, 237–251 (2019). https://doi.org/10.1016/j.socscimed.2019.05.035

15. Kotis, K.I., Pliatsios, A., Goumopoulos, C., Kotis, K.: A review on IoT frameworks supporting multi-level interoperability-the semantic social network of things framework. Article in Int. J. Advanc. Intern. Technol. **13**(1), 46–64 (2020). [Online]. Available: http://www.iariajournals.org/internet_technology/
16. Santos, K.S.S., Pinheiro, L.B.L., Maciel, R.S.P.: Interoperability Types Classifications: A Tertiary Study. In: ACM International Conference Proceeding Series (2021). https://doi.org/10.1145/3466933.3466952
17. Fernandes, J., Graciano Neto, V.V., dos Santos, R.P.: An Approach Based on Conceptual Modeling to Understand Factors that Influence Interoperability in Systems-of-Information Systems, pp. 1–10 (2021). https://doi.org/10.1145/3493244.3493275
18. Vila, M., Sancho, M.R., Teniente, E., Vilajosana, X.: Semantics for connectivity management in IoT sensing. In: Lecture Notes in Computer Science (including subseries Lecture Notes in Artificial Intelligence and Lecture Notes in Bioinformatics), vol. 13011 LNCS, no. ISBN 978-3-030-89021-6, pp. 297–311 (2021). https://doi.org/10.1007/978-3-030-89022-3_24
19. Amara, F.Z., Hemam, M., Djezzar, M., Maimour, M.: Semantic web technologies for internet of things semantic interoperability. In: Lecture Notes in Networks and Systems, vol. 357 LNNS, pp. 133–143 (2022). https://doi.org/10.1007/978-3-030-91738-8_13
20. Widell, N., Keranen, A., Badrinath, R.: What is Semantic interoperability in IoT and why is it important? Information Services and Use **39**(3), 205–213 (2019). https://doi.org/10.3233/ISU-190045
21. Deshmukh, R.A., Jayakody, D., Schneider, A., Damjanovic-behrendt, V.: Data spine: a federated interoperability enabler for heterogeneous iot platform ecosystems. Sensors **21**(12), 1–28 (2021). https://doi.org/10.3390/s21124010
22. Davies, J., Fisher, M.: Data platforms: interoperability and insight. The Internet of Things 37–50 (2020). https://doi.org/10.1002/9781119545293.ch4
23. Balakrishna, S., Solanki, V.K., Gunjan, V.K., Thirumaran, M.: A survey on semantic approaches for IoT data integration in smart cities. In: ICICCT 2019 – System Reliability, Quality Control, Safety, Maintenance and Management, pp. 827–835 (2020). https://doi.org/10.1007/978-981-13-8461-5_94
24. Ganzha, M., Paprzycki, M., Pawłowski, W., Solarz-Niesłuchowski, B., Szmeja, P., Wasielewska, K.: Semantic interoperability. Internet of Things 133–165 (2021). https://doi.org/10.1007/978-3-030-82446-4_5
25. Gruber, T.R.: A translation approach to portable ontology specifications. Knowledge Acquisition **5**(2), 199–220 (1993). https://doi.org/10.1006/knac.1993.1008
26. IDA: What is the difference between rdf-, triple and quad store? (2021). https://instructionaldesign.com.au/the-semantic-web/ accessed 22 Jun 2021
27. OWL Working Group: Web Ontology Language (OWL) (2012). https://www.w3.org/OWL/ accessed 22 Jun 2021
28. Ontotext: What is RDF, Why Bother with RDF (2021). https://www.ontotext.com/knowledgehub/fundamentals/what-is-rdf/ accessed 21 Jun. 2021
29. Wylot, M., Hauswirth, M., Cudre-Mauroux, P., Sakr, S.: RDF data storage and query processing schemes A survey. ACM Computing Surveys **51**(4) (2018). https://doi.org/10.1145/3177850
30. Le-Tuan, A., Hayes, C., Hauswirth, M., Le-Phuoc, D.: Pushing the scalability of rdf engines on IOT edge devices. Sensors (Switzerland) **20**(10), 15–18 (2020). https://doi.org/10.3390/s20102788
31. Lehmann, F.: Semantic networks. Comp. Maths. Appli. **23**(2–5), 1–50 (1992). https://doi.org/10.1016/0898-1221(92)90135-5

32. Salem, A.-B.M., Alfonse, M.: Ontology versus semantic networks for medical knowledge representation. In: Proceedings of the 12th WSEAS international Conference on COMPUTERS, no. 6, pp. 769–774 (2008). [Online] Available: http://www.wseas.us/e-library/conferences/2008/crete/Computers/120-computers.pdf
33. Ullah, F., Habib, M.A., Farhan, M., Khalid, S., Durrani, M.Y., Jabbar, S.: Semantic interoperability for big-data in heterogeneous IoT infrastructure for healthcare. Sustainable Cities and Society **34**, 90–96 (2017). https://doi.org/10.1016/j.scs.2017.06.010
34. Balakrishna, S., Solanki, V.K., Gunjan, V.K., Thirumaran, M.: A survey on semantic approaches for IoT data integration in smart cities. In: Gunjan, V., Garcia Diaz, V., Cardona, M., Solanki, V., Sunitha, K. (eds.) ICICCT 2019 – System Reliability, Quality Control, Safety, Maintenance and Management. ICICCT 2019. Springer, Singapore (2020). https://doi.org/10.1007/978-981-13-8461-5_94
35. Rahman, H., Hussain, I.: LiO-IoT : A Light-weight Ontology to provide Semantic Interoperability in Internet of Things', no. April, pp. 571–575 (2019)
36. Zong Chen, A.J.I., Smy, S.: Interoperability improvement in internet of things using fog assisted semantic frame work. Journal of Trends in Computer Science and Smart Technology **2**(1), 56–68 (2020). https://doi.org/10.36548/jtcsst.2020.1.006
37. Swar, A., Khoriba, G., Belal, M.: A unified ontology-based data integration approach for the internet of things. Int. J. Electri. Comp. Eng. **12**(2), 2097–2107 (2022). https://doi.org/10.11591/ijece.v12i2.pp2097-2107
38. Weeve: Data Interoperability has proven a challenge for companies trying to transform and understand data from various systems while still generating business value (2021). https://www.iotforall.com/data-to-dollars-a-data-interoperability-challenge accessed 16 Jun. 2021

An Intelligent Method Based on Dissimilarity Measure Picture Fuzzy and Apply to Supplier Selection

Nguyen Xuan Thao and Quang-Dung Pham[✉]

Faculty of Information Technology, Vietnam National University of Agriculture, Hanoi, Vietnam
{nxthao,pqdung}@vnua.edu.vn

Abstract. Supplier selection is the MCDM problem determining the optimal choice among the possible alternatives. This problem had inaccurate and vague information on alternatives. It may also be due to an inadequate understanding of the group of decision-makers on the criteria or alternatives. Classical techniques are known to have difficulty dealing with such situations. The picture fuzzy set theory is one of the adaptive intelligent tools that can be used to handle such situations. This article proposes the MCDM method called picture fuzzy TOPSIS. In which, the value of each alternative and the weight of criteria is evaluated by the decision-maker over the linguistic variables whose values are picture fuzzy numbers. Finally, a numerical example is given illustrating the proposed method.

Keywords: Picture fuzzy set · TOPSIS · MCDM

1 Introduction

Because of the uncertainty and complexity of the nature of decision-making, so it often uses fuzzy techniques. Using the MCDM methods, we can identify the best alternative from the set of alternatives based on some criteria. In recent decades, supplier selection has increasingly played a vital role in both academia and industry. Having many MCDM techniques were investigated for this problem. [2, 8, 22]. However, the methods above are restricted to using the crisp set theory. Therefore, it is difficult to treat uncertainty or an incomplete environment. So having the proposed MCDM methods using fuzzy set theory or intuitionistic fuzzy set to supplier selection is introduced. [3, 7–15, 21, 22, 24].

Fuzzy set theory [23] is a useful method to study the problems of imprecision and uncertainty. After that, a lot of having many to deal with imprecision and uncertainty new theories have been introduced. For instance, Intuitionistic fuzzy set [1] is a generalization of the notion of a fuzzy set. It gives two components (a degree of membership and a degree of non-membership) in the fuzzy set only gives one degree of membership. The intuitionistic fuzzy set has many applications in different practical problems. The intuitionistic fuzzy set has many applications in different practical problems [1, 3, 22, 24, 29, 30, 33]. The picture fuzzy set [4] has three memberships (a degree of positive membership, a degree of negative membership, and a degree of neutral membership of

N.-T. Nguyen et al. (Eds.): ICIT 2022, LNDECT 148, pp. 32–43, 2022.
https://doi.org/10.1007/978-3-031-15063-0_3

an element) is an extension of the fuzzy set. This approach is widely used by researchers in both theory and application [4, 5, 7, 16–18, 25–28, 31, 32]. Hoa and Thong [7] used picture fuzzy sets modified fuzzy clustering algorithms and applied them for geographic data clustering. Singh investigated the correlation coefficient for picture fuzzy sets and applied them in type clustering and pattern recognition [14]. Son applied of picture fuzzy set in the clustering problems [16–18]. Garg [6] and Wei [20] introduced some picture fuzzy aggregation operators and their application to multiple attribute decision making. The picture fuzzy database is investigated by Van Dinh et al. [19]. Dinh et al. [5] defined some distance and dissimilarity measures of picture fuzzy set apply to pattern recognition problems. Le et al. [26] pointed out the limitations of the dissimilarity measure of Dinh et al. in [5] and proposed a new dissimilarity measure. But the proposed measure of Le et al. also has limitations in many different cases. Therefore, we found the need to improve the dissimilarity measurements to overcome the limitations of existing measures.

Moreover, the dissimilarity measure based picture fuzzy TOPSIS is not studied. It is the motivation for us to propose a new dissimilarity measure and construct the picture fuzzy TOPSIS model based on dissimilarity measures. The main difference between the picture fuzzy sets and the previous fuzzy sets is in their concept. Picture fuzzy sets have the advantage over fuzzy sets in that fuzzy sets have three components: positive, negative, and neutral functions of an element for a given set. Thus, the picture fuzzy set has a complementary function that measures the degree of neutralization (whereas the fuzzy set has only a membership function and intuitionistic fuzzy set have also two parts of the function and the non-integral function of a given set of fuzzy sets). Thus, an experts' assessments of a given product in addition to evaluating the good (positive) or bad (negative) aspects of a product, the expert adds another dimension. In contrast, the intuitionistic fuzzy set and fuzzy set do not reflect this.

In this research, a new dissimilarity measure between picture fuzzy sets is defined to overcome the limitation of the previous measures. Then, it is used to give TOPSIS method and apply it to solve the supplier selection problems. The next of the paper is as follows: In Sect. 2, we rewrite some basic concepts relative to picture fuzzy sets. In Sect. 3, we propose a new TOPSIS method, which uses picture fuzzy sets, named the picture fuzzy TOPSIS method. Finally, we apply the picture fuzzy TOPSIS method for supplier selection in Sect. 4.

2 Basic Concepts

In this section, we recall the concept of picture fuzzy set. That is used to construct the TOPSIS model in this paper.

2.1 Picture Fuzzy Set

Definition 1 [4]. A Picture fuzzy set (PFS) A on the universal set U is $A = \{(u, \mu_A(u), \eta_A(u), \gamma_A(u))|u \in U\}$ where $\mu_A(u), \eta_A(u)$ and $\gamma_A(u)$ are the "degree of positive membership, degree of neutral membership and $\gamma_A(x)(\in [0, 1])$ $\gamma_A(x)(\in [0, 1])$ degree of negative membership of u in A, respectively" where $\mu_A(u), \eta_A(u)$ μ_A, γ_A μ_A, γ_A $\gamma_A(u) \in [0, 1]$ η_A η_A satisfy the following condition:

$$0 \le \mu_A(u) + \eta_A(u) + \gamma_A(u) \le 1, \forall u \in U. \tag{1}$$

The collection of all picture fuzzy set in U is denoted by PFS(U).

Definition 2. The picture fuzzy set $B = \{(u, \mu_B(u), \eta_B(u), \gamma_B(u)) | u \in U\}$ is called the subset of the picture fuzzy set $A = \{(u, \mu_A(u), \eta_A(u), \gamma_A(u)) | u \in U\}$ iff $\mu_B(u) \leq \mu_A(u)$, $\eta_B(u) \leq \eta_A(u)$ and $\gamma_B(u) \geq \gamma_A(u)$ for all $u \in U$.

Definition 3. The   complement   of   picture   fuzzy   set $A = \{(u, \mu_A(u), \eta_A(u), \gamma_A(u)) | u \in U\}$ is

$$A^C = \{(u, \gamma_A(u), 1 - \mu_A(u) - \eta_A(u) - \gamma_A(u), \mu_A(u)) | u \in U\} \tag{2}$$

For convenience, we consider $\alpha = (\mu_\alpha, \eta_\alpha, \gamma_\alpha)$ with $\mu_\alpha \geq 0, \eta_\alpha \geq 0, \gamma_\alpha \geq 0$ and $\mu_\alpha + \eta_\alpha + \gamma_\alpha \leq 1$ is a picture fuzzy number.

Definition 4. Let α, β be two picture fuzzy sets on U. Then

$$\text{(i)} \quad \alpha \oplus \beta = \left(\mu_\alpha + \mu_\beta - \mu_\alpha.\mu_\beta, \eta_\alpha.\eta_\beta, \gamma_\alpha.\gamma_\beta\right) \tag{3}$$

$$\text{(ii)} \quad \alpha \otimes \beta = \left(\mu_\alpha.\mu_\beta, \eta_\alpha.\eta_\beta, \gamma_\alpha + \gamma_\beta - \gamma_\alpha.\gamma_\beta\right) \tag{4}$$

$$\text{(iii)} \quad \lambda\alpha = \left(1 - (1 - \mu_\alpha)^\lambda, \eta_\alpha^\lambda, \gamma_\alpha^\lambda\right) \text{ where} \lambda > 0 \tag{5}$$

$$\text{(iv)} \quad \alpha^\lambda = \left(\mu_\alpha^\lambda, 1 - (1 - \eta_\alpha)^\lambda, 1 - (1 - \gamma_\alpha)^\lambda\right) \text{ where} > 0. \tag{6}$$

Definition 5 [6]. Let $\{\alpha_1, \alpha_2, \ldots, \alpha_N\}$ be the picture fuzzy numbers. The picture fuzzy weighted averaging operator is

$$PFWA_\delta(\alpha_1, \alpha_2, \ldots, \alpha_N) = (1 - \prod_{t=1}^{N}(1 - \mu_{\alpha_t})^{\delta_t}, \prod_{t=1}^{N}(\eta_{\alpha_t})^{\delta_t}, \prod_{t=1}^{N}(\gamma_{\alpha_t})^{\delta_t} \tag{7}$$

where $\delta = (\delta_1, \delta_2, \ldots, \delta_N)$ with $\delta_t > 0 (t = 1, 2, \ldots, N)$ and $\sum_{t=1}^{N} \delta_t = 1$.

2.2 Dissimilarity Between Two Picture Fuzzy Sets

Suppose A, B are two picture fuzzy sets on $U = \{u_1, u_2, \ldots, u_N\}$, the dissimilarity measure of two picture fuzzy sets A and B defined as follow:

Definition 5. A function $d : PFS(U) \times PFS(U) \rightarrow [0, 1]$ is a dissimilarity measure between two picture fuzzy sets A, B if it satisfies follow properties:

(d1) $0 \leq d(A, B) \leq 1$
(d2) $d(A, B) = 0$ if only if $A = B$
(d3) $d(A, B) = d(B, A)$
(d4) if $A, B, C \in PFS(U)$ and $A \subseteq B \subseteq C$, then $d(A, C) \geq \max\{d(A, B), d(B, C)\}$.

Theorem 1. Given $U = \{u_1, u_2, \ldots, u_N\}$ is a universal set. For $A, B \in PFS(U)$, then

$$d_T(A, B) = \frac{1}{3N} \sum_{k=1}^{N} \left[|\mu_A(u_k) - \mu_B(u_k)| + \left(1 - e^{-|\eta_A(u_k) - \eta_B(u_k)|}\right) + \left(1 - e^{-|\gamma_A(u_k) - \gamma_B(u_k)|}\right) \right] \quad (8)$$

is a dissimilarity measure of two picture fuzzy sets.

Proof. We show that d_H satisfies definition 5. Indeed, we have $(d2)$, $(d3)$ is obviously.
   + (d1). Evidently $d_H(A, B) \geq 0$.
   Moreover, since $|\mu_A(u_k) - \mu_B(u_k)|, |\eta_A(u_k) - \eta_B(u_k)|, |\gamma_A(u_k) - \gamma_B(u_k)| \leq 1$ then $d_H(A, B) \leq 1$.
   + (d4). From definition 2 and $A \subseteq B \subseteq C$, we have $\mu_A(u_k) \leq \mu_B(u_k) \leq \mu_C(u_k)$ so that $|\mu_A(u_k) - \mu_C(u_k)| \geq \max\{|\mu_A(u_k) - \mu_B(u_k)|, |\mu_B(u_k) - \mu_C(u_k)|\}$, for all $u_k \in U, k = 1, 2, \ldots, N$. Similarity, $|\eta_A(u_k) - \eta_C(u_k)| \geq \max\{|\eta_A(u_k) - \eta_B(u_k)|, |\eta_B(u_k) - \eta_C(u_k)|\}$ and $|\gamma_A(u_k) - \gamma_C(u_k)| \geq \max\{|\gamma_A(u_k) - \gamma_B(u_k)|, |\gamma_B(u_k) - \gamma_C(u_k)|\}$ for all $u_k \in U, k = 1, 2, \ldots, N$. Hence $d(A, C) \geq \max\{d(A, B), d(B, C)\}$. $\qquad\square$

3 Proposed Picture Fuzzy TOPSIS Method

Here, we introduce a new TOPSIS method using the distance measure of picture fuzzy set. Suppose that $A = \{A_1, A_2, \ldots, A_m\}$ is a set of alternatives and $C = \{C_1, C_2, \ldots, C_n\}$ is a set of criteria. Let $W = \{w_1, w_2, \ldots, w_n\}$ is the weight vector assigned for the criteria $C = \{C_1, C_2, \ldots, C_n\}$, which determined by the decision makers.
   We construct the procedure of picture fuzzy TOPSIS process, which is as follows:

Step 1. Determine the weight of decision makers
Assume that there is a decision group having p decision makers $DM = \{DM_1, DM_2, \ldots, DM_p\}$. In which, the importance of the decision makers are considered as linguistic terms expressed in picture fuzzy numbers and $DM_t = (\mu_{dm_t}, \eta_{dm_t}, \gamma_{dm_t})$ is a picture fuzzy number express the rating of tth decision maker $(t = 1, 2, \ldots, p)$. Then we obtain the weight δ_t of tth decision maker $(t = 1, 2, \ldots, p)$ as follows:

$$\delta_t = \frac{1 - \frac{\mu_{dm_t}^2 + \eta_{dm_t}^2 + \gamma_{dm_t}^2}{3}}{\sum_{t=1}^{p} \left[1 - \frac{\left(\mu_{dm_t}^2 + \eta_{dm_t}^2 + \gamma_{dm_t}^2\right)}{3} \right]} \quad (9)$$

for all $t = 1, 2, \ldots, p$ and $\sum_{t=1}^{p} \delta_t = 1$.

Step 2. Construction of aggregated picture fuzzy decision matrix respect to the decision maker.

Let $D^{(t)} = \left(d_{ik}^{(t)}\right)_{m\times n} = \left(\left(\mu_{d_{ik}^t}, \eta_{d_{ik}^t}, \gamma_{d_{ik}^t}\right)\right)_{m\times n}$ be the picture fuzzy matrix of tth decision maker. Then, the decision matrix D of decision makers defined by

$$
\begin{array}{c}
\quad C_1\ C_2\ \ldots\ C_n \\
\begin{array}{c} A_1 \\ A_2 \\ \vdots \\ A_m \end{array}
\left(\begin{array}{cccc}
d_{11} & d_{12} & \ldots & d_{1n} \\
d_{21} & d_{22} & \ldots & d_{2n} \\
\vdots & \vdots & \ldots & \vdots \\
d_{m1} & d_{m2} & \cdots & d_{mn}
\end{array}\right)
\end{array}
\tag{10}
$$

where $d_{ik} = \left(\mu_{d_{ik}}, \eta_{d_{ik}}, \gamma_{d_{ik}}\right) = PFWA_\delta\left(d_{ik}^{(1)}, d_{ik}^{(2)}, \ldots, d_{ik}^{(p)}\right)$, in which with $\delta = (\delta_1, \delta_2, \ldots, \delta_p)$, δ_t is calculus by Eq. (9), specifically

$$
\mu_{d_{ik}} = 1 - \prod_{t=1}^{p}\left(1 - \mu_{d_{ik}^t}\right)^{\delta_t}, \eta_{d_{ik}} = \prod_{t=1}^{p}\left(\eta_{d_{ik}}\right)^{\delta_t}, \gamma_{d_{ik}} = \prod_{t=1}^{p}\left(\gamma_{d_{ik}^t}\right)^{\delta_t}.
\tag{11}
$$

Step 3. Determine the weights of criteria
Each decision maker has different decision weights for each criterion. To determine the grouped weight of each criterion, we need to aggregate the weight of the decision makers for each criterion. Assume that weights of criteria are denoted by $W = (w_1, w_2, \ldots, w_n)$ where $w_k = \left(\mu_{w_k}, \eta_{w_k}, \gamma_{w_k}\right)$ is the weight of kth criteria $C_k (k = 1, 2, \ldots, n)$. Let $w_k^t = \left(\mu_{w_k^t}, \eta_{w_k^t}, \gamma_{w_k^t}\right)$ be a picture fuzzy number expressing criteria $C_k (k = 1, 2, \ldots, n)$ by the tth decision maker. Then $w_k = PFWA_\delta\left(w_k^1, w_k^2, \ldots, w_k^p\right)$ with $= (\delta_1, \delta_2, \ldots, \delta_p)$, δ_t is calculus by Eq. (9). Hence, we have

$$
\mu_{w_k} = 1 - \prod_{t=1}^{p}\left(1 - \mu_{w_k^t}\right)^{\delta_t}, \eta_{w_k} = \prod_{t=1}^{p}\left(\eta_{w_k^t}\right)^{\delta_t}, \gamma_{w_k} = \prod_{t=1}^{p}\left(\gamma_{w_k^t}\right)^{\delta_t}
\tag{12}
$$

Step 4. Construction of aggregated weighted picture fuzzy decision matrix with respect to criteria
The aggregated weight decision matrix $D^* = \left(d_{ik}^*\right)_{m\times n}$ where $d_{ik}^* = w_k \otimes d_{ik} = \left(\mu_{d_{ik}^*}, \eta_{d_{ik}^*}, \gamma_{d_{ik}^*}\right)$ is computed based on Eq. (3)

$$
\mu_{d_{ik}^*} = \mu_{w_k}\cdot\mu_{d_{ik}}, \eta_{d_{ik}^*} = \eta_{w_k}\cdot\eta_{d_{ik}}\ \gamma_{d_{ik}^*} = \gamma_{w_k} + \gamma_{d_{ik}} - \gamma_{w_k}\cdot\gamma_{d_{ik}}.
\tag{13}
$$

Step 5. Calculation picture fuzzy–positive ideal solution (PF-PIS) and picture fuzzy–negative ideal solution (PF-NIS)
Now, we define the PF-PIS $d_k^{max} = \left(\mu_{d_k^{max}}, \eta_{d_k^{max}}, \gamma_{d_k^{max}}\right)$ and $d_k^{min} = \left(\mu_{d_k^{min}}, \eta_{d_k^{min}}, \gamma_{d_k^{min}}\right)$ in $D^* = \left(d_{ik}^*\right)_{m\times n}$. Let G_1 be a collection of benefit criteria and G_2 be a collection cost criteria. Based on Eq. (13) we defined PF-PIS and PF-NIS as follows:

$$
\mu_{d_k^{max}} = \begin{cases} \max_i \mu_{d_{ik}^*}, ifA_i \in G_1 \\ \min_i \mu_{d_{ik}^*}, ifA_i \in G_2 \end{cases}, \gamma_{d_k^{max}} = \begin{cases} \min_I \gamma_{d_{ik}^*}, ifA_i \in G_1 \\ \max_i \gamma_{d_{ik}^*}, ifA_i \in G_2 \end{cases}
$$

$$\eta_{d_k^{max}} = \begin{cases} \max_i \{min(\eta_{d_{ik}^*}, 1 - \mu_{d_k^{max}} - \gamma_{d_k^{max}})\}, ifA_i \in G_1 \\ \min_i \eta_{d_{ik}^*}, ifA_i \in G_2 \end{cases}, \qquad (14)$$

and

$$\mu_{d_k^{min}} = \begin{cases} \min_i \mu_{d_{ik}^*}, ifA_i \in G_1 \\ \max_i \mu_{d_{ik}^*}, ifA_i \in G_2 \end{cases}, \gamma_{d_k^{min}} = \begin{cases} \max_I \gamma_{d_{ik}^*}, ifA_i \in G_1 \\ \min_i \gamma_{d_{ik}^*}, ifA_i \in G_2 \end{cases}$$

$$\eta_{d_k^{min}} = \begin{cases} \min_i \eta_{d_{ik}^*}, ifA_i \in G_1 \\ \max_i \{min(\eta_{d_{ik}^*}, 1 - \mu_{d_k^{max}} - \gamma_{d_k^{max}})\}, ifA_i \in G_2 \end{cases}, \qquad (15)$$

Step 6. Determination of the dissimilarity measure from PF-PIS and PF-NIS to each alternative.
The dissimilarity of each alternative $d_{ik}^* = \left(\mu_{d_{ik}^*}, \eta_{d_{ik}^*}, \gamma_{d_{ik}^*}\right)$ from PF-PIS $d_k^{max} = \left(\mu_{d_k^{max}}, \eta_{d_k^{max}}, \gamma_{d_k^{max}}\right)$ for $(i = 1, 2, \ldots, m)$ computed by

$$s_i^+ = \frac{1}{3m} \sum_{k=1}^n \left[\left|\mu_{d_{ik}^*} - \mu_{d_k^{max}}\right| + \left|\eta_{d_{ik}^*} - \eta_{d_k^{max}}\right| + \left|\gamma_{d_{ik}^*} - \gamma_{d_k^{max}}\right|\right] \qquad (16)$$

The dissimilarity measure of each alternative $d_{ik}^* = \left(\mu_{d_{ik}^*}, \eta_{d_{ik}^*}, \gamma_{d_{ik}^*}\right)$ from PF-NIS $d_k^{min} = \left(\mu_{d_k^{min}}, \eta_{d_k^{min}}, \gamma_{d_k^{min}}\right)$ for $(i = 1, 2, \ldots, m)$ computed by

$$s_i^- = \frac{1}{3m} \sum_{k=1}^n \left[\left|\mu_{d_{ik}^*} - \mu_{d_k^{mim}}\right| + \left|\eta_{d_{ik}^*} - \eta_{d_k^{mim}}\right| + \left|\gamma_{d_{ik}^*} - \gamma_{d_k^{min}}\right|\right] \qquad (17)$$

Step 7. Determine the closeness coefficient
The relative closeness coefficient of each alternative A_i to PF-NIS is defined as follows

$$CC_i = \frac{S_i^+}{S_i^- + S_i^+} (i = 1, 2, \ldots, m). \qquad (18)$$

Step 8. Determine the rank of alternatives
We have the ranking of the decrements of alternatives according to $CC_i (i = 1, 2, \ldots, m)$.
$A_i \succ A_j$ if only if $CC_i \geq CC_j$ for all $i, j = 1, 2, \ldots, m$.

4   Application to Supplier Selection

Assume that there is a group of four decision makers $\{DM_1, DM_2, DM_3, DM_4\}$ invited to choose the best option from five supplier alternatives $(A_i, i = 1, 2, \ldots, 5)$ with respect to four performance criteria such that Quality (C_1), Relationship closeness (C_2), Delivery

Table 1. Importance weight as linguistic variables

Linguistic terms	Picture fuzzy sets
Very important	$(0.9, 0.1, 0)$
Important	$(0.8, 0.15, 0.02)$
Medium	$(0.6, 0.15, 0.15)$
Unimportant	$(0.2, 0.15, 0.6)$
Very unimportant	$(0, 0.1, 0.9)$

performance (C_3), and Price (C_4). Their decision powers are considered as the linguistic term in which the values of a linguistic term is a picture fuzzy set that expressed in Table 1.

The decision makers use weighted form linguistic variables (in Table 2) to evaluate the performance of each criterion. The information of weights provided to the four criteria by the four decision makers presented in Table 3. The weighted values of these criteria are given by the company or organization that invited the team of experts to help them make a final assessment of the alternatives. The value of the criteria for alternatives is given by the linguistic set shown in Table 4.

Now, we apply the picture fuzzy TOPSIS method was proposed in Sect. 3 to this supplier selection problem.

Step 1. Using Eq. (9), we have the weight of decision maker with the linguistic term in the Table 5.

Step 2. By using Eq. (10), we obtained the aggregated picture fuzzy decision matrix with respect to the decision maker in Table 6.

Step 3. The weights of criteria were computed based on Eq. (12), the results are shown in Table 7.

Step 4. To construct the aggregated weighted picture fuzzy decision matrix with respect to criteria, we use the Eq. (13) and give it in Table 8.

Step 5. Calculate the picture fuzzy –positive ideal solution (PF-PIS) and picture fuzzy– negative ideal solution (PF-NIS) using Eq. (14) and Eq. (15), the result given in Table 9.

Step 6. Determination of the distance measure from PF-PIS and PF-NIS to each alternative based on Eq. (16) and Eq. (17), those shown in Table 10.

Step 7. We determine the closeness coefficient of all alternatives by using Eq. (18) and give it in Table 11.

Step 8. According to the results calculated here, the ranking order of alternatives is $A_2 \succ A_1 \succ A_3 \succ A_4 \succ A_5$ as in the last column of Table 11.

The proposed method here overcomes the limitations of subjective characteristics of decision-makers by integrating weights determined by four decision-makers. The difference between the method here and most of the other MCDMs is that the weights are also the picture fuzzy numbers aggregated from the formal language variables. The information for decision-making is integrated from the decision matrix with formal variables and defined by four decision-makers. This is a unique feature of this study.

5 Conclusion

In this paper, we introduce an intelligent method based on new dissimilarity measures between picture fuzzy sets. This proposed measure overcomes the limitation of some existing dissimilarity measures. Also, we extended the TOPSIS method which is the familiar method in MCDM problems, and apply it to supplier selection. In the future, we can study picture fuzzy TOPSIS methods as the adaptive intelligent methods using other measures or new aggregation operators; also the proposed approach can be used for dealing with decision-making problems such as personal selection in academia, project evaluation, and many other areas of management systems.

Table 2. Linguistic terms to rate the importance of alternative

Linguistic terms	Picture fuzzy sets
Extremely good (EG)/extremely high (EH)	$(1, 0, 0)$
Very very good (VVG)/very very high (VVH)	$(0.9, 0.1, 0)$
Very good (VG)/ very high (VH)	$(0.8, 0.15, 0.05)$
Good (G)/high (H)	$(0.7, 0.1, 0.15)$
Medium good (MG)/ medium high (MH)	$(0.6, 0.25, 0.1)$
Medium (M)/Fair (F)	$(0.5, 0.4, 0.05)$
Medium bad (MB)/medium low (ML)	$(0.4, 0.15, 0.35)$
Bad (B)/low(L)	$(0.3, 0.1, 0.6)$
Very bad (VB)/very low (VL)	$(0.2, 0.25, 0.5)$
Very very bad (VVB)/very very low (VVL)	$(0.1, 0.1, 0.8)$
Extremely bad (EB)/extremely low (EL)	$(0, 0, 1)$

Table 3. The importance weights of criteria.

Criteria	DM_1	DM_2	DM_3	DM_4
C_1	VI	VI	I	M
C_2	I	M	VI	I
C_3	VI	M	VI	M
C_4	M	I	M	VI

Table 4. The rating of the alternatives

Criteria	Supplier	Decision makers			
		DM_1	DM_2	DM_3	DM_4
C_1 (Quality)	A_1	G	VG	G	MG
	A_2	MG	VG	G	G
	A_3	M	VG	G	M
	A_4	G	VG	M	MG
	A_5	VVG	VG	G	MG
C_2 (Relationship closeness)	A_1	VB	MG	G	MG
	A_2	G	G	VG	MG
	A_3	MG	VG	G	G
	A_4	M	VG	G	MG
	A_5	G	VG	M	MG
C_3 (Delivery performance)	A_1	VVG	VG	G	G
	A_2	VB	MG	G	MB
	A_3	G	G	VG	M
	A_4	M	VG	G	M
	A_5	G	VG	M	MG
C_4 (Price)	A_1	VH	H	F	H
	A_2	VH	VH	F	H
	A_3	H	H	VH	ML
	A_4	H	VL	F	VL
	A_5	H	F	F	ML

Table 5. Importance of decision maker and their weights

	DM1	DM2	DM3	DM4
Linguistic term	M	VI	I	UI
δ_t	0.2678	0.225	0.2412	0.266

Table 6. The aggregated picture fuzzy decision matrix respect to decision maker

	C_1	C_2	C_3	C_4
A_1	(0.7044, 0.1398, 0.1052)	(0.584, 0.1748, 0.1542)	(0.7797, 0.1398, 0)	(0.6956, 0.1557, 0.0858)
A_2	(0.7042, 0.14, 0.1051)	(0.7063, 0.1407, 0.1033)	(0.584, 0.1748, 0.1542)	(0.7221, 0.1706, 0.067)
A_3	(0.6403, 0.2296, 0.0652)	(0.7042, 0.14, 0.1051)	(0.7063, 0.1407, 0.1033)	(0.6729, 0.1228, 0.1442)
A_4	(0.6566, 0.1953, 0.0807)	(0.6403, 0.2296, 0.0652)	(0.6403, 0.2296, 0.0652)	(0.4507, 0.2191, 0.2078)
A_5	(0.7797, 0.1398, 0)	(0.6566, 0.1953, 0.0807)	(0.6566, 0.1953, 0.0807)	(0.5423, 0.2126, 0.1126)

Table 7. The weights of criteria

Criteria	Weights
C_1	(0.8291, 0.1228, 0)
C_2	(0.8022, 0.136, 0)
C_3	(0,8025, 0.122, 0)
C_4	(0.7633, 0.1347, 0)

Table 8. The aggregated weighted picture fuzzy decision matrix

	C_1	C_2	C_3	C_4
A_1	(0.584, 0.0172, 0.1052)	(0.4685, 0.0238, 0.1542)	(0.6257, 0.0171, 0)	(0.531, 0.021, 0.0858)
A_2	(0.5839, 0.0172, 0.1051)	(0.5666, 0.0191, 0.1033)	(0.4687, 0.0213, 0.1542)	(0.5512, 0.023, 0.067)
A_3	(0.5309, 0.0282, 0.0652)	(0.0.5649, 0.019, 0.1051)	(0.5668, 0.0172, 0.1033)	(0.6729, 0.1228, 0.1442)
A_4	(0.5444, 0.024, 0.0807)	(0.5136, 0.0312, 0.0652)	(0.5138, 0.028, 0.0652)	(0.344, 0.0295, 0.2078)
A_5	(0.6464, 0.0172, 0)	(0.5267, 0.0266, 0.0807)	(0.5269, 0.0238, 0.0807)	(0.4139, 0.0286, 0.1126)

Table 9. PF-PIS and PF-NIS

Criteria	PF-PIS	PF-NIS
C_1	(0.6464, 0.0282, 0)	(0.5309, 0.0172, 0.1052)
C_2	(0.5666, 0.0312, 0.0652)	(0.4685, 0.019, 0.1542)
C_3	(0.6257, 0.028, 0)	(0.4687, 0.0171, 0.1542)
C_4	(0.344, 0.0165, 0.2078)	(0.5512, 0.0295, 0.067)

Table 10. The dissimilarity measure from PF-PIS and PF-NIS to each alternative.

Alternatives	S+	S−
A_1	0.0548	0.0447
A_2	0.0635	0.0442
A_3	0.0558	0.0482
A_4	0.047	0.0569
A_5	0.0364	0.0523

Table 11. The closeness coefficient

Alternatives	CC_i	Rank
A_1	0.5508	2
A_2	0.6008	1
A_3	0.5365	3
A_4	0.4524	4
A_5	0.4104	5

References

1. Atanassov, K.T.: Intuitionistic fuzzy sets. Fuzzy sets and Systems **20**(1), 87–96 (1986)
2. Bhutia, P.W., Phipon, R.: Application of AHP and TOPSIS method for supplier selection problem. IOSR Journal of Engineering **2**(10), 43–50 (2012)
3. Boran, F.E., Genç, S., Kurt, M., Akay, D.: A multi-criteria intuitionistic fuzzy group decision making for supplier selection with TOPSIS method. Expert Systems with Applications **36**(8), 11363–11368 (2009)
4. Cuong, B.C., Kreinovich, V.: Picture Fuzzy Sets-a new concept for computational intelligence problems. In: Information and Communication Technologies (WICT), 2013 Third World Congress on, pp. 1–6. IEEE (Dec 2013)
5. Dinh, N.V., Thao, N.X., Chau, N.M.: Some dissimilarity measures of picture fuzzy set. Fair **10**(2017), 104–109 (2017)
6. Garg, H.: Some picture fuzzy aggregation operators and their applications to multicriteria decision-making. Arabian Journal for Science and Engineering 1–16 (2017)
7. Hoa, N.D., Thong, P.H.: Some improvements of fuzzy clustering algorithms using picture fuzzy sets and applications for geographic data clustering. VNU J. Sci. Comp. Sci. Commu. Eng. **32**(3) (2017)
8. Jadidi, O., Firouzi, F., Bagliery, E.: TOPSIS method for supplier selection problem. World Academy of Science, Engineering and Technology **47**, 956–958 (2010)
9. Kavita, S.P., Kumar, S.: A multi-criteria interval-valued intuitionistic fuzzy group decision making for supplier selection with TOPSIS method. Lecture Notes in Computer Science **5908**, 303–312 (2009)
10. Khishtandar, S., Zandieh, M., & Dorri, B. (2016). A multi criteria decision making framework for sustainability assessment of bioenergy production technologies with hesitant fuzzy linguistic term sets: The case of Iran. *Renewable and Sustainable Energy Reviews.*

11. Maldonado-Macías, A., Alvarado, A., García, J.L., Balderrama, C.O.: Intuitionistic fuzzy TOPSIS for ergonomic compatibility evaluation of advanced manufacturing technology. The Int. J. Advan. Manuf. Technol. **70**(9–12), 2283–2292 (2014)
12. Pérez-Domínguez, L., Alvarado-Iniesta, A., Rodríguez-Borbón, I., Vergara-Villegas, O.: Intuitionistic fuzzy MOORA for supplier selection. Dyna **82**(191), 34–41 (2015)
13. Omorogbe, D.E.: A review of intuitionistic fuzzy topsis for supplier selection. AFRREV STECH: An Int. J. Sci. Technol. **5**(2), 91–102 (2016)
14. Singh, P.: Correlation coefficients for picture fuzzy sets. J. Intelli. Fuzzy Sys. **28**(2), 591–604 (2015)
15. Solanki, R., Gulati, G., Tiwari, A., Lohani, Q.D. A correlation based Intuitionistic fuzzy TOPSIS method on supplier selection problem. In: Fuzzy Systems (FUZZ-IEEE), 2016 IEEE International Conference on, pp. 2106–2112. IEEE (July 2016)
16. Son, L.H.: DPFCM: A novel distributed picture fuzzy clustering method on picture fuzzy sets. Expert systems with applications **42**, 51–66 (2015)
17. Son, L.H.: Generalized picture distance measure and applications to picture fuzzy clustering. Applied Soft Computing **46**(C), 284–295 (2016)
18. Son, L.H.: Measuring analogousness in picture fuzzy sets: from picture distance measures to picture association measures. Fuzzy Optimization and Decision Making 1–20 (2017).
19. Van Dinh, N., Thao, N.X., Chau, N.M.: On the picture fuzzy database: theories and application. J. Sci **13**(6), 1028–1035 (2015)
20. Wei, G.: Picture fuzzy aggregation operators and their application to multiple attribute decision making. Journal of Intelligent & Fuzzy Systems **33**(2), 713–724 (2017)
21. Yayla, A.Y., Yildiz, A., Ozbek, A.: Fuzzy TOPSIS method in supplier selection and application in the garment industry. Fibres & Textiles in Eastern Europe (2012)
22. Yildiz, A., Yayla, A.Y.: Multi-criteria decision-making methods for supplier selection: a literature review. South African J. Indu. Eng. **26**(2), 158–177 (2015)
23. Zadeh, L.A.: Fuzzy sets. Information and control **8**(3), 338–353 (1965)
24. Zeng, S., Xiao, Y.: TOPSIS method for intuitionistic fuzzy multiple-criteria decision making and its application to investment selection. Kybernetes **45**(2), 282–296 (2016)
25. Thao, N.X., Dinh, N.V.: Rough picture fuzzy set and picture fuzzy topologies. J. Comp. Sci. Cyberneti. **31**(3), 245–253 (2015)
26. Le, N.T., Van Nguyen, D., Ngoc, C.M., Nguyen, T.X.: New dissimilarity measures on picture fuzzy set and applications. J. Comp. Sci. Cyberneti. **34**(3), 219–231 (2018)
27. Nguyen, X.T., Nguyen, V.D., Nguyen, D.D.: Rough fuzzy relation on two universal sets. Int. J. Intelli. Sys. Appl. **6**(4), 49–55 (2014)
28. Thao, N.X., Cuong, B.C., Ali, M., Lan, L.H.: Fuzzy equivalence on standard and rough neutrosophic sets and applications to clustering analysis. In: Information Systems Design and Intelligent Applications, pp. 834–842. Springer, Singapore (2018)
29. Xuan Thao, N.: A new correlation coefficient of the intuitionistic fuzzy sets and its application. Journal of Intelligent & Fuzzy Systems **35**(2), 1959–1968 (2018)
30. Thao, N.X., Ali, M., Smarandache, F.: An intuitionistic fuzzy clustering algorithm based on a new correlation coefficient with application in medical diagnosis. Journal of Intelligent & Fuzzy Systems **36**(1), 189–198 (2019)
31. Van Dinh, N., Thao, N.X.: Some measures of picture fuzzy sets and their application in multi-attribute decision making. Int. J. Mathem. Sci. Comp. (IJMSC) **4**(3), 23–41 (2018)
32. Nguyen, X.T.: Evaluating water reuse applications under uncertainty: a novel picture fuzzy multi criteria decision making medthod. Int. J. Info. Eng. Electro. Bus. **10**(6), 32–39 (2018)
33. Nguyen, X.T.: Support-intuitionistic fuzzy set: a new concept for soft computing. Int. J. Intelli. Sys. Appl. **7**(4), 11–16 (2015)

A New Intelligent Computing Method for Scheduling and Crashing Projects with Fuzzy Activity Completion Times

Nguyen Hai Thanh$^{(\boxtimes)}$ ⃝iD

Faculty of Applied Sciences, VNU International School, Hanoi, Vietnam
nhthanh.ishn@isvnu.vn

Abstract. Program evaluation and review technique is a powerful computing tool in project management when the project activity completion times assume crisp values. In the era of digital economy, these time data are not always crisp, and it is necessary to develop intelligent computing methods to support the project management board when the project activity completion times assume some kinds of uncertainty, such as fuzziness or randomness. A lot of research has been recently conducted to study project scheduling and project crashing, where the completion times of project activities are fuzzy numbers. However, there is a need to improve modeling and computing aspects of Program evaluation and review technique to deal with fuzzy project activity completion times more sufficiently. This paper proposes a new intelligent computing fuzzy linear programming-based method for project scheduling, i.e. for determining project's critical activities and critical path, and for project crashing, i.e. for shortening project completion time with the minimum total crashing cost where project activity completion times are modeled by fuzzy left triangular numbers. As a result, new linear programming-based algorithm frames have been constructed to efficiently support project analysis and management with Program evaluation and review technique.

Keywords: Program evaluation and review technique · Fuzzy linear programming · Project scheduling · Project crashing

1 Introduction

Program evaluation and review technique (PERT) is a powerful computing tool in project management, in general, project scheduling and project crashing, in particular. Along with the development of big data management tools and modern computing tools, PERT is increasingly being improved in terms of modeling and scientific computing to create a powerful tool for the efficient analysis and management of complex projects with input data that include both crisp data and uncertain data, which can assume fuzziness and randomness [3, 5, 8, 9, 11, 18, 19, 22, 24]. In Vietnam, PERT is taught in undergraduate and graduate degree programs for business, management, finance, technology students, etc. [1, 12, 14, 21], and is applied in practice for implementing and managing real-world projects [4, 17]. However, most of this literature only presents classical

© The Author(s), under exclusive license to Springer Nature Switzerland AG 2022
N.-T. Nguyen et al. (Eds.): ICIT 2022, LNDECT 148, pp. 44–57, 2022.
https://doi.org/10.1007/978-3-031-15063-0_4

computational procedures based on the PERT network diagram but has not yet applied a powerful computational tool such as linear programming for project scheduling and project crashing. Recently, there have been research papers on PERT where the completion times of project activities are modeled by triangular or trapezoidal fuzzy numbers. Most of these papers only focus on project scheduling, including determining the total project completion time and critical path of the PERT network on the basis of fuzzy logic application [10, 20, 23] or applying operations with fuzzy numbers using interval numbers or α - cuts [2, 6] or using simulation to simulate different scenarios of project activity completion time data [7]. As it can be seen, these research papers have presented quite diverse methods in handling PERT with fuzzy activity completion times. However, the computational procedures are often quite complex and have not been generalized into detailed algorithms. Moreover, to my best knowledge, there are no works related to the application of fuzzy linear programming to schedule and crash projects with fuzzy activity completion times.

This paper proposes a new intelligent computing fuzzy linear programming-based method for project scheduling, i.e. for determining project's critical activities and critical path, and for crashing project's activities, i.e. for shortening total project completion time with the minimum total crashing cost where project activity completion times are modeled by fuzzy left triangular numbers. As a result, new linear programming-based algorithm frames have been constructed to efficiently support project analysis and management with PERT. The remaining sections of this paper are organized as follows: First, in Sect. 2, a linear programming approach to scheduling and crashing projects with crisp activity completion times, as it has recently been proposed by the author of this paper, will be reviewed using generalized notations. Next, in Sect. 3, a fuzzy linear programming approach will be investigated for scheduling projects with fuzzy activity completion times. Sect. 4 then will present the results of applying the fuzzy linear programming approach for crashing projects with fuzzy activity completion times. Finally, concluding observations will be made in Sect. 5.

2 Linear Programming Approach to Project Scheduling and Crashing

PERT with conventional crisp activity completion times may be referred as classical PERT. Also, PERT with fuzzy activity completion times may be referred as fuzzy PERT. Basic concepts of classical PERT can be reviewed based on the textbooks and lecture notes that are used in Vietnam for teaching undergraduate and graduate courses [1, 12, 14, 21]. In this paper, notation $A = \{1, 2, ..., n\}$ is used to indicate the set of all n project's activities. Then activities i and j are said to satisfy the precedence relation R defined in the set A if activity i is an immediate predecessor of activity j, or in other words, if activity j is an immediate successor of activity i. Also, to mean that activity i is an immediate predecessor of activity j, we simply write $(i, j) \in R$. For each activity j, denote: t_j = completion time; ES_j = earliest start time; EF_j = earliest finish time; LS_j = latest start time; LF_j = latest finish time; $pred(j) = \{i \in A | (i, j) \in R\}$ and $succ(j) = \{i \in A | (j, i) \in R\}$. Also denote T = the project completion time, which is the minimum total time required to complete the project. For an activity j, j $\in$ A, the slack

$s_j = LS_j - ES_j = LF_j - EF_j$. Activity j is called critical activity if its slack is 0, or, if $LS_j = ES_j$ and $LF_j = EF_j$. Hence, critical activities require the utmost attention in the sense that they should be conducted with a tight schedule. A project having n activities can be presented by a project network with $n + 2$ nodes, where except start node and finish node, each node presents an activity. Node i and node j are connected by an arrow if $(i, j) \in R$. Start node (node 0) is connected by an outgoing arrow with activity j when $pred(j) = \emptyset$, finish node (node $n + 1$) is connected by an ingoing arrow with activity j when $succ(j) = \emptyset$. A critical path is a path in the project network consisting of the start node, critical activities, and finish node, which are connected by corresponding arrows.

2.1 Linear Programming Approach to Project Scheduling (see [16])

Consider a project having n activities whose completion times are given. The extended set of project's activities is denoted by $A^+ = \{0, 1, 2, ..., n, n + 1\}$, where activity 0 is the start activity and activity $n + 1$ is the finish activity. The extended precedence relation in A^+ is denoted by R^+ and is defined by:

$$R^+ = R \cup \{(0, j)| j \in A \, and \, pred(j) = \emptyset\} \cup \{(i, n + 1)| i \in A \, and \, succ(i) = \emptyset\},$$

where R is the precedence relation given in A: $R = \{(i, j)| j \in A, i \in pred(j)\}$.

Denote the finish time of activity j by x_j, obviously, $EF_j \leq x_j \leq LF_j$, for $j = \overline{1, n}$. Set, by convention, $x_0 = 0$ and $x_{n+1} = T$. Then the following linear programming problem (LPP) can be considered to find activity finish times and the shortest project completion time T.

Problem 1: Min $z = x_{n+1}$, subject to: $x_j - x_i \geq t_j$ for $(i, j) \in R^+$; $x_0 = 0$, $x_j \geq 0$ and integer for $j = \overline{1, n + 1}$; where t_j is the completion time of activity j, $j = \overline{1, n}$, and, by convention, $t_{n+1} = 0$.

To find activity earliest finish times EF_j, $j = \overline{1, n}$, the following LPP shall be considered:

Problem 2: Min $z = x_1 + x_2 + ... + x_n$, subject to the same constraints of Problem 1 and an additional constraint: $x_{n+1} = T$; where T is the shortest project completion time as found in Problem 1.

Activity latest finish times LF_j, $j = \overline{1, n}$, shall be found by solving the following LPP:

Problem 3: Max $z = x_1 + x_2 + ... + x_n$, subject to the same constraints of Problem 2.

Lemma: The optimal solution of Problem 2 provides the activity earliest finish times EF_j, for $j = \overline{1, n}$. The optimal solution of Problem 3 provides the activity latest finish times LF_j, for $j = \overline{1, n}$. (For proof see [16]).

Table 1. Project's activities and their start and finish times

Activity	Im. pred. activities	Completion time (weeks)	ES	EF	LS	LF	Slack	Note
(1)		3	0	3	0	3	0	Critical
(2)	1	3	3	6	6	9	3	
(3)	1	2	3	5	3	5	0	Critical
(4)	2	3	6	9	9	12	3	
(5)	3	7	5	12	5	12	0	Critical
(6)	2, 3	3	6	9	15	18	9	
(7)	4, 5	6	12	18	12	18	0	Critical
(8)	3	2	5	7	16	18	11	

Illustrative example 1: For the input data of a project as given in the first three columns of Table 1, the optimal solution of Problem 1 is $(x_0, x_1, x_2, x_3, x_4, x_5, x_6, x_7, x_8, x_9) = (0, 3, 9, 5, 12, 12, 12, 18, 7, 18)$. The optimal solutions of Problem 2 and Problem 3 are shown in 5th and 7th columns. Activity earliest and latest start times can be calculated by: $ES_j = EF_j - t_j$, $LS_j = LF_j - t_j$, for $j = \overline{1, 8}$, and are shown in 4th and 6th columns. The total time to complete the project $T = 18$ and critical path is $(1) \rightarrow (3) \rightarrow (5) \rightarrow (7)$.

2.2 Linear Programming Approach to Project Crashing (see [16])

Illustrative example 2: Reconsider Illustrative example 1. The completion time for this project is 18 weeks. Which activities should be crashed, and by how many weeks, to complete the project in 16 weeks with a minimum total crash cost?

To answer this question, it is "a must" to ask for the following information for each activity j, $j = \overline{1, 8}$: i) Activity cost under the normal or expected activity time; ii) Time to complete the activity under maximum crashing (i.e., the shortest possible activity time); iii) Activity cost under maximum crashing. It is assumed that the normal time to complete an activity, t_j, which can be met at a normal cost, c_j, can be crashed to a reduced time, t_j', under maximum crashing for an increased cost, c_j'. Then the activity j's maximum time reduction, r_j, is calculated by: $r_j = t_j - t_j'$. It is assumed that its cost per unit reduction, k_j, is linear and can be calculated by: $k_j = (c_j' - c_j)/r_j$. Suppose that all these necessary input data are known and summarized in Table 2 wherein it is assumed $k_j = 0$ if $r_j = 0$.

For generalization, consider a project with extended set of activities A^+ wherein precedence relation R^+ is defined as discussed in Sect. 2.1. All input data for project crashing: $t_j, c_j, t_j', c_j', r_j, k_j$ are given, $j = \overline{1, n}$. Denote the finish time of activity j by x_j, $j = \overline{1, n}$, and by convention, set $x_0 = 0$ and $x_{n+1} = T'$, which is the project completion time after crashing. Also denote y_j the amount of time activity j is crashed, for $j = \overline{1, n}$ and, by convention, set $y_{n+1} = 0$. To decide what activities to crash and how much to crash activity times, the following LPP is considered:

Table 2. Normal and crash times and costs

Activity	Normal time t_j	Normal cost c_j	Crash time t_j'	Crash cost c_j'	Time reduction r_j	Crash cost/day k_j
(1)	3	20	2	30	1	10
(2)	3	100	2	150	1	50
(3)	2	50	2	50	0	0
(4)	3	250	3	250	0	0
(5)	7	180	5	300	2	60
(6)	3	30	2	40	1	10
(7)	6	100	4	140	2	20
(8)	2	150	2	150	0	0

Problem 4: Min $z = k_1 y_1 + k y_2 + \ldots + k_n y_n$, subject to: $x_j - x_i + y_j \geq t_j$ for $(i, j) \in R^+$; $y_j \leq r_j$ for $j = \overline{1, n}$; $x_0 = 0$, $x_{n+1} = T'$, $y_{n+1} = 0$, x_j & $y_j \geq 0$ and integer for $j = \overline{1, n+1}$; wherein t_j is the completion time of activity j, $j = \overline{1, n}$, and, $t_0 = 0$, $t_{n+1} = 0$.

Illustrative example 2 (contd.): With the data given in Table 2 and T' = 16, the optimal solution of Problem 4 is $(x_0, x_1, x_2, x_3, x_4, x_5, x_6, x_7, x_8, x_9, y_1, y_2, y_3, y_4, y_5, y_6, y_7, y_8) = (0, 2, 5, 4, 11, 11, 8, 16, 6, 16, 1, 0, 0, 0, 0, 0, 1, 0)$, which gives $y_1 = 1$, $y_2 = 0$, $y_3 = 0$, $y_4 = 0$, $y_5 = 0$, $y_6 = 0$, $y_7 = 1$, $y_8 = 0$, and $z_{min} = 30$. It is meant that to crash the total completion time of the project by 2 weeks with the minimum total crash cost of 30, each of the two activities 2 and 7 should be crashed by 1 week. Now, with the revised activity times: $t_1 = 2$, $t_2 = 3$, $t_3 = 2$, $t_4 = 3$, $t_5 = 7$, $t_6 = 3$, $t_7 = 5$, $t_8 = 2$, the approach as described in Problem 1, Problem 2 and Problem 3 can be considered to calculate the activity earliest and latest start and finish times and to find the critical path $(1) \rightarrow (3) \rightarrow (5) \rightarrow (7)$.

3   Fuzzy PERT: Finding Critical Activities and Critical Path

3.1   Concepts of Flexible Fuzzy Linear Programming

Consider the following classical LPP:

Problem 5: $Min\, z = \sum_{j=1}^{n} c_j x_j$, subject to : $\sum_{j=1}^{n} a_{ij} x_j \geq b_i (i = \overline{1, m})$; $x_j \geq 0 (j = \overline{1, n})$.

Fuzzy flexible constraints: Consider i^{th} constraint: $\sum_{j=1}^{n} a_{ij} x_j \geq b_i$, then a membership function may be created regarding decision vector $\mathbf{x} = (x_1, x_2, \ldots, x_n) \in R^n$ to indicate the level with which $\mathbf{x}$ belongs to the fuzzy set C_i of all decision vectors essentially satisfying the fuzzy flexible constraint [13, 15]:

$$\mu_{Ci}(x) = \begin{cases} 1 & \text{if } \sum_{j=1}^{n} a_{ij} x_j \geq b_i \\ \frac{\sum_{j=1}^{n} a_{ij} x_j - (b_i - p_i)}{p_i} & \text{if } b_i > \sum_{j=1}^{n} a_{ij} x_j \geq b_i - p_i \\ 0 & \text{if } b_i - p_i > \sum_{j=1}^{n} a_{ij} x_j \end{cases}$$

wherein p_i is some (positive) left spread of b_i, as specified by the decision maker.

The meaning of this membership is to essentially satisfy the crisp constraint $\sum_{j=1}^{n} a_{ij}x_j \geq b_i$ by considering a fuzzy flexible constraint $\sum_{j=1}^{n} a_{ij}x_j \tilde{\geq} \tilde{b}_i$ where $\tilde{b}_i$ is a fuzzy left triangular number or a fuzzy left threshold of the type $(b_i, p_i)_{LL}$, b_i is the reference point and p_i is the left spread for the fuzzy number $\tilde{b}_i$. In fact, the flexible fuzzy inequality is a proposition that is satisfied at a membership level between 0 and 1, i.e. its truth value may be any value between 0 and 1.

Problem 5 can now be rewritten as follows:

Problem 6: $Minz = \sum_{j=1}^{n} c_j x_j$, subject to : $\sum_{j=1}^{n} a_{ij}x_j \tilde{\geq} \tilde{b}_i (i = \overline{1, m}); x_j \geq 0 (j = \overline{1, n})$.

Fuzzy flexible goal: A fuzzy flexible goal for objective function $Minz = \sum_{j=1}^{n} c_j x_j$ in Problem 6 may be constructed as follows. First, the upper limit z_u must be found by solving the following LPP (*Step 1*):

$$Minz = \sum_{j=1}^{n} c_j x_j, \text{ subject to : } \sum_{j=1}^{n} a_{ij}x_j \geq b_i (i = \overline{1, m}); x_j \geq 0 (j = \overline{1, n}).$$

Then, the lower limit z_l can be found by solving the following LPP (*Step 2*):

$$Minz = \sum_{j=1}^{n} c_j x_j, \text{ subject to : } \sum_{j=1}^{n} a_{ij}x_j \geq b_i - p_i (i = \overline{1, m}); x_j \geq 0 (j = \overline{1, n}).$$

Based on the values z_l and z_u, the fuzzy set G w.r.t the objective function is determined by the following membership function [13, 15]:

$$\mu_G(x) = \begin{cases} 1 & if \ \sum_{j=1}^{n} c_j x_j \leq z_l \\ \frac{\sum_{j=1}^{n} c_j x_j - z_u}{z_l - z_u} & if \ z_l < \sum_{j=1}^{n} c_j x_j \leq z_u \\ 0 & if \ z_u < \sum_{j=1}^{n} c_j x_j \end{cases}$$

The meaning of this membership is to essentially minimize the value of the objective function $\sum_{j=1}^{n} c_j x_j$. In other words, the fuzzy flexible goal as specified above can be understood as a fuzzy flexible constraint $\sum_{j=1}^{n} c_j x_j \tilde{\leq} \tilde{z}$ where $\tilde{z}$ is a fuzzy right triangular number or a fuzzy right threshold of the type $(z_l, z_u - z_l)_{RR}$, z_l is the reference point and $z - z_l$ is the right spread of $\tilde{z}$.

Having denoted C_i the fuzzy set determined by the ith fuzzy flexible constraints, $i = 1, 2, ..., m$, and G the fuzzy set determined by the fuzzy flexible goal, now the following fuzzy optimization problem is formulated [13, 15] to find the optimal solution $\mathbf{x} = (x_1, x_2, ..., x_n)$ of Problem 6:

Problem 7: $Max \ \lambda = \mu_D(\mathbf{x}) = min \ \{\mu_G(x), \mu_{C1}(x), \mu_{C2}(x), ..., \mu_{Cm}(x)\}$, subject to: $(\mathbf{x}, \mu_D(\mathbf{x})) \in D = (G) \cap (C_1) \cap (C_2) \cap ... \cap (C_m)$; where $\mathbf{x} = (x_1, x_2, ..., x_n)$ and D is the fuzzy intersection of fuzzy sets $G, C_1, C_2, ..., C_m$.

This problem is equivalent to the following LPP (*Step 3*):

Problem 8: *Max* λ, *subject to:* $\lambda(z_l - z_u) - \sum_{j=1}^{n} c_j x_j \geq -z_u$; $-\lambda p_i + \sum_{j=1}^{n} a_j x_{ij} \geq b_i - p_i (i = \overline{1, m})$; $0 \leq \lambda \leq 1$; $x_j \geq 0 (j = \overline{1, n})$.

The three step solution procedure for solving fuzzy flexible LPPs

Note that to solve Problem 6, it is necessary to perform *the three step solution procedure* consisting of Step 1 (to find z_u), Step 2 (to find z_l) and Step 3 (Problem 8).

3.2 Calculating the Project Completion Time with Fuzzy Activity Completion Times

For finding the project completion time when activity completion times are fuzzy left triangular numbers or fuzzy left thresholds of the type $\tilde{t}_j = (b_j, p_j)_{LL}$, $j = \overline{1, n}$, the following fuzzy LPP can be considered:

Problem 9: Min $z = x_{n+1}$, subject to: $x_j - x_i \geq \tilde{t}_j$ for $(i, j) \in R^+$; $x_0 = 0$, $x_j \geq 0$ and integer for $j = \overline{1, n+1}$; where $x_j =$ finish time for activity j, $\tilde{t}_j$ is the fuzzy completion time of activity j, $j = \overline{1, n}$, and, by convention, $\tilde{t}_{n+1} = (0, 0)_{LL}$.

Illustrative example 1 (extended): Applying the Fuzzy Delphy method, a consensus of expert opinions about activity completion times has been derived as summarized in Table 3.

In Table 3, the fact, that 3 left spreads (for the fuzzy completion times of activities 3, 4 and 8) have value of 0, can be understood as if they assume a positive small value ε. Also, fuzzy completion time $(b_j, p_j)_{LL}$ for activity j, for $j = \overline{1, 8}$, is modeled by a fuzzy left triangular number or a fuzzy left threshold. For example, consider the fuzzy completion time for activity 1: $(b_1, p_1)_{LL} = (3, 1)_{LL}$, it is meant that under normal conditions activity 1 can be completed within $b_1 = 3$ weeks and, under most optimistic conditions, it can be completed within $b_1 - p_1 = 2$ weeks. So the fuzziness inherent in this threshold represents vagueness in the project manager's evaluation of the completion time for activity 1.

Table 3. Fuzzy activity completion times

Activity	Fuzzy completion time $\tilde{t}_j =$ $(b_j, p_j)_{LL}$		Activity	Fuzzy completion time $\tilde{t}_j =$ $(b_j, p_j)_{LL}$	
	b_j	p_j		b_j	p_j
(1)	3	1	(5)	7	2
(2)	3	1	(6)	3	1
(3)	2	0	(7)	6	2
(4)	3	0	(8)	2	0

With the precedence relation as specified in Table 1 and the data given in Table 3, Problem 9 assumes the form:

Problem 9a: Min z $= x_9$, subject to: $x_1 - x_0 \overset{\sim}{\geq} (3,1)_{LL}$; $x_2 - x_1 \overset{\sim}{\geq} (3,1)_{LL}$; $x_3 - x_1 \overset{\sim}{\geq}$ $(2,0)_{LL}$; $x_4 - x_2 \overset{\sim}{\geq} (3,0)_{LL}$; $x_5 - x_3 \overset{\sim}{\geq} (7,2)_{LL}$; $x_6 - x_2 \overset{\sim}{\geq} (3,1)_{LL}$; $x_6 - x_3 \overset{\sim}{\geq} (3,1)_{LL}$; $x_7 - x_4 \overset{\sim}{\geq} (6,2)_{LL}$; $x_7 - x_5 \overset{\sim}{\geq} (6,2)_{LL}$; $x_8 - x_3 \overset{\sim}{\geq} (2,0)_{LL}$; $x_9 - x_6 \overset{\sim}{\geq} (0,0)_{LL}$; $x_9 - x_7 \overset{\sim}{\geq}$ $(0,0)_{LL}$; $x_9 - x_8 \overset{\sim}{\geq} (0,0)_{LL}$; $x_0 = 0$, $x_j \geq 0$ and integer for $j = \overline{1, 9}$; wherein $x_j =$ finish time of activity for $j = \overline{1, 8}$, and x_9 is the project completion time.

Now, it is needed to apply *the three step solution procedure* to solve this LPP. First, to find the value $z_u = 18$, and $z_l = 13$, the following two LPPs can be considered:

Problem 9b: *(Step 1)* . Min z $= x_9$, subject to: $x_1 - x_0 \geq 3$; $x_2 - x_1 \geq 3$; $x_3 - x_1 \geq 2$; $x_4 - x_2 \geq 3$; $x_5 - x_3 \geq 7$; $x_6 - x_2 \geq 3$; $x_6 - x_3 \geq 3$; $x_7 - x_4 \geq 6$; $x_7 - x_5 \geq 6$; $x_8 - x_3 \geq$ 2; $x_9 - x_6 \geq 0$; $x_9 - x_7 \geq 0$; $x_9 - x_8 \geq 0$; $x_0 = 0$, $x_j \geq 0$ and integer for $j = \overline{1, 9}$.

Problem 9c: *(Step 2)*. Min z $= x_9$, subject to: $x_1 - x_0 \geq 2$; $x_2 - x_1 \geq 2$; $x_3 - x_1 \geq 2$; $x_4 - x_2 \geq 3$; $x_5 - x_3 \geq 5$; $x_6 - x_2 \geq 2$; $x_6 - x_3 \geq 2$; $x_7 - x_4 \geq 4$; $x_7 - x_5 \geq 4$; $x_8 - x_3 \geq 2$; $x_9 - x_6 \geq 0$; $x_9 - x_7 \geq 0$; $x_9 - x_8 \geq 0$; $x_0 = 0$, $x_j \geq 0$ and integer for $j = \overline{1, 9}$.

Since $z_u = 18$ and $z_l = 13$, $z_l - z_u = -5$. Applying Problem 8 and using the data given in Table 3, the following LPP can be solved to find finish times for activity j, $j = \overline{1, 8}$:

Problem 9d: *(Step 3)*. Max z $= x_{10}$, subject to: $-x_9 - 5x_{10} \geq -18$; $x_1 - x_0 - x_{10} \geq 2$; $x_2 - x_1 - x_{10} \geq 2$; $x_3 - x_1 \geq 2$; $x_4 - x_2 \geq 3$; $x_5 - x_3 - 2x_{10} \geq 5$; $x_6 - x_2 - x_{10} \geq 2$; $x_6 - x_3 - x_{10} \geq 2$; $x_7 - x_4 - 2x_{10} \geq 4$; $x_7 - x_5 - 2x_{10} \geq 4$; $x_8 - x_3 \geq 2$; $x_9 - x_6 \geq 0$; $x_9 - x_7 \geq 0$; $x_9 - x_8 \geq 0$; $x_0 = 0$, $0 \leq x_{10} \leq 1$; $x_j \geq 0$ and integer for $j = \overline{1, 9}$; where x_{10} is identical to variable λ in Problem 8.

The optimal solution of this LPP is $\mathbf{x} = (x_0, x_1, x_2, x_3, x_4, x_5, x_6, x_7, x_8, x_9, x_{10}) =$ $(0, 3, 6, 5, 10, 11, 9, 16, 7, 16, 0.4)$. The total project completion time is $x_9 = T = 16$ and $x_{10} = \lambda = \mu_D(\mathbf{x^o}) = 0.4$ is the maximum level of membership function describing the level with which decision vector $\mathbf{x^o} = (x_0, x_1, x_2, x_3, x_4, x_5, x_6, x_7, x_8, x_9)$ belongs to the fuzzy set of decision vectors / solutions of Problem 9a.

To find the optimal solution $\mathbf{x^o} = (x_0, x_1, x_2, x_3, x_4, x_5, x_6, x_7, x_8, x_9)$ of fuzzy LPP, Problem 9a, so that $\mu_D(\mathbf{x^o}) = 0.4$ and x_9 assumes the smallest possible value, we now solve the following LPP:

Problem 9e: Min z $= x_9$, subject to the same constraints as specified in Problem 9d and an additional constraint: $x_{10} = 0.4$.

In this example, the optimal solution of Problem 9e is fortunately the same optimal solution of Problem 9d as obtained above. In particular, $x_9 = 16$ and $\mu_D(\mathbf{x^o}) = 0.4$ remain the same as obtained when solving Problem 9d.

3.3 Finding Critical Activities and Critical Path with Fuzzy Activity Completion Times

Using the approach as shown in Problem 2 and Problem 3, the earliest and latest finish times of activity j, for $j = 1, 2, \ldots, n$, can be found.

Illustrative example 1 (extended, contd.): We now shall use the data as given in Table 3 to find the earliest finish time EF_j and latest finish time LF_j of activity j, for j = 1, 2, ..., 8, by solving the following 02 LPPs, respectively:

Problem 9f: Min $z = x_1 + x_2 + ... + x_8$, subject to the same constraints as specified in Problem 9e and an additional constraint: $x_9 = 16$; and

Problem 9g: Max $z = x_1 + x_2 + ... + x_8$, subject to the same constraints as specified in Problem 9f.

The optimal solution of Problem 9f is $\mathbf{x} = (x_0, x_1, x_2, x_3, x_4, x_5, x_6, x_7, x_8, x_9, x_{10})$ = (0, 3, 6, 5, 9, 11, 9, 16, 7, 16, 0.4), $x_9 = 16$ is the completion time and $x_{10} = \mu_D(\mathbf{x^o})$ = 0.4. The optimal solution of Problem 9 g is $\mathbf{x} = (x_0, x_1, x_2, x_3, x_4, x_5, x_6, x_7, x_8, x_9,$ $x_{10}) = (0, 3, 8, 5, 11, 11, 16, 16, 16, 16, 0.4)$, $x_9 = 16$ is the completion time and x_{10} = $\mu_D(\mathbf{x^o}) = 0.4$. Since $EF_j = LF_j$, for j = 1, 3, 5, 7, activities 1, 3, 5, 7 are critical, and the critical time path is (1) → (3) → (5) → (7). Crisp / defuzzified completion times for these activities may be calculated and understood as reference time values when conducting the project: $t_1 = x_1 = 3$, $t_3 = x_3 - x_1 = 2$, $t_5 = x_5 - x_3 = 6$, $t_7 = x_7 - x_5 = 5$. Based on the precedence relation, crisp / defuzzified completion times for non-critical activities may also be calculated using formula: $t_j = Min_{i \in pred(j)}\{x_j - x_i\}$, where x_j = EF_j for j = $\overline{1, 8}$. Hence $t_2 = Min\{x_2 - x_1\} = 3$, $t_4 = Min\{x_4 - x_2\} = 3$, $t_6 = Min\{x_6 - x_2, x_6 - x_3\} = 3$, $t_8 = Min\{x_8 - x_3\} = 2$. The results of the above computing are presented in Table 4.

Table 4. Project's activities and their integer start and finish times with fuzzy activity completion times

Activity	ES	EF	LS	LF	t	Slack	Activity	ES	EF	LS	LF	t	Slack
(1)	0	3	0	3	3	0	(5)	5	11	5	11	6	0
(2)	3	6	5	8	3	2	(6)	6	9	13	16	3	7
(3)	3	5	3	5	2	0	(7)	11	16	11	16	5	0
(4)	6	9	8	11	3	2	(8)	5	7	14	16	2	9

Note 1: Obviously, optimal solutions of Problem 9f and Problem 9g, as obtained above, are optimal solutions of Problem 9e. Therefore, these two optimal solutions and the optimal solution of Problem 9e, are *optimal solutions* of Problem 9d and of the fuzzy LPP, Problem 9a. All these three optimal solutions provide decision vectors having the maximum level of membership function $\mu_D(.)$ = 0.4 describing the level with which they belong to the fuzzy set of decision vectors / solutions of Problem 9a.

4　Fuzzy PERT: Crashing the Total Project Completion Time

4.1　Crashing Fuzzy Activity Completion Time Procedure

To decide where and how much to crash activity completion times for a project, the following fuzzy LPP (a fuzzified version of Problem 4) can be considered:

Problem 10: Min $z = k_1y_1 + k_2y_2 + \ldots + k_ny_n$, subject to: $x_j - x_i + y_j \geq \tilde{t}_j$ for (i, j) $\in R^+$; $y_j \leq r_j$, for $j = \overline{1, n}$; $x_0 = 0$, $x_{n+1} = T'$, $y_{n+1} = 0$, $x_j \geq 0$ and integer, $y_j \geq 0$ and integer, for $j = \overline{1, n+1}$; where x_j = finish time for activity j, y_j = amount of time activity j to be crashed, k_j = given cost per unit reduction, r_j = given maximum time reduction, $\tilde{t}_j$ = fuzzy completion time of activity j, for $j = \overline{1, n}$, and, by convention, $\tilde{t}_{n+1}$ $= (0, 0)_{LL}$. Also, T' = the project completion time after crashing.

Illustrative example 2 (extended): Consider activity fuzzy completion times $\tilde{t}_j =$ $(b_j, p_j)_{LL}$ for $j = \overline{1, 8}$ as given in Table 3. As it is known, the three decision vectors / optimal solutions of fuzzy LPP, Problem 9a (see Note 2), provide the project completion time $x_9 = T = 16$. Now, the question is which activities should be crashed and by how many weeks to complete the project within, say, 14 weeks, so that the total crash cost is minimum. With the data as given in Table 3 and Table 4, Problem 10 becomes:

Problem 10a: Min $z = 10y_1 + 50y_2 + 60y_5 + 10y_6 + 20y_7$, subject to: $x_1 - x_0 + y_1$ $\gtrsim (3,1)_{LL}$; $x_2 - x_1 + y_2 \gtrsim (3,1)_{LL}$; $x_3 - x_1 + y_3 \gtrsim (2,0)_{LL}$; $x_4 - x_2 + y_4 \gtrsim (3,0)_{LL}$; x_5 $- x_3 + y_5 \gtrsim (7,2)_{LL}$; $x_6 - x_2 + y_6 \gtrsim (3,1)_{LL}$; $x_6 - x_3 + y_6 \gtrsim (3,1)_{LL}$; $x_7 - x_4 + y_7 \gtrsim$ $(6,2)_{LL}$; $x_7 - x_5 + y_7 \gtrsim (6,2)_{LL}$; $x_8 - x_3 + y_8 \gtrsim (2,0)_{LL}$; $x_9 - x_6 + y_9 \gtrsim (0,0)_{LL}$; $x_9 - x_7$ $+ y_9 \gtrsim (0,0)_{LL}$; $x_9 - x_8 + y_9 \gtrsim (0,0)_{LL}$; $y_1 \leq 1$; $y_2 \leq 1$; $y_3 = 0$; $y_4 = 0$; $y_5 \leq 2$; $y_6 \leq$ 1; $y_7 \leq 2$; $y_8 = 0$; $x_0 = 0$, $x_9 = 14$, $y_9 = 0$, $x_j \geq 0$ and integer, $y_j \geq 0$ and integer, for j $= \overline{1, 9}$; wherein x_j = finish time for activity j, and y_j = amount of time activity j to be crashed, for $j = \overline{1, 8}$.

It is necessary to apply *the three-step solution procedure* to solve this fuzzy LPP. By performing Step 1 (to solve a LPP, called problem 10b) and Step 2 (to solve a LPP, called problem 10c), it can be found that $z_u = 110$ and $z_l = 0$. Then, Step 3 can be applied for finding an optimal solution with the minimum total crash cost by solving the following LPP:

Problem 10d: *(Step 3):* Max $z = x_{10}$, subject to: $- (10y_1 + 50y_2 + 60y_5 + 10y_6 +$ $20y_7) - 110x_{10} \geq -110$; $x_1 - x_0 + y_1 - x_{10} \geq 2$; $x_2 - x_1 + y_2 - x_{10} \geq 2$; $x_3 - x_1 + y_3$ ≥ 2; $x_4 - x_2 + y_4 \geq 3$; $x_5 - x_5 + y_5 - 2x_{10} \geq 5$; $x_6 - x_2 + y_6 - x_{10} \geq 2$; $x_6 - x_3 + y_6$ $- x_{10} \geq 2$; $x_7 - x_4 + y_7 - 2x_{10} \geq 4$; $x_7 - x_5 + y_7 - 2x_{10} \geq 4$; $x_8 - x_3 + y_8 \geq 2$; $x_9 - x_6$ $+ y_9 \geq 0$; $x_9 - x_7 + y_9 \geq 0$; $x_9 - x_8 + y_9 \geq 0$; $y_1 \leq 1$; $y_2 \leq 1$; $y_3 = 0$; $y_4 = 0$; $y_5 \leq 2$; $y_6 \leq 1$; $y_7 \leq 2$; $y_8 = 0$; $x_0 = 0$, $x_9 = 14$, $y_9 = 0$, $x_j \geq 0$ and integer, $y_j \geq 0$ and integer, for $j = \overline{1, 9}$.

The optimal solution of the above LPP is $\mathbf{x} = (x_0, x_1, x_2, x_3, x_4, x_5, x_6, x_7, x_8, x_9,$ $x_{10}, y_1, y_2, y_3, y_4, y_5, y_6, y_7, y_8) = (0, 2, 6, 4, 10, 10, 8, 14, 6, 14, 0.5, 1, 0, 0, 0, 0,$ 1, 1, 0). The total project completion time is $x_9 = 14$, and $x_{10} = \mu_D(\mathbf{x^o}) = 0.5$ is the maximum level of membership function with which decision vector $\mathbf{x^o} = (x_1, x_2, x_3,$ $x_4, x_5, x_6, x_7, x_8, x_9, y_1, y_2, y_3, y_4, y_5, y_6, y_7, y_8) = (2, 6, 4, 10, 10, 8, 14, 6, 14, 1, 0,$ 0, 0, 0, 1, 1, 0) belongs to the fuzzy set of decision vectors / solutions of Problem 10a. After the crashing, the revised activity completion times are: $t_1 = 2$, $t_2 = 3$, $t_3 = 2$, t_4 $= 3$, $t_5 = 7$, $t_6 = 3$, $t_7 = 5$, $t_8 = 2$, and the revised project completion time is T' = 14. The minimum total crash cost is $10y_1 + 50y_2 + 60y_5 + 10y_6 + 20y_7 = 10 \times 1 + 50$ $\times 0 + 60 \times 0 + 10 \times 1 + 20 \times 1 = 40$.

To find the optimal solution of fuzzy LPP, Problem 10a, so that $\mu_D(x^o) = 0.5$ and x_9 assumes the smallest possible value, we now solve the following LPP:

Problem 10e: Min $z = x_9$, subject to the same constraints as specified in Problem 10d and additional constraints: $y_1 = 1$, $y_2 = 0$, $y_3 = 0$, $y_4 = 0$, $y_5 = 0$, $y_6 = 1$, $y_7 = 1$, $y_8 = 0$ and $x_{10} = 0.5$.

In this example, the optimal solution of Problem 10e is different from the optimal solution of Problem 10d as obtained above: $x = (x_1, x_2, x_3, x_4, x_5, x_6, x_7, x_8, x_9, x_{10}, y_1, y_2, y_3, y_4, y_5, y_6, y_7, y_8) = (2, 6, 4, 10, 10, 8, 14, 6, 14, 0.5, 1, 0, 0, 0, 0, 1, 1, 0)$, but $x_9 = 14$ and $x_{10} = \mu_D(x^o) = 0.5$ remain the same as obtained when solving Problem 9d.

4.2   Finding Critical Path with Fuzzy Activity Times Under Crashing

Using the approach as shown in Problem 2 and Problem 3, the earliest and latest finish times of activity j, for $j = \overline{1, n}$, can be found.

Illustrative example 2 (extended, contd.): We now shall use the data as given in Table 2 and Table 3 to find the earliest finish time EF_j and latest finish time LF_j of activity j under crushing, for $j = \overline{1, 8}$, by solving the following 02 LPPs, respectively:

Problem 10f: Min $z = x_1 + x_2 + x_3 + x_4 + x_5 + x_6 + x_7 + x_8$, subject to the same constraints of Problem 10e and the additional constraint: $x_9 = 14$; and

Problem 10g: Max $z = x_1 + x_2 + x_3 + x_4 + x_5 + x_6 + x_7 + x_8$, subject to the same constraints of Problem 10f.

Table 5. Project's activities and their integer start and finish times under crashing with fuzzy activity completion times

Activity	ES	EF	LS	LF	t	Slack	Activity	ES	EF	LS	LF	t	Slack
(1)	0	2	0	2	2	0	(5)	5	10	5	10	6	0
(2)	3	5	5	7	3	2	(6)	6	7	13	14	2	7
(3)	3	4	3	4	2	0	(7)	11	14	11	14	4	0
(4)	6	8	8	10	3	2	(8)	5	6	14	14	2	9

The optimal solution of Problem 10f is: $x_1 = 2$, $x_2 = 5$, $x_3 = 4$, $x_4 = 8$, $x_5 = 10$, $x_6 = 7$, $x_7 = 14$, $x_8 = 6$, $x_9 = 14$, $x_{10} = 0.5$, $y_1 = 1$, $y_2 = 0$, $y_3 = 0$, $y_4 = 0$, $y_5 = 0$, $y_6 = 1$, $y_7 = 1$, $y_8 = 0$, and $x_{10} = \mu_D(x) = 0.5$. The optimal solution of Problem 10 g is: $x_1 = 2$, $x_2 = 7$, $x_3 = 4$, $x_4 = 10$, $x_5 = 10$, $x_6 = 14$, $x_7 = 14$, $x_8 = 14$, $x_9 = 14$, $x_{10} = 0.5$, $y_1 = 1$, $y_2 = 0$, $y_3 = 0$, $y_4 = 0$, $y_5 = 0$, $y_6 = 1$, $y_7 = 1$, $y_8 = 0$. The total project completion time is $x_9 = 14$, and $x_{10} = \mu_D(x) = 0.5$. Since $EF_j = LF_j$, for $j = 1, 3, 5, 7$, activities 1, 3, 5, 7 are critical, and the critical time path is $(1) \rightarrow (3) \rightarrow (5)$

$\rightarrow$ (7). Crisp / defuzzified completion times for these activities are: $t_1 = x_1 = 2$, $t_3 = x_3 - x_1 = 2$, $t_5 = x_5 - x_3 = 6$, $t_7 = x_7 - x_5 = 4$. Based on the precedence relation, crisp / defuzzified completion times for non-critical activities can be calculated using formula: $t_j = Min_{i \in pred(j)}\{x_j - x_i\}$; where $x_j = EF_j$ for $j = \overline{1, 8}$. Hence $t_2 = Min\{x_2 - x_1\} = 3$, $t_4 = Min\{x_4 - x_2\} = 3$, $t_6 = Min\{x_6 - x_2, x_6 - x_3\} = 2$, $t_8 = Min\{x_8 - x_3\} = 2$. The results of the above computing are presented in Table 5.

Note 2: Obviously, optimal solutions of Problem 10f and Problem 10g, as obtained above, are optimal solutions of Problem 10e. Therefore, these two optimal solutions and the optimal solution of Problem 10e, are *optimal solutions* of Problem 10d and of the fuzzy LPP, Problem 10a. All these three optimal solutions / decision vectors have level of membership function $\mu_D(.) = 0.5$ describing the level with which they belong to the fuzzy set of decision vectors / solutions of Problem 10a.

5 Concluding Observations

This paper has reviewed some basic concepts of classical PERT and a linear programming approach, as it has recently been proposed, to scheduling and crashing projects with crisp activity completion times using generalized notations. The main result of the research as described in this paper is the proposal of a new intelligent computing fuzzy linear programming-based method for scheduling and crashing projects with fuzzy activity completion times which are modeled by fuzzy left triangular numbers.

New linear programming-based algorithm frames have been constructed to efficiently support project analysis and management with PERT. The series of LPPs (problems 9a, 9b, 9c, 9d, 9e, 9f, 9 g), as proposed in Sect. 3, is a linear programming-based algorithm frame for scheduling projects with fuzzy activity completion times, that is, for finding the total project completion time, critical path, and critical activities. Similarly, the series of LPPs (problems 10a, 10b, 10c, 10d, 10e, 10f, 10 g), as proposed in Sect. 4, is a linear programming-based algorithm frame for crashing projects with fuzzy activity completion times, that is, for finding critical path with fuzzy activity completion times under crashing to shorten the total project completion time with the minimum total crashing cost. Based on these linear programming-based algorithm frames, an intelligent software package may be created to support project management board.

The research has also shown that linear programming is an immensely powerful modeling and computational tool for project analysis and management. Especially, fuzzy linear programming can be efficiently applied for solving project scheduling as well as project crashing problems with fuzzy activity completion times. Based on the results that have been achieved in this research, some research directions may be proposed on applying linear programming for solving project management problems, including scheduling, crashing, continuously reviewing, controlling, rescheduling projects where activity completion times assume various kinds of uncertainty, fuzziness, and randomness.

References

1. Anderson, D., Sweeney, D., Williams, T., Camm, J., Cochran, J.: Quantitative methods for business, 13th edn. Cengage Learning, South-Western (2015)

2. Dubois, D., Fortin, J., Zielinski, P.: Interval PERT and its fuzzy extension. In: Kahraman, C., Yavuz, M. (eds.) Production engineering and management under fuzziness, Studies in fuzziness and soft computing 252, 179–199. Springer, Berlin (2010). https://doi.org/10.1007/978-3-642-12052-7

3. Ehsani, E., Kazemi, N., Udoncy Olugu, E., Grosse, E.H., Schwindld, K.: Applying fuzzy multi-objective linear programming to a project management decision with non-linear fuzzy membership functions, Neural computing and applications **28**, 2193–2206 (2017)

4. EVN Central Power Corporation: PC3-INVEST: An application of PERT in scheduling and controlling hydropower plant Da Krong construction project (2015)

5. Gouda, A., Monhor, D., Szántai, T.: Stochastic Programming Based PERT Modeling. In: Coping with Uncertainty. Lecture Notes in Economics and Mathematical Systems, vol 581. Springer, Berlin, Heidelberg (2006). https://doi.org/10.1007/3-540-35262-7_14

6. Hsiau, H.J., Lin, C.W.R.: A fuzzy PERT approach to evaluate plant construction project scheduling risk under uncertain resources capacity. J. indus. Eng. Manage. **2**(1), 31–47 (2009)

7. Tenjo-García, J.S., Figueroa-García, J.C.: Simulation-based fuzzy PERT problems. In: 2019 IEEE Colombian conference on applications in computational intelligence proceedings, pp. 1–5 (2019)

8. Kim, J.Y., Kang, C.W., Hwang, I.K.: A practical approach to project scheduling: considering the potential quality loss cost in the time–cost tradeoff problem. Int. J. Proj. Manage. **30**(2), 264–272 (2012)

9. Karmaker, C.L., Halder, P.: Scheduling project crashing time using linear programming approach: Case study. Int. J. Res. Indus. Eng. **6**(4), 283–292 (2017)

10. Mete, M., Ali Fuat, G.: CPM PERT and project management with fuzzy logic technique and implementation on a business. Procedia - social and behavioral sciences **210**, 348–357 (2015)

11. Ramadhanti, L.C., Yenradee, P.: Analysis and improvement of late completion of aircraft engine maintenance using fuzzy PERT/CPM with limited resources. Int. J. Knowl. Sys. Sci. (IJKSS) **12**(4), 1–25 (2021)

12. Thanh, N.H.: Applied mathematics, Postgraduate textbook. University of Education Publishing House (2005)

13. Thanh, N.H.: Optimization in fuzzy-stochastic environment and its applications in industry and economics. Internalization studies **1**, 131–143 (2012)

14. Thanh, N.H.: Quantitative methods for management and finance, Lecture notes for master's in financial management program. VNU International School (2019)

15. Thanh, N.H.: Applications of fuzzy programming for solving portfolio optimization problems: Some aspects of modeling and computing. In: Proceedings of 2020 12th International conference on knowledge and systems engineering, pp. 236–242 (2020)

16. Thanh, N.H.: Application of linear programming to project scheduling and project crashing. Vietnam J. Agri. Sci. **19**(11), 1499–1508 (2021)

17. Phong, N.T., Van, L.T.: Quantitative methods for business and project management in construction industry. Construction Publishing House (2015)

18. Elmarbrouk, O.M.: A linear programming technique for the optimization of the activities in maintenance projects. Int. J. Eng. Technol. **11**(1), 24–29 (2011)

19. Ballesteros-Perez, P., Elamrousy, K.M., Gonzalez-Cruz, M.C.: Non-linear time-cost trade-off models of activity crashing: Application to construction scheduling and project compression with fast-tracking. Automation in construction **97**, 229–240 (2019)

20. Ramo, R.M.: Fuzzy PERT for project management. Int. J. adv. Eng. Technol. **7**(4), 1150–1160 (2014)

21. Taha, A. H.: Operations research: An introduction, 10th edition. Prentice Hall (2017)

22. Wallace, A.: Project planning and scheduling using PERT and CPM techniques with linear programming: case study. Int. J. Sci. Technol. Res. **4**(8), 222–227 (2015)

23. Shakenova, Y.: Using fuzzy logic to obtain PERT three-time estimates in oil and gas projects. Adv. Eng. Technol. Appl. **5**(2), 29–34 (2016)
24. Nemaa, Z.K., Aswed, G.K.: Forecasting construction time for road projects and infrastructure using the fuzzy PERT method. 2021 IOP Conf. Ser.: Mater. Sci. Eng. **1076**, 012123 (2021)

A Study of Filter-Based Feature Selection in Software Fault Prediction

Ha Thi Minh Phuong[1], Le Thi My Hanh[2], and Nguyen Thanh Binh[1(✉)] [iD]

[1] The University of Danang - Vietnam - Korea University of Information
and Communication Technology, Danang, Vietnam
{htmphuong,ntbinh}@vku.udn.vn
[2] The University of Danang - University of Science and Technology,
Danang, Vietnam
ltmhanh@dut.udn.vn

Abstract. Software fault prediction (SFP) assists developers in diagnosing the potential defects in the early stage. In SFP, software metrics have strong influence on the performance of a predictive model. However, high dimensional data impacts negatively on the predictive accuracy. As a solution, feature selection provides a process of selecting the optimal features that combine with machine learning techniques to build SFP models. For feature selection, filter selection is a way of addressing the high dimensionality, reducing computation time and improving prediction performance. In this research, we investigate a comparative analysis to review how different of nine filter feature selection methods on both datasets in PROMISE repository, namely CM1 and KC1. The experimental results show that the performances of classifiers are varying on different datasets, especially, in the CM1 dataset, Gain Ratio and Relief based on XGBoost (XGB) and Extra Trees (ET) achieved the highest accuracy and AUC values. In KC1, Gain Ratio and Mutual Information presented the greatest performance among nine methods.

Keywords: Feature selection · Filter · Machine learning algorithms · Fault prediction

1 Introduction

Fault prediction is one of the most crucial stages that enhance the performance and reliability of systems in software development process. The activities of software fault prediction help developers in anticipating potential errors and coordinating their testing resources. Several practice researchers have investigated machine learning techniques to identify which software modules or code areas are likely to be faulty. Most of these techniques focus on metrics (e.g. object-oriented metrics, process metrics) that reflect the characteristics of software modules. However, the large number of features are usually high, thereby

L. T. M. Hanh and N. T. Binh—These authors contributed equally to this work.

primarily leading to a high-dimensional data problem that causes high time complexity and low performance. In addition, some of the metrics are uncorrelated to the faults and some others consist of noisy, redundant or irrelevant features [1]. Feature selection is an approach to address these problems to reduce computational time, minimize the negative effect of dimensional dataset and improve the predictive accuracy [2]. The main idea of feature selection is to obtain the useful metrics which can be described as the input data of the predictive models. Therefore, feature selection is an important pre-processing stage that enhances the data quality and consequently improves the prediction accuracy [3].

Recent studies have concluded that some methods achieve the best performance [4] while some others argued that the efficacy of methods for selecting the optimal metrics has no significance in software fault prediction [5]. In this paper, we examine the comparison of different filtering methods for the feature elimination. The outcome of this study assist researchers with selection of appropriate filtering methods for software defect prediction. The structure of the paper is presented as follows. In Sect. 2, we introduce a background of feature selection methods. Section 3 highlights filtering methods in feature elimination, learners, datasets and performance metrics. In Sect. 4, experimental results are illustrated by a conclusion in Sect. 5.

2 Related Work

In this section, we introduce the previous research on filter feature selection approach. Gao et al. [6] proposed a hybrid feature elimination approach by combining feature ranking and attribute subset selection to build a defect predictor. They have summarized that the proposed model performs well and there are no significant reference on the performance of models if we reduce the irrelevant metrics. Rodriguez et al. [7] applied both filter and wrapper methods on various datasets. They also concluded that the reduced metrics achieves high accuracy

Table 1. List of filter methods for feature elimination

	Feature selection	Criterion
Filter feature ranking methods	Information Gain (IG)	Ranking
	Gain Ratio (GR)	Ranking
	Chi-Square	Ranking
	Fisher Score	Ranking
	Relief	Ranking
	Mutual Information (MI)	Ranking
	Correlation-based Feature Selection (CFS)	Ranking
	Pearson Correlation	Ranking
	Lap-score	Ranking

of defect prediction than the full dataset. Gayatri et al. [8] proposed a predictive model which integrated decision tree classier with feature selection and trained the model on 18 classifiers. They concluded that the proposed model obtains the high performance.

3   Background

Some data have irrelevant and redundant features that cause the curse of dimensionality and overfitting issues. Feature ranking technique scores each individual feature by some decisive factor and then selects appropriate variables for each dataset. Otherwise, the feature elimination technique uses the search strategy for finding optimal subsets. According to Zheng [9], Filter, Wrapper and Embedded methods are three types of feature selection.

(a) Filter methods perform as prepossessing data to sort the variables wherein the top-ranked features are obtained and applied for predictive models. A criterion is used to sort the metric values with a specific threshold hence the redundant variables are removed from the dataset. Filter methods measure the relevance of a feature to the output or data, hence it does not require complexity computation. Table 1 presents a lists of filter ranking selection methods.

(b) Wrapper methods are different from filter methods as they apply a classifier to generate a subset that gains the best predictive performance. According to Chandrashekar [2], there are two categories Wrapper methods, namely Sequential Feature Selection and Heuristic Search. The Sequential Feature Selection will add features to empty set or remove features from full set in the training dataset by using some evaluation functions. The Heuristic Search Algorithm is referred to as an evolutionary algorithm that optimizes the objective function by evaluating different subsets.

(c) Embedded methods reclassify different subsets which are done in Wrapper methods to decrease the computation time [2]. Lasso regression [10], and elastic net [11] are some embedded methods which are usually use in feature selection.

4   Methodology

4.1   Feature Selection by Filtering

The filtering methods proposed several metrics for evaluating the information value of each variable. In this section, we present a general description and the comparison of nine feature elimination methods by filtering.

(a) Chi-Square technique aids in the solution of statistical and feature selection problems by assessing the independence of two categorical variables in order to determine the quality of fit between the expected value E and the observed value O. If these two categories are independent, these O and E values will be close, but if they are related, the Chi-squared value will be high. The calculation of Chi-Square is presented as follows:

$$X^2_{(D,t,c)} = \Sigma_{e_t \in 0,1} \Sigma_{e_c \in 0,1} \frac{(N_{e_t e_c} - E_{e_t e_c})^2}{E_{e_t e_c}} \tag{1}$$

where e_t and e_c can have one of two values: 0 and 1, N describes the value that is observed in D and E defines the value that is expected.

(b) Information Gain is a parameter measured as the reduction in entropy induced by dividing a dataset based on a particular value of a random variable. It is computed by comparing the dataset's entropy before and after a split. After computing the information gain of each feature, we can simply choose the feature with the highest information gain score. The calculation of Information gain is the following [12]:

$$G(S, A) = I(S) - \sum_{v \in Values(A)} \frac{|S_v|}{|S|} I(S_v) \tag{2}$$

where, $Values(A)$ represents the set of all possible values for attribute A; S_v is the subset of S, where the value of A is v and $I(S)$ is the entropy which is described as follows:

$$I(S) = -\sum_{i=1}^{k} P(value_i) * log_2(P(value_i)) \tag{3}$$

wherein, $(value_i)$ shows each possible(i^{th}) value of X and $P(value_i)$ is the probability of getting the possible(i^{th}) value from the set

(c) Gain Ratio is basically the information gain divided by the split feature value. Then the intrinsic information are larger, the value of feature will decrease [13].

$$Gainratio(Feature) = \frac{Gain(Feature)}{Intrinsicinfo(Feature)} \tag{4}$$

(d) Pearson Correlation [14] describes the simply "correlation" between two variables. The value ranges between -1 and 1 that represents how closely two variables are connected linearly. If the value is near to zero, it indicates a lesser connection (precise zero meaning no correlation). If it is near to 1, two variables have a greater positive association. Otherwise, if it is near to -1, two variables have a greater negative association.

(e) Mutual Information [15] is a measure of the two variables mutual dependency, or the reduction in uncertainty for one variable given a known value of the other variable. Mutual information is inextricably tied to the concept of entropy and information gain.

(f) Fisher score is the usual feature elimination to gain the subset of variables. The formulation of Fisher Score on the $j - th$ feature is illustrated by [16]:

$$F(x^j) = \frac{\sum_{k=1}^{c} n_k(\mu_k^j - \mu^j)^2}{(\sigma^i)^2} \tag{5}$$

where $(\sigma^i)^2 = \sum_{k=1}^c n_k(\sigma_k^i)^2$

μ_k is the mean vector and n_k is size of the $k - th$ class in the reduced data space. The top variables have highest scores will be chosen by computing the Fisher score value of each metric.

(g) Relief assigns a feature score to each feature which is applied to measure and select the top-ranking variables. The following calculation define the weight of a given feature:

$$W = \frac{(W - diff(x_{xj}), nearhit_{ij}^2 + diff(x_{ij}, nearmiss_{ij})^2)}{m} \qquad (6)$$

where m is the sample size which is randomly choosen j distance from a subset of the training data), (x_{xj} describes the value of metric that randomly chosen j distance, $nearhit_{ij}^2$ is the value of the feature within the closest training set in the same class, $diff(x_{xj}, nearhit_{ij}^2)$ is the distinction between values of feature within randomly chosen j range and $nearhit ij2$ value of attribute within the closest data point in the same class. The xij and $nearmiss ij$ values for a beneficial characteristic are supposed to be very close to each other.

(h) CFS simply presents a metric to assess the effectiveness of the feature subset. It examines a subset of characteristics, taking into consideration each feature's distinct prediction capacity as well as the degree of duplication among them.

(i) Lap-score [17] is a measure to assess the feature importance based on its capability to preserve locality. This approach entails embedding the data on the closest neighbour network using an arbitrary distance metric, followed by creating a weight matrix. A laplacian score is then computed for each feature, with the smallest values corresponding to the most essential dimensions.

4.2 Learners

In this study, various distinctively different classifiers were applied for evaluating the efficiency of the FS methods by filtering, namely Decision Tree (DT),Random Forest (RF), KNeighborsClassifier(KNN), Multilayer Perceptron (MLP), Adaboost (Ada), Gradient Boosting (GB), Hist Gradient Boosting (HGB), XGBoost (XGB), Extra Trees (ET), Bagging. The classification models were selected to review how filter methods work on different classifiers with different characteristics such as DT, Ada, GB, HGB, XGB, ET for Tree-based, KNN for instance based and RF, Bagging for ensemble-based methods.

4.3   Experimental Setup

Firstly, we apply the nine filter-based methods which are introduced in Sect. 4.1 in the full defect dataset. Each filter method will rank the software metric values based on their respective characteristics. In this study, we apply $log_2 N$ to filter the number of the top-ranked features which N is the total features in each software defect dataset [6]. In the next step, ten above learners were applied to examine the efficiency of the reduced features and the original features in software fault prediction.

4.4   Datasets

We conduct the experiment on CM1 and KC1 datasets which extracted from the PROMISE Software Engineering Repository to examine the efficacy among various filter methods. The PROMISE repository has been employed in many researches due to its reliability on the classification results.

4.5   Performance Metrics

A set of performance metrics to assess the performance of filter feature section methods is often utilized. We use two performance metrics, namely Accuracy and Area under Curve (AUC). Accuracy define a ratio of correctly predicted samples to the total samples. The range of accuracy value is from 0 to 1. Area Under the ROC Curve (AUC) measure the capability of a learner to differentiate between classes.

5   Experimental Results

The section discusses the experimental results of applying nine feature elimination techniques on two datasets based on ten classifiers demonstrated in Sect. 4.2. The Accuracy and AUC results are evaluated on both original datasets and reduced datasets. The highest values for each dataset were marked in bold font to indicate the best accuracy performance. From the results in Table 2, on the CM1 dataset, the Gain Ratio method achieves the highest accuracy value on the predictive performance of RF, AdaBoost, Gradient Boosting and XGB with 88.50, 85.75, 86.66 and 88.39 respectively. Moreover, HG and ET based on Relief reached the highest accuracy values of 90.00 and 91.05. In addition, Gain Ratio and Relief methods obtain the best AUC values on KNN, HGB and ET in Table 3. This clearly shows that Gain Ratio and Relief reach a positive effect on the predictive models on the CM1 dataset.

As shown in Table 4, we give the results of comparing the accuracy performance of the ten above filter methods with the objective to find the subset has the maximum accuracy values on the KC1 dataset. We can see that the Gain Ratio and Mutual Information methods obtain the high accuracy performance from KNN, RF, HG and ET with the percentage of 82.56, 82.70,81.78 and 84.01

Table 2. Accuracy values of filter methods on CM1 dataset

	KNN	DT	RF	MLP	AB	GB	HG	BG	XGB	ET
Without FS	0.8614	0.7965	0.8851	0.8877	0.7649	0.8105	0.8228	0.7836	0.8807	0.9093
Chi-Square	0.7895	0.7746	0.8395	0.8132	0.7263	0.7649	0.8667	0.7509	0.8333	0.8754
Information gain	0.8615	0.8188	0.8565	0.8647	0.8263	0.8507	0.8439	**0.8882**	0.8462	0.8816
Gain ratio	0.8900	0.8234	**0.8850**	0.8581	**0.8575**	**0.8666**	0.8750	0.8881	**0.8839**	0.8773
Pearson correlation	0.7982	0.8395	0.8816	0.7246	0.7877	0.7877	0.8754	0.8158	0.8649	0.9053
Mutual information	0.8399	0.7879	0.8497	0.7647	0.7457	0.7612	0.8736	0.7640	0.8314	0.8777
Fisher score	0.7965	0.7500	0.8474	0.8316	0.7316	0.7842	0.8474	0.7842	0.8386	0.8895
Relief	**0.8965**	0.8184	0.8553	**0.8921**	0.7561	0.8018	**0.9000**	0.8614	0.8825	**0.9105**
CFS	0.8035	0.7798	0.8728	0.8614	0.7140	0.7667	0.8807	0.8298	0.8439	0.9175
Lap-score	0.8158	0.7596	0.8351	0.8509	0.7596	0.7632	0.8860	0.7877	0.8544	0.8825

Table 3. AUC values of filter methods on CM1 dataset

	KNN	DT	RF	MLP	AB	GB	HG	BG	XGB	ET
Without FS	0.9124	0.8281	0.9588	0.9276	0.8477	0.8640	0.9559	0.8732	0.9435	0.9659
Chi-Square	0.9177	0.8934	0.9215	0.9923	**0.9612**	0.9430	0.9686	0.9880	0.8890	0.9420
Information gain	0.8613	0.8859	0.9503	0.9931	0.9612	0.9430	0.9610	0.9696	0.9619	0.9686
Gain ratio	**0.9745**	**0.9147**	**0.9750**	**0.9970**	0.9096	**0.9582**	0.9612	**0.9921**	**0.9714**	0.9487
Pearson correlation	0.9304	0.8785	0.9461	0.9165	0.7998	0.8611	0.9547	0.8841	0.9255	0.9562
Mutual information	0.9214	0.8605	0.9223	0.8286	0.8168	0.8356	0.9254	0.8104	0.8942	0.9403
Fisher score	0.9192	0.8084	0.9265	0.8975	0.7754	0.8143	0.9282	0.8605	0.9053	0.9515
Relief	0.8963	0.8483	0.9285	0.9341	0.8488	0.8504	**0.9612**	0.9185	0.9271	**0.9661**
CFS	0.9167	0.8283	0.9393	0.9057	0.7757	0.8145	0.9417	0.8895	0.9241	0.9651
Lap-score	0.9241	0.8296	0.9224	0.9055	0.7711	0.8015	0.9411	0.8613	0.9191	0.9453

respectively. The same also is observed for the AUC in Table 5, the percentage value of Gain Ratio and Mutual Information methods are better than remaining methods. Therefore, we assume that Gain Ratio and Mutual Information gain the best efficacy among various predictive models. It was conclusively filter feature selection methods help reduce the redundant features for building a good classification model which obtains high-performance prediction.

Based on the work of Aljamaan and Hamoud [18] in Table 6, they conducted an experiment to compare the predictive performance of Tree-based ensembles which combined feature selection using Gain Ratio and data balance with the SMOTE technique. In their experiment, the best accuracy was achieved by using Random Forest for CM1 and Extra Tree for KC1. The same also is observed for our experiment result, the good performance among the classifiers on Gain Ratio was obtained on both CM1 and KC1 datasets. Moreover, the prediction accuracy

Table 4. Accuracy values of filter methods on KC1 dataset

	KNN	DT	RF	MLP	AB	GB	HG	BG	XGB	ET
Without FS	0.8372	0.7138	0.7767	0.6743	0.5926	0.6398	0.7733	0.6599	0.6887	0.8044
Chi-Square	0.7428	0.6643	0.6953	0.6274	0.6196	0.6035	0.6899	0.6254	0.6479	0.7250
Information gain	0.7702	0.7322	0.7602	0.7545	0.7212	**0.7674**	0.7165	0.7939	0.7724	0.7366
Gain ratio	0.7523	0.6558	0.7057	**0.7560**	**0.7277**	0.7667	0.5974	**0.7946**	0.7036	0.7007
Pearson correlation	0.7399	0.6856	0.7212	0.6579	0.6265	0.6415	0.7250	0.6524	0.7123	0.7405
Mutual information	**0.8256**	**0.7892**	**0.8270**	0.7521	0.7203	0.7299	**0.8178**	0.7422	**0.7790**	**0.8401**
Fisher score	0.7266	0.6651	0.7155	0.6196	0.5960	0.6220	0.6887	0.6277	0.6357	0.7221
Relief	0.7831	0.6723	0.7474	0.6329	0.6173	0.6461	0.7204	0.6449	0.6691	0.7739
CFS	0.7514	0.6781	0.7089	0.6387	0.6370	0.6485	0.7129	0.6306	0.6686	0.7405
Lap-score	0.7123	0.7109	0.7336	0.6283	0.6467	0.6530	0.7474	0.6248	0.7135	0.7659

Table 5. AUC values of filter methods on KC1 dataset

	KNN	DT	RF	MLP	AB	GB	HG	BG	XGB	ET
Without FS	0.8072	0.7342	0.8453	0.7394	0.6326	0.6856	0.8089	0.6957	0.7649	0.8691
Chi-Square	0.7428	0.6987	0.7645	0.6830	0.6592	0.6697	0.7517	0.6663	0.7198	0.7828
Information gain	0.8120	0.7729	0.8148	0.9538	0.7995	**0.8325**	0.7985	0.8575	0.7198	0.7826
Gain Ratio	0.7894	0.6809	0.7600	**0.9541**	**0.8229**	0.7412	0.7372	0.8499	0.7615	0.7566
Pearson correlation	0.7936	0.7240	0.7781	0.7044	0.6452	0.6743	0.7849	0.6799	0.7566	0.7829
Mutual information	**0.8576**	**0.8590**	**0.8965**	0.8244	0.7649	0.8061	**0.8178**	**0.8956**	**0.8605**	**0.8949**
Fisher score	0.8062	0.7177	0.7814	0.6768	0.6370	0.6703	0.7556	0.6653	0.7076	0.7963
Relief	0.7831	0.7288	0.8126	0.6744	0.6358	0.6779	0.7928	0.6809	0.7295	0.8325
CFS	0.8005	0.7212	0.7765	0.6976	0.6608	0.7021	0.7721	0.6839	0.7355	0.7992
Lap-score	0.7122	0.7612	0.7924	0.6771	0.6915	0.7110	0.8066	0.6605	0.7747	0.8134

Table 6. Comparison performance between tree-based ensemble learners on CM1 and KC1 datasets

	CM1		KC1	
	Accuracy	AUC	Accuracy	AUC
DT	82.28	0.82	69.1	0.69
Ada	82.28	0.82	71.23	0.77
RF	**91.23**	0.97	72.55	0.79
ET	90.53	**0.98**	**73.07**	**0.8**
GB	90	0.95	69.04	0.76
HGB	89.82	0.97	70.83	0.77
XGB	88.25	0.95	67.55	0.74
CAT	88.95	0.96	67.66	0.74

and AUC of classifiers on Mutual Information methods is higher than the results of the work of Aljamaan and Hamoud on the KC1 dataset.

6 Conclusion

The selection of optimal metrics from the input dataset is focused to obtain a powerful model in fault prediction. Several feature elemination methods have been proposed but it is uncertain to identify which the method achieved the best performance. In this paper, we investigated the comparative analysis of feature selection based on filter approach. From the experimental results, we found that feature selection method performance varies from one dataset to the other. Especially, Gain Ratio and Relief obtained the highest performance over other filter methods on CM1 dataset. In the case of KC1, the high prediction performance was gained on Gain Ratio and Mutual Information. We conclusively summarized that there are a difference in the performance for feature elimination methods the used datasets and the selection of classifiers. Further research could be focused to create an ensemble model which combine feature selection and ensemble techniques.

References

1. Jimoh, R., Balogun, A., Bajeh, A., Ajayi, S.: A promethee based evaluation of software defect predictors. J. Comput. Sci. Appl. **25**(1), 106–119 (2018)
2. Chandrashekar, G., Sahin, F.: A survey on feature selection methods. Comput. Electr. Eng. **40**(1), 16–28 (2014)
3. Balogun, A.O., Basri, S., Abdulkadir, S.J., Hashim, A.S.: Performance analysis of feature selection methods in software defect prediction: a search method approach. Appl. Sci. **9**(13), 2764 (2019)
4. Ghotra, B., McIntosh, S., Hassan, A.E.: A large-scale study of the impact of feature selection techniques on defect classification models. In: 2017 IEEE/ACM 14th International Conference on Mining Software Repositories (MSR), pp. 146–157. IEEE (2017)
5. Xu, Z., Liu, J., Yang, Z., An, G., Jia, X.: The impact of feature selection on defect prediction performance: an empirical comparison. In: 2016 IEEE 27th International Symposium on Software Reliability Engineering (ISSRE), pp. 309–320. IEEE (2016)
6. Gao, K., Khoshgoftaar, T.M., Wang, H., Seliya, N.: Choosing software metrics for defect prediction: an investigation on feature selection techniques. Softw. Pract. Exp. **41**(5), 579–606 (2011)
7. Rodriguez, D., Ruiz, R., Cuadrado-Gallego, J., Aguilar-Ruiz, J., Garre, M.: Attribute selection in software engineering datasets for detecting fault modules. In: 33rd EUROMICRO Conference on Software Engineering and Advanced Applications (EUROMICRO 2007), pp. 418–423. IEEE (2007)
8. Gayatri, N., Nickolas, S., Reddy, A., Reddy, S., Nickolas, A.: Feature selection using decision tree induction in class level metrics dataset for software defect predictions. In: Proceedings of the World Congress on Engineering and Computer Science, vol. 1, pp. 124–129. Citeseer (2010)

9. Zheng, Z., Wu, X., Srihari, R.: Feature selection for text categorization on imbalanced data. ACM Sigkdd Explor. Newsl. **6**(1), 80–89 (2004)

10. Khanji, C., Lalonde, L., Bareil, C., Lussier, M.-T., Perreault, S., Schnitzer, M.E.: Lasso regression for the prediction of intermediate outcomes related to cardiovascular disease prevention using the transit quality indicators. Med. Care **57**(1), 63–72 (2019)

11. Zou, H., Hastie, T.: Regularization and variable selection via the elastic net. J. Roy. Stat. Soc. Ser. B (Stat. Methodol.) **67**(2), 301–320 (2005)

12. Hall, M.A., Smith, L.A.: Practical feature subset selection for machine learning. In: Proceedings of the 21st Australasian Computer Science Conference ACSC 1998, PP. 181–191 (1998)

13. Han, J., Kamber, M., Pei, J.: 3 - data preprocessing. In: Han, J., Kamber, M., Pei, J. (eds.) Data Mining. The Morgan Kaufmann Series in Data Management Systems, 3rd edn., pp. 83–124. Morgan Kaufmann, Boston (2012)

14. Kirch, W.: Pearson's correlation coefficient. Encyclopedia of Public Health, pp. 1090–1091 (2008)

15. Walters-Williams, J., Li, Y.: Estimation of mutual information: a survey. In: Wen, P., Li, Y., Polkowski, L., Yao, Y., Tsumoto, S., Wang, G. (eds.) RSKT 2009. LNCS (LNAI), vol. 5589, pp. 389–396. Springer, Heidelberg (2009). https://doi.org/10.1007/978-3-642-02962-2_49

16. Duda, R.O., Hart, P.E., Stork, D.G.: Pattern Classification, 2nd edn. Wiley-Interscience, Hoboken (2000)

17. He, X., Cai, D., Niyogi, P.: Laplacian score for feature selection. Adv. Neural Inf. Process. Syst. **18**, 1–8 (2005)

18. Aljamaan, H., Alazba, A.: Software defect prediction using tree-based ensembles. In: Proceedings of the 16th ACM International Conference on Predictive Models and Data Analytics in Software Engineering, pp. 1–10 (2020)

Applying Lecturer Expectation and Learning Retention to an Adaptive Learning System

Tien Vu-Van[1,2], Huy Tran[1,2], Duy Tran Ngoc Bao[1,2], Hoang-Anh Pham[1,2(✉)], and Nguyen Huynh Tuong[1,2]

[1] Ho Chi Minh City University of Technology (HCMUT), 268 Ly Thuong Kiet Street, District 10, Ho Chi Minh City, Vietnam
{vvtien,tranhuy,duytnb,anhpham,htnguyen}@hcmut.edu.vn
[2] Vietnam National University Ho Chi Minh City (VNU-HCM), Linh Trung Ward, Thu Duc District, Ho Chi Minh City, Vietnam

Abstract. Adaptive learning systems (ALS) have become popular because they encourage learners' self-study discipline by recommending various learning materials that adapt to their knowledge. However, previous studies showed a gap in considering learning retention in study programming. The new knowledge becomes familiar to the learner by learning repetitively. Additionally, it lacks the lecturer's expectations in determining the learner's knowledge. This study tries to resolve these two above problems for ALS. First, we propose six parameters that help teachers define their expectations and assess learners' abilities throughout the learning process. Second, we adopt an algorithm to generate a reasonable appearing probability for each question and select a question randomly based on its probability, giving the learners a chance to practice questions in the past again. Repeated questions help learners learn by doing and mastering programming skills rather than memorizing a solution to a problem.

Keywords: Adaptive learning system · Lecturer expectations · Learning retention · Probability · Clustering

1 Introduction

The pandemic COVID-19 has a negative impact on studying at school. The learners can't go to school and have face-to-face absorbing the knowledge from the instructor. Conversely, the instructor can't help the learners with their questions directly. The learning process has shipped to another form by leveraging a learning management system (LMS) to ensure learning while avoiding large numbers of learners have gathered in one place. An LMS is a convenient environment where instructors can share lecture videos, slides, and exercises, and learners can learn and do the homework at any time. In programming courses, there are some demands on practice coding exercises for learners who need to be familiar with writing code to resolve a problem, debugging, and fixing bugs

© The Author(s), under exclusive license to Springer Nature Switzerland AG 2022
N.-T. Nguyen et al. (Eds.): ICIT 2022, LNDECT 148, pp. 68–77, 2022.
https://doi.org/10.1007/978-3-031-15063-0_6

independently without consulting some instructors' help. General exercises published in the LMS may not be sufficient for learners because each person has prior personal knowledge, learning style, and learning pace. The LMS should provide and suggest appropriate exercises to learners based on their current abilities.

To support the need for adaptive learning that allows learning without time and classroom restrictions, adaptive learning systems (ALS) have been developed to deliver learners with personalized learning content based on their learning models. In other words, learning materials in ALS are constructed based on learners' behavior to adapt to each learner individually. In [1], the authors state that a well-developed ALS can produce the same effectiveness as classroom learning or even better.

In our study, we investigated some existing ALSs with two concerning questions: (1) how does an ALS appropriately describe the learners' characteristics, and (2) how does it recommend the materials to learners. In addition, we also consider about lecturer's engagement in determining learners' characteristics.

Learners' characteristics are the information extracted from learners that can help bring them appropriate questions. The ALS will check these characteristics with some conditions to see if the learner qualified for a higher degree. One of the ways to check these conditions is to ask learners to take tests, but it may annoy learners during practice if it happens frequently. Instead, the ALS should provide an open and unconstrained place to practice. On the other hand, if there is lecturers' engagement, learners need to do some explicit tests as the systems in [2,3]. If there is no lecturers' engagement, the system can implicitly measure learners' characteristics as systems in [4–6]. Lecturers have much more experience than learners, so they should have reasonable expectations for testing learners. They should be involved in the testing process and provide the conditions to require learners to achieve specific knowledge. In our approach, we propose some conditional parameters to define if a learner stays in the current difficulty degree or moves up to a higher degree. The lecturers provide these parameters to present their expectations for the learning process of learners. We reuse our previous clustering model [7] to classify questions into three different difficulty groups: easy, medium, and hard. If learners pass the lecturer's condition of a certain difficulty group, the system will update their ability to a higher difficulty and bring them questions in a harder group.

Regarding recommending process, the systems in [2,4] suggest materials based on a group of learners have high relevance to the target learner. Meanwhile, the systems in [5,6] indicate that learners need to redo the questions on the topic that they make many mistakes. These studies pay less attention to learning retention. Learning retention is the procedure of transferring new knowledge into long-term memory so that you have understood the knowledge and can recognize it in the future. Programming is a subject that learners need to practice to gain observation, problem-solving, and debugging skills. Correctly answered questions could not guarantee that the learner has mastered that question. The author in [8] points out that programming is more practical than theoretical and must be learned "by doing" rather than memorizing. Our proposed system will generate appearing probabilities for all questions based on the last time each appears to a learner to support learning retention. Then, the system will randomly choose a question and give it to the learner. Therefore, an accomplished question can have a chance to appear to the learner again.

2   Proposed Approach

Figure 1 depicts the overview of our system, in which there are two types of users participating in the system such as lecturers and learners. Lecturers can manage courses and exercises using the "Learning Contents Management" module. A question bank is also provided to separate questions from courses for re-usability, i.e., a question can be used in two distinguished exercises or courses. Lecturers can add prepared questions into an exercise that can be an **assigned exercise** or a **practice one**.

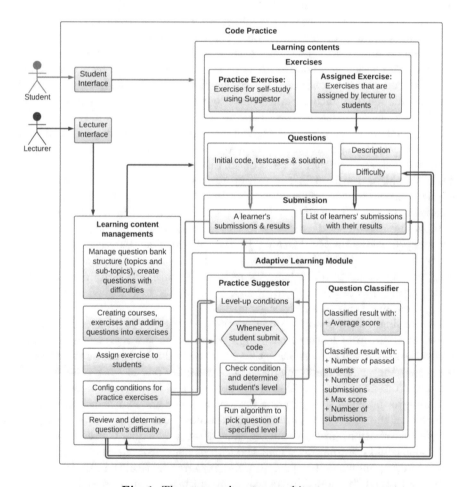

Fig. 1. The proposed system architecture

An assigned exercise will limit the maximum number of submissions and expired time. Lecturers must set up these constraints for each question in a specific exercise. Meanwhile, a practice exercise does not have any constraints. Instead, the system will automatically suggest appropriate questions with the learner's characteristics. Lecturers need to configure some conditional parameters to represent their expectations. More difficult questions will be provided once the learner is qualified for these parameters. Moreover, a **Question Classifier** in our proposed system will help the lecturers review the difficulty of questions. This process will help them see the difficulty difference according to the learner's result and lecturer estimation. Lecturers can then adjust the question's difficulty based on their expectations.

Learners need to work hard to achieve good scores under the question's constraints. Therefore, the **Question Classifier** will fetch learners' submissions to assigned exercises as they are reliable data. Regarding practice exercises, each time learners submit answers, the **Practice Suggestor** will retrieve all their submissions, extract interesting information, and compare them to conditional parameters, which are level-up conditions in Fig. 2, and determine appropriate difficulty for the learner. The system will run an algorithm to generate appearing probabilities of all questions belonging to determined difficulty; then it randomly gives a new question to the learner.

2.1   Question Classifier

This module refers to our previous work in [7] to implement the question difficulty classifier. However, we do not make statistics on classification results by topic like the prior study. In this study, we instead provide the classification results to lecturers for reviewing and editing question difficulty. Again, the classification model uses submissions data from assigned exercises. We consider five difficulty-related factors that can be extracted from learners' submissions. Then, the system cluster the question difficulty on two following models:

- Model 1 has only one feature that is the average scores
- Model 2 has four features including the number of passed learners, the number of passed submissions, the max score of submissions, and the number of submissions.

After the question is classified, the system presents the results of these two models to the lecturer to review and edit the difficulty level if needed. This process brings two advantages. First, lecturers take references from the learners' perspective to consider about question's difficulty. Second, they have full control over the setup question's difficulty based on their expectations despite the results from the question classifier. As an example shown in Fig. 2, the question at number 2 (Deep Copy) has two consistently difficult of Medium from the Question Classifier, while the current level is Easy. Lecturers can adjust the current level value to the Medium level. However, the question at number 3, named Most Occurrence Character, is classified as Hard, while the lecturer has expected it as Medium level. The lecturer can leave it without adjusting its current level value.

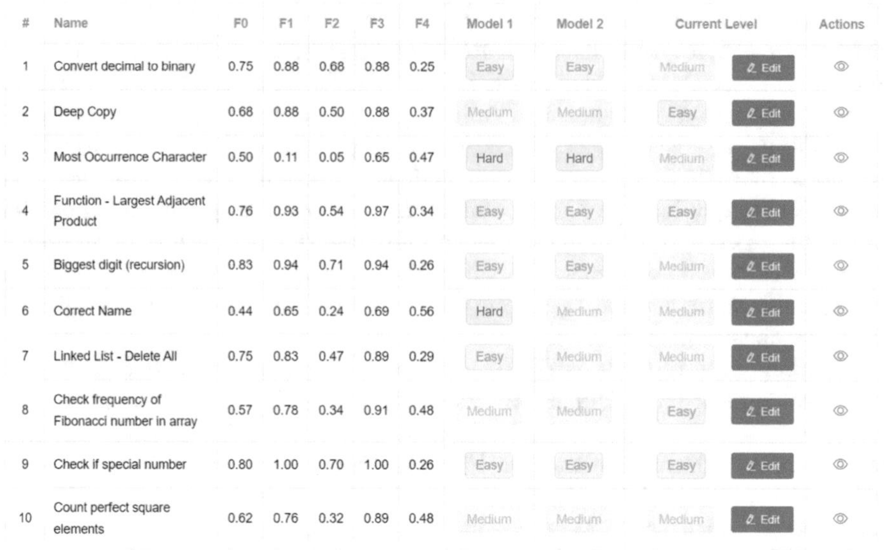

#	Name	F0	F1	F2	F3	F4	Model 1	Model 2	Current Level		Actions
1	Convert decimal to binary	0.75	0.88	0.68	0.88	0.25	Easy	Easy	Medium	✎ Edit	◉
2	Deep Copy	0.68	0.88	0.50	0.88	0.37	Medium	Medium	Easy	✎ Edit	◉
3	Most Occurrence Character	0.50	0.11	0.05	0.65	0.47	Hard	Hard	Medium	✎ Edit	◉
4	Function - Largest Adjacent Product	0.76	0.93	0.54	0.97	0.34	Easy	Easy	Easy	✎ Edit	◉
5	Biggest digit (recursion)	0.83	0.94	0.71	0.94	0.26	Easy	Easy	Medium	✎ Edit	◉
6	Correct Name	0.44	0.65	0.24	0.69	0.56	Hard	Medium	Medium	✎ Edit	◉
7	Linked List - Delete All	0.75	0.83	0.47	0.89	0.29	Easy	Medium	Medium	✎ Edit	◉
8	Check frequency of Fibonacci number in array	0.57	0.78	0.34	0.91	0.48	Medium	Medium	Easy	✎ Edit	◉
9	Check if special number	0.80	1.00	0.70	1.00	0.26	Easy	Easy	Easy	✎ Edit	◉
10	Count perfect square elements	0.62	0.76	0.32	0.89	0.48	Medium	Medium	Medium	✎ Edit	◉

Fig. 2. Question Classifier interface

2.2 Determine the Appropriate Difficulty Level for Learners

When a learner practices for the first time, he will be assigned an easy question. Once the learner submits the answer, the system considers the following conditional parameters to determine whether the learner will be on the current difficulty level or go to a higher difficulty level.

- **The minimum number of questions that learners need to do** (mq): Learners need to be exposed to several questions at a specific difficulty degree. When doing a question, learners will read, understand the topic and think about how to do it. Learners can come up with a failure or success solution. After being exposed to many questions, learners will be familiar with the usual problem types and solutions; then, they can understand how to program with the current difficulty.
- **The minimum of good questions** (mgq): A question is considered "good" when the number of submissions is low, and the maximum score is high. To check these two conditions, the lecturer will set up "gns" (a good number of submissions), the is the number of submissions that students are not allowed to be exceeded. Simultaneously, the lecturer will set up "gms" (good max score), a question's score that the student's result must be over.
- **The minimum score needed for** ns **last submissions** (mls): Once learners are familiar with the current difficulty, we expect learners to achieve a relatively good score on consecutive last questions. It requires that learners have understood the description. Then, they have a solution to the question and have implemented an actual code. In addition, learners need to prove that they don't pass this condition by fortunate.

The lecturer's expectations are presented through six above conditional parameters: $(mq, mgq, gns, gms, ns, mls)$. The lecturer needs to provide two parameters groups: easy-to-medium and medium-to-hard parameters.

2.3   Choose Question to Suggest to Learners

The process of selecting questions is primarily using the random selection method. However, unlike the normal random process, we also consider learning retention in the process of suggesting. Learners may not completely understand the question they did, so the questions that appear again will help learners practice more on the question itself. We propose modifying the question's appearing probability: the farther learners did the questions, the higher the probability of appearing, and a new question will have a high probability of appearing.

Let p_{max} is the maximum appearing probability of questions that learners did. Let p_{new} is the probability of new questions. We set $p_{new} = k * p_{max}$, then we can adjust $k > 1$ if we want new questions to appear more or $k < 1$ if we want old questions to appear more to learners.

The suggestion algorithms are presented in Algorithm 1 and Algorithm 2.

Algorithm 1: Calculate appearing probabilities of questions.

Input:

   ques_times[]: An array of submitting time of last submission of questions. If learner didn't do a question, its submitting time is -1.0.

Output:

   An array of appearing probabilities of questions.

begin

   $N \longleftarrow ques\_scores.length$;
   $currentTime \longleftarrow GET\_CURRENT\_TIME()$;
   $s[N]$;   /* create new array s of size N */
   for $i = 0$ **to** $N - 1$ **do**
   | **if** $ques\_times[i] == -1$ **then** $s[i] \longleftarrow -1$;
   | **else** $s[i] \longleftarrow currentTime - ques\_times[i]$;
   end
   $maxS \longleftarrow MAX(s)$;
   for $i = 0$ **to** $N - 1$ **do**
   | **if** $s[i] == -1$ **then** $s[i] = maxS * k$;   /* Unassigned question */
   end
   /* Calculate the probabilities   */;
   $sumS \longleftarrow SUM(s)$;
   $p[N]$;
   for $i = 0$ **to** $N - 1$ **do** $p[i] \longleftarrow s[i]/sumS$;
   return p;

end

Algorithm 2: Choose question to suggest.

Input:
  p[]: An array of appearing probabilities of questions.
Output:
  Index of suggested question.
begin
  $cumSumP[N]$;
  $cumSumP[0] \longleftarrow p[i]$;
  for $i = 1$ **to** $N - 1$ **do** $cumSumP[i] \longleftarrow cumSumP[i - 1] + p[i]$;
  $idx \longleftarrow 0$;
  /* Randomly generates a floating-point in $[0 - 1)$ */
  $x = RANDOM\_FLOAT(0, 1)$;
  while $x >= cumSumP[idx]$ **do** $idx + +$;
  return idx
end

---

3   Implementation and Evaluation

3.1   Implementation Results

We have also successfully implemented our proposed approach as an ALS system named Code Practice, which provides lecturers with three main functions such as managing learning contents (Exercise), managing group of learners (Group), and Question Classifier, as shown in Fig. 2. The grouping function will not be presented because it is out of the context of our current study. As aforementioned, lecturers can manage learning content such as exercise creation and adjustment. For example, in Fig. 3, lecturers can assign constraints including start time, end time, and the number of submissions for each question before adding it to an assigned exercise.

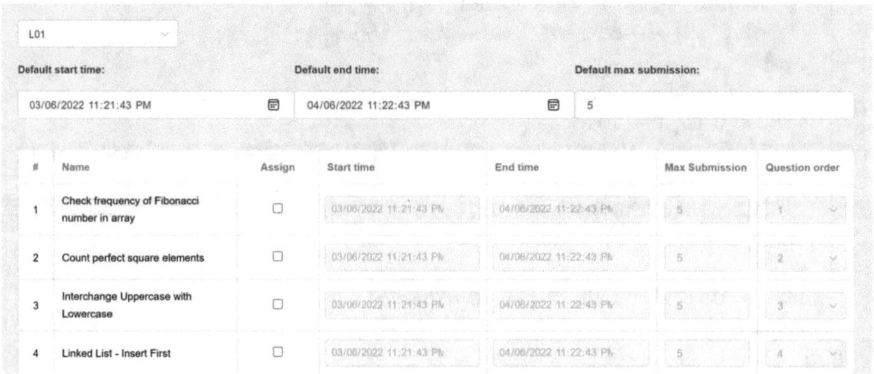

Fig. 3. Configuration of assigned exercises

Students are also provided with two different interfaces for doing the exercises corresponding to the two types of exercises configured by lecturers. Regarding assigned exercises, as shown in (Fig. 4), students can see constraints such as deadline and number of submissions. Meanwhile, students no longer see constraints in practice exercises, as shown in (Fig. 5). Instead, they can see the displayed question difficulty corresponding to their current level.

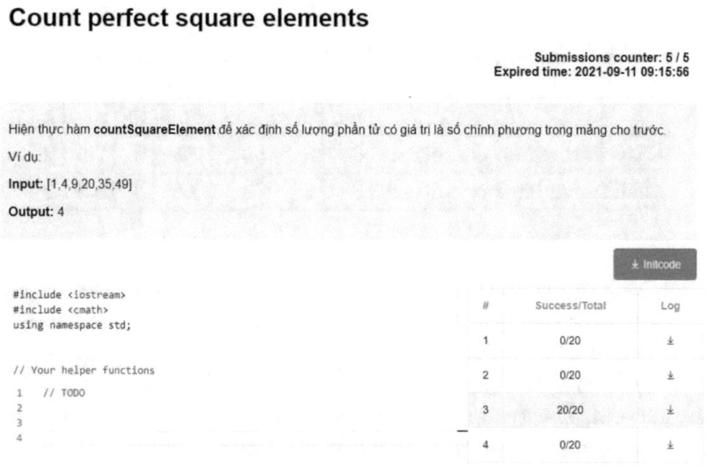

Fig. 4. An assigned exercise for learners

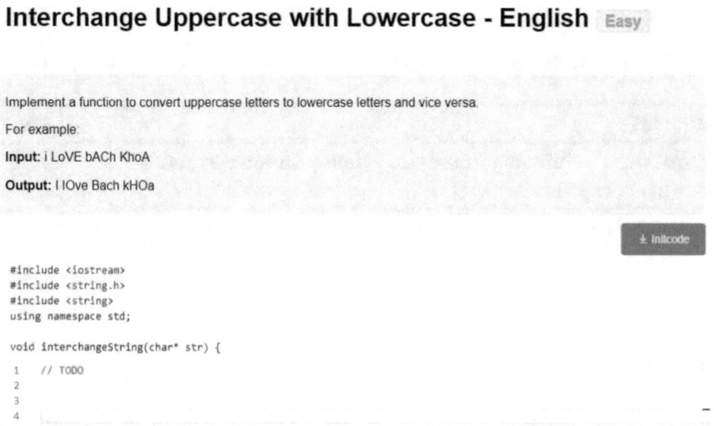

Fig. 5. A practice exercise for learners

3.2   Preliminary Evaluation

We arrange a practice exercise in the system to test the usability and have a straightforward look at the effectiveness of the adaptive learning process. There are 40 learners under the same context of reviewing their knowledge before the final exam of the Programming Fundamentals course offered by HCMUT in the 2nd semester of the 2020–2021 year. The practice exercise consists of 5 Easy, 5 Medium, and 5 Hard questions about the Array topic as summarised in Table 1.

Table 1. Setup parameters for the adaptive learning process

	mq	mgq	gns	gms	ns	mls	k
Easy-to-medium conditions	3	3	3	7.5	3	7.5	2
Medium-to-hard conditions	3	3	3	8.5	3	9.0	2

The questionnaire consists of two questions that focus on the impacts of the adaptive learning process. Each question is rated from 1 (i.e., strongly disagree) to 5 (i.e., strongly agree). Evaluation results are aggregated by averaging the scores given by users. Details of the statements and scores are summarised in Table 2. In general, the system got a relatively high score for interesting statements. However, the suitability of question difficulty suggested to learners is not excellent. The reason may be the lack of the number of questions in our system, which leads to the difficulty distribution being inaccurate. We are building a more significant question bank to examine the system's efficiency.

Table 2. Student feedback on the adaptive learning process

#	Statement	Mean score
1	The question's difficulty increases when you answer more questions correctly	4.31
2	The suggested difficulty level of questions is suitable for you	4.15

4   Conclusion

We presented an approach for an adaptive learning support system, unlike the existing works that mainly concentrate on relevant properties but less novelty and retention of learning materials. Our approach uses automatically recorded thresholds to determine the learners' ability and bring more challenging questions. It also applies probability to ensure further questions and faraway questions will often appear to learners. We have conducted experiments and got positive feedback from both learners and teachers. However, it will take time to collect the learners' interactions and measure the impact of our approach.

Acknowledgement. This research is funded by Vietnam National University Ho Chi Minh City (VNU-HCM) under grant number DS2022-20-07. We acknowledge the support of time and facilities from Ho Chi Minh City University of Technology (HCMUT), VNU-HCM for this study.

References

1. Anindyaputri, N.A., Yuana, R.A., Hatta, P.: Enhancing students' ability in learning process of programming language using adaptive learning systems: a literature review. Open Eng. **10**, 820–829 (2020)
2. Gavrilović, N., Arsić, A., Domazet, D., Mishra, A.: Algorithm for adaptive learning process and improving learners' skills in Java programming language. Comput. Appl. Eng. Educ. **26**(5), 1362–1382 (2018)
3. Chrysafiadi, K., Virvou, M., Sakkopoulos, E.: Optimizing programming language learning through student modeling in an adaptive web-based educational environment. In: Virvou, M., Alepis, E., Tsihrintzis, G.A., Jain, L.C. (eds.) Machine Learning Paradigms. ISRL, vol. 158, pp. 205–223. Springer, Cham (2020). https://doi.org/10.1007/978-3-030-13743-4_11
4. Karagiannis, I., Satratzemi, M.: Implementation of an adaptive mechanism in Moodle based on a hybrid Dynamic User Model. In: Auer, M.E., Tsiatsos, T. (eds.) ICL 2018. AISC, vol. 916, pp. 377–388. Springer, Cham (2020). https://doi.org/10.1007/978-3-030-11932-4_36
5. Chookaew, S., Panjaburee, P., Wanichsan, D., Laosinchai, P.: A personalized e-learning environment to promote student's conceptual learning on basic computer programming. Procedia. Soc. Behav. Sci. **116**, 815–819 (2014)
6. Hsiao, I.H., Sosnovsky, S., Brusilovsky, P.: Guiding students to the right questions: adaptive navigation support in an e-learning system for Java programming. J. Comput. Assist. Learn. **26**, 270–283 (2010)
7. Tran, H., Tien, V.V., Hoang, N.V., Duy, T.N.B., Thinh, T.N., Thanh-Van, L.: An efficient approach to measure the difficulty degree on practical programming exercises based on student performances. In: 2nd Symposium on Computer Science and Engineering (SCSE), p. 4 (2021)
8. Vizcaíno, A., Contreras, J., Favela, J., Prieto, M.: An adaptive, collaborative environment to develop good habits in programming. In: Gauthier, G., Frasson, C., VanLehn, K. (eds.) ITS 2000. LNCS, vol. 1839, pp. 262–271. Springer, Heidelberg (2000). https://doi.org/10.1007/3-540-45108-0_30

An Intelligent Approach for Multi-criterial Decision Making Using Similarity of Intuitionistic Fuzzy Sets

Tran Duc Quynh[1]([⊠])(iD), Nguyen Xoan Thao[2], Nguyen Doan Dong[1], and Nguyen Quang Thuan[1]

[1] Faculty of Applied Sciences - International School, Vietnam National University, Hanoi, Vietnam
{ducquyn,dongnd,thuannq}@vnu.edu.vn
[2] Faculty of Information Technology, Vietnam National University of Agriculture, Hanoi, Vietnam
nxthao@vnua.edu.vn

Abstract. We are living in the digital era with many modern technologies assisting people's lives. However, making decisions with massive information still very difficult. In Multi-criterial decision-making (MCDM) problems, the information is usually uncertain and ambiguous. In such problems, applying fuzzy set seems to be appropriate but limited in information representation. The entropy based on similarity measures of intuitionistic fuzzy sets (IFS) can be applied to tackle such issues. In this work, we introduce a formula which can satisfy all the conditions of entropy. We also propose an algorithm applying a new entropy and similarity measures for portfolio selection problems. We demonstrate the proposed approach to rank assets in stock markets in Vietnam. The experimental results on the Vietnamese stocks data set show that our approach is very useful. The entropy and the similarity measures of IFS can be an alternative tool for solving portfolio selection problems.

Keywords: Intuitionistic fuzzy set · Similarity measures · Vietnamese stock markets · Stocks ranking

1 Introduction

Making decisions involving multiple criteria or multi-criteria decision-making (MCDM), in our daily-life, always is much strenuous because of incomplete data, time pressure, ambiguous information. There are many solutions for these problems in order to assist the decision-maker to have an optimum choice. The conventional methods are used mathematical approximations but the results is often poor in performance. Another solution is applying the fuzzy set with many work applied in various fields such as banking, education, etc. However, the fuzzy set approach do not represent much information, so the decision is able to be incorrect. In order to increase the accuracy and ambiguity, the IFS is a solution [16]. Compare to the classical fuzzy set, the IFS adds a membership function and two functions in the IFS have the sum [1] smaller than or equal 1. Additionally, the topic in the IFS entropy and the similarity were widely used [7] and

N.-T. Nguyen et al. (Eds.): ICIT 2022, LNDECT 148, pp. 78–86, 2022.
https://doi.org/10.1007/978-3-031-15063-0_7

applied in many fields of real problems such as pattern recognition [9], selecting target market [21], etc.

In finance management, one of the most important problem is the portfolio selection. In the stock markets, the goal of the investors is to select good assets with not only low risks but also high returns. There are some indicators applied to measure risks and some to indicate profits or returns. To indicate the risks of the assets, there were many publishes in the past. In [13], the Markowitz's model was introduced for evaluation of the risk by applying the variance in the model. In [24], an indicator so-called Sharpe ratio (by combining the return and standard deviation) was used to maximize. The authors in [10, 12] used constraints of cardinality to diversify the portfolio.

In Vietnam, investors tend to choose very few assets namely one or two assets to invest, so to build a good strategic investment is much more difficult when applying optimization models. As a result, ranking the assets and selecting the good ones are optimal solutions.

Ranking alternatives can be done using the MCDM models thanks to the combination of multiple criteria. A criterion has influences on the ranking result due to its weight. This weight is evaluated by using the entropy of each criterion to measure the uncertainty of information [2,5,6,11,14,17]. There are some variant formulations of entropy measures based on the classical one as in [3,8,19,26], sine or cosine functions as in [20,25] but it is not intuitive. Besides, the similarity also can be used to rank the alternatives [16,23]. Using an entropy measures of IFS derived from the existing similarity measures as in [22] are much more easier and intuitive. Therefore, this is the our motivation to utilize the MCDM model with the similarity of IFS in order to rank assets.

A single or multi criteria has been utilized in many work recently to rank the assets. The most well-known approaches can be the AHP, Simple Additive Weight (SAW), p-VIKOR, Estimation of Weight, p-TOPSIS method [4]. The distances on R^n are applied in these approaches, so the performances of the results are poor because the indicators of assets are not certainty.

In this paper, we use the similarity on IFSs to rank the assets. From returns in history, we compute these asset indicators. The criteria set can be seen as a universal set and an asset can be seen as an intuitionistic fuzzy set. Based on those assets the similarity between them can be calculated. After that, we calculate perfect elements. Finally, the results of ranking assets is yielded based on the similarities.

The organization of the paper is as follows: The preliminary is introduced in Sect. 2; Sect. 3 recalls an entropy of IFS; Sect. 4 discusses the applications; finally, the conclusion is in Sect. 5.

2   Background

In this part, the concepts of the intuitive fuzzy sets (IFS) are discussed.

Definition 1 (Atanassov). *On the universal set U, an IFS is defined as follows:*

$$A = \{(u, \mu_A(u), v_A(u)) | u \in U\}$$

where $\mu_A(u) \in [0,1]$, $v_A(u) \in [0,1]$ are the membership and non-membership functions of u in A, respectively, and

$$\mu_A(u) + v_A(u)) \leq 1$$

for all u ∈ U.

$\pi_A(u) = 1 - \mu_A(u) - v_A(u)$ is defined as the hesitant of u in U.
Denoted that $IFS(U)$ is the family of IFS in the universal U.
With two IFSs $A, B \in IFS(U)$, it can be obtained that:

- $A \subset B$ iff $\mu_A(u) \leq \mu_B(u)$ and $v_A(u) \geq v_B(u)$.
- $A = B$ iff $\mu_A(u) = \mu_B(u)$ and $v_A(u) = v_B(u)$.
  For all $u \in U$
- The complement of $A \in IFS(U)$ is

$$C(A) = \{(u, v_A(u), \mu_A(u)) | u \in U\}$$

- For all $\alpha > 0$, it is defined that:

$$A = \{(u, (\mu_A(u))^\alpha, 1 - (1 - v_A(u))^\alpha) | u \in U\} \tag{1}$$

Definition 2. *A real function $S : IFS(U) \times IFS(U) \to [0, 1]$ is a measure of similarity in the IFSs if it satisfies the axioms as follows:*

(S1) $0 \leq S(A, B) \leq 1$ for all $A, B \in IFS(U)$,
(S2) $S(A, C(A)) = 0$ if A is crisp set,
(S3) $S(A, A) = 1$ for all $A \in IFS(U)$,
(S4) $S(A, B) = S(B, A)$ for all $A, B \in IFS(U)$,
(S5) If $A, B, C \in IFS(U)$ such that $A \subset B \subset C$ then $S(A, C) \leq min\{S(A, B), S(B, C)\}$.

In our approach, the finite universal set $U = \{u_1, u_2, ..., u_n\}$ is considered. A real number $\omega_i \in [0, 1]$ is so-called a weight of each element u_i if it satisfies the condition $\sum_{i=1}^n \omega_i = 1$.
Let $A = \{(u, \mu_A(u), v_A(u)) | u \in U\}, B = \{(u, \mu_B(u), v_B(u)) | u \in U\}$ be two IFSs on U.

Example 1. There have been several proposed measures of similarity:

1. In [18], similarity measure is defined as:

$$S_S(A, B) = \sum_{i=1}^n \frac{\omega_i}{3} \{2\sqrt{\mu_A(u_i)\mu_B(u_i)} + 2\sqrt{v_A(u_i)v_B(u_i)} + \sqrt{\pi_A(u_i)\pi_B(u_i)}$$
$$+ 2\sqrt{(1 - \mu_A(u_i))(1 - \mu_B(u_i))} + 2\sqrt{(1 - v_A(u_i))(1 - v_B(u_i))}\} \tag{2}$$

2. In [15], similarity measure is defined as:

$$S_Q(A, B) = \frac{1}{n} \sum_{i=1}^n S_Q^i(A, B) \tag{3}$$

where

$$S_Q^i(A, B) = \begin{cases} 1, & \text{if } \mu_A(u_i) = \mu_B(u_i) \text{ and } v_A(u_i) = v_B(u_i) \\ \frac{3 + min\{\mu_A(u_i), \mu_B(u_i)\} - max\{v_A(u_i), v_B(u_i)\} - |\mu_A(u_i) - \mu_B(u_i)| - |v_A(u_i) - v_B(u_i)|}{4}, & \text{other cases} \end{cases}$$

for all $i = 1, 2, ..., n$.

3   Entropy of IFSs Based on Similarity Measures

The concept of IFS entropy was proposed in [19]. The idea of the method is to measure the IFS intuition.

Definition 3. *An entropy of $A \in IFS(U)$ is a function $E : IFS(U) \to [0,1]$ such that: For all $u_i \in U$*

(e1) $E(A) = 0$ if A having $\mu_A(u_i) = 1$ or $v_A(u_i) = 1$,
(e2) $E(A) = 1$ if $\mu_A(u_i) = v_A(u_i)$,
(e3) $E(A) = E(C(A))$,
(e4) If $A, B \in IFS(U)$, such that:
   $\mu_A(u_i) \le \mu_B(u_i) \le v_B(u_i) \le v_A(u_i)$
   or $\mu_A(u_i) \ge \mu_B(u_i) \ge v_B(u_i) \ge v_A(u_i)$,
   then $E(A) \le E(B)$.

The entropy $E : IFS(U) \to [0,1]$ is determined as the expression as follows:

Theorem 1. $E(A) = S(A, C(A))$ $(A \in IFS(U))$.

From Theorem 1, there are some new entropies of IFS as follows:
Denoted that the weight of u_i is $\omega_i = \frac{1}{n}$, for all $i = 1, 2, ..., n$. It is obtained that:

1. Based on the Eq. (2), it is obtained that:

$$E_S(A) = \frac{1}{3n} \sum_{i=1}^{n} \{ 4\sqrt{\mu_A(u_i) v_A(u_i)} + 1 - \mu_A(u_i) - v_A(u_i) + 2\sqrt{(1 - \mu_A(u_i))(1 - v_A(u_i))} \}$$

$$(4)$$

2. Based on the Eq. (3), it is obtained that:

$$E_Q(A) = \sum_{i=1}^{n} E_Q^i(A) \tag{5}$$

where

$$E_Q^i(A) = \begin{cases} 1, \text{ if } \mu_A(u_i) = v_A(u_i) \\ \frac{3 + \min\{\mu_A(u_i), v_A(u_i)\} - \max\{\mu_A(u_i), v_A(u_i)\} - 2|\mu_A(u_i) - v_A(u_i)|}{4}, & \text{other cases} \end{cases}$$

for all $i = 1, 2, ..., n$.
The feasibility of these entropies is shown in the [22].

4   Application of Proposed Measure in Ranking Vietnamese Stocks

4.1   Problem Definitions

The investor aims to choose an asset to invest, so the problem of portfolio selection in ranking stocks in Vietnam are considered to demonstrate our approach. Suppose that we have m assets and the history of data from [0 to , T].
   We denoted that the at the period t:

– The price of asset i is p_{it}.
– And the expression of the returns r_{it} of the asset i is as $r_{it} = \dfrac{p_{it} - p_{i(t-1)}}{P_{i(t-1)}}$.

There are several indicators of finance for a asset and they are classified into two groups. One group indicates the asset profitability. The other indicate the risk of the asset.

We consider 5 indicators for the asset i:

ER - The calculation of Expected Return (ER): $ER_i = \frac{1}{T} \sum\limits_{t=1}^{T} r_{it}$

Std - The calculation of the standard deviation: $Std_i = \sqrt{\frac{1}{T} \sum\limits_{t=1}^{T} (r_{it} - ER_i)^2}$

SR - The calculation of the Sharpe Ratio (SR): $SR_i = \frac{ER_i}{Std_i}$

PR - The calculation of the Positive Rate (PR): $PR_i = \frac{1}{T} \sum\limits_{t=1}^{T} \eta(r_{it})$ where $\eta(r_{it}) = 1$
if $r_{it} > 0$, 0 otherwise
ML - The calculation of the Maximum Loss:

$$ML_i = \begin{cases} |\min\limits_{t} r_{it}|, \text{if there is } r_{it} < 0 \\ 0, \text{otherwise} \end{cases}$$

The ranking of the assets based on those indicators can assists the investor to choose optimal assets to invest.

It is difficult for this task when working on the real values data. In our approach, a vector of the indicator for each asset is converted from the data to an IFS. Then ranking those assets can be done easily thanks to the measure of the similarity.

4.2   Solution Method

Considering that $A_1, A_2, ..., A_m$ and $C_1, C_2, ..., C_n$ for finance are m assets and n indicators for finance, respectively. We denote that the universal set is $C = \{C_1, C_2, ..., C_n\}$. The objective is to define the membership and non-membership functions of the asset $A_i (i = 1, 2, ..m)$. If the indicator j for the asset i, denoted by $u_{i,j} (j = 1, 2, .., n)$ is computed based on the history data, so two functions mentioned above can be calculated as follows:
The membership function:

$$\mu_{ij} = \mu_{A_i}(C_j) = \frac{u_{ij} - lb_j}{ub_j - lb_j + \varepsilon} \tag{6}$$

The non-membership function:

$$v_{ij} = v_{A_i}(C_j) = \frac{ub_j - u_{ij}}{ub_j - lb_j + \varepsilon} \tag{7}$$

where $lb_j = \min\limits_{i}(u_{ij})$, $ub_j = \max\limits_{i}(u_{ij})$ and ε is a positive constant. It is obvious that our defined functions, the membership function μ and the non-membership function v, are still satisfied the conditions. In the next step, we develop the intuitionistic fuzzy model to select portfolio, a MCDM model, applying to rank assets. The new entropy and measure of IFS similarity are used in our approach.

The approach are described as following:

For all $i = 1, 2, ..., m$ and $(j = 1, 2, ..., n)$ with m is the number of alternatives and n is the number of criteria.

1. Determining $C = \{C_1, C_2, ..., C_n\}$ as the criteria for assessment
2. Identifying $A = \{A_1, A_2, ..., A_m\}$ as alternatives for assessment
3. Determining the IFSs $A_i = \{(C_j, \mu_{ij}, v_{ij}) | C_j \in C\}$.
   Denoted that A_i is an IFS of C
4. Determining ω_j is the weight of the criteria C_j.
   In which, $C_j = \{(A_i, \mu_{ij}, v_{ij}) | A_i \in A\}$ is as the IFS on the universal set A.
   The expression to calculate the weight of C_j can be defined as following:

$$\omega_j = \frac{1 - e_j}{n - \sum_{i=1}^{n} e_j} \qquad (8)$$

   with $e_j = E(C_j)$.
5. Choosing A_p (the perfect alternative - an IFS on universal set C):
   For the non-benefit (or so-called cost criteria) are in range $(0, 1)$ and the benefit criteria are also in range $(1, 0)$.

6. Calculating $S(A_b, A_i)$ as the expression stated in Example 2.
7. Ranking the options by applying the measure of similarity.
   With two alternative A_i, A_k, it is said that A_i is ranked higher than A_k once $S(A_i, A_b) > S(A_k, A_b)$.

5   Experiment and Results

We apply the described approach on the Vietnamese stock price dataset from 2020 to 2021. The data has different number of stocks. There are some missing values in the dataset; however, we use the interpolated approach to fill those missing values. The brief description of the dataset is on the Table 1.

Table 1. Vietnamese stocks dataset from 2020 to 2021

No.	Column	Count	Non-Nul	Data type
0	Symbol	428105	Non-null	string
1	Open price	428105	Non-null	number
2	Close price	428105	Non-null	number
3	High price	428105	Non-null	number
4	Low price	428105	Non-null	number
5	Trading date	428105	Non-null	datetime

As presented in Sect. 4, Standard Deviation (Std), Positive Rate (PR), Expected Return (ER), Max Loss (ML), Sharpe Ratio (SR) are used as indicators in our method. The formulas of these indicators are provided in the Sect. 4. The value of 0.001 was chosen for epsilon in our experimental settings.

The results were ranked by SR indicator and Similarity indicator for the Vietnamese stocks datasets after calculating as in Table 2 to 3. As stated previously, in the results, we use the SR indicator because the SR is one of the most important indicators due to the balance between the return and the standard deviation. We show on the report only top ten assets ranking on Similarity and Sharpe Ratio.

As in the results, it is obvious that:

– All the top ten assets ranking by Similarity have ML with value of 0.0
– The ER or Expected Return of top assets is positive
– Assets with symbol BTR, HFS, VTI appear in both results raking by Similarity and SR.
– There are differences between the Similarity ranking and SR ranking. However, the SR has a heavy influence on the similarity ranking because the SR values are quite high.

It also can be seen that the similarity measure combines many indicators therefore, the results ranking by similarity is better than SR ranking. Also, the similarity reflect the comparison of the assets because of the membership and non-membership functions. From the results of ranking assets in stock market, it can be helpful for the investors to select assets to invest based on the ranking the assets. As a result, the similarity of IFSs applied to rank the assets can be seen as a new approach for portfolio selection problems.

Table 2. Results using similarity indicator to rank Vietnamese stocks

Name	ER	PR	SR	STD	ML	Similarity	Similarity ranking	SR ranking
BTR	0.003356	0.024457	0.146215	0.022955	0.0	0.390227	1	2
HFS	0.002054	0.014599	0.105251	0.019520	0.0	0.387163	2	9
HAB	0.001817	0.011952	0.102290	0.017763	0.0	0.386997	3	12
VTI	0.002976	0.010526	0.103142	0.028857	0.0	0.384555	4	10
CEC	0.000438	0.005865	0.061915	0.007069	0.0	0.383124	5	57
XDH	0.001733	0.007968	0.079351	0.021841	0.0	0.382396	6	21
BNW	0.000259	0.003984	0.055665	0.004651	0.0	0.382175	7	70
HPU	0.001601	0.005650	0.075378	0.021245	0.0	0.381490	8	24
BXT	0.000019	0.001992	0.044677	0.000421	0.0	0.380126	9	111
VLP	0.001041	0.003984	0.063186	0.016480	0.0	0.379957	10	54

Table 3. Results using SR indicator to rank Vietnamese stocks

Name	ER	PR	SR	STD	ML	Similarity	Similarity ranking	SR ranking
PTO	0.005457	0.063745	0.163315	0.033413	0.162304	0.133712	279	1
BTR	0.003356	0.024457	0.146215	0.022955	0.000000	0.390227	1	2
TVG	0.005029	0.025896	0.141962	0.035423	0.141304	0.131424	282	3
CYC	0.004452	0.085657	0.132511	0.033598	0.166667	0.115867	289	4
IME	0.002955	0.031873	0.131066	0.022548	0.176471	0.113053	292	5
SPH	0.003122	0.035857	0.127800	0.024428	0.171671	0.113502	291	6
ISG	0.003393	0.037849	0.127407	0.026628	0.173913	0.112886	294	7
T12	0.002999	0.035857	0.109913	0.027288	0.175299	0.110222	296	8
HFS	0.002054	0.014599	0.105251	0.019520	0.000000	0.387163	2	9
VT1	0.002976	0.010526	0.103142	0.028857	0.000000	0.384555	4	10

6 Conclusions

In this paper, we demonstrate an intelligent approach for MCMD problem using entropy on IFSs. We have applied the new entropy and the measure of similarity to rank Vietnamese stocks. The experimental results on stock datasets in Vietnamese market from 2020–2021 have show that our proposed approach can be seen as an alternative to solve problems in portfolio selection. In the future, we will continue to investigate this approach in order to apply on other domains.

References

1. Atanassov, K.T.: Intuitionistic fuzzy sets. Fuzzy Sets Syst. **20**(1), 87–96 (1986). https://doi.org/10.1016/S0165-0114(86)80034-3. https://www.sciencedirect.com/science/article/pii/S0165011486800343
2. Batyrshin, I.: On fuzzinesstic measures of entropy on Kleene algebras. Fuzzy Sets Syst. **34**(1), 47–60 (1990)
3. Burillo, P., Bustince, H.: Entropy on intuitionistic fuzzy sets and on interval-valued fuzzy sets. Fuzzy Sets Syst. **78**(3), 305–316 (1996)
4. Chou, S., Duong, T.T., Xuan Thao, N.: Renewable energy selection based on a new entropy and dissimilarity measure on an interval-valued neutrosophic set. J. Intell. Fuzzy Syst. **40**(6), 11375–11392 (2021)
5. De Luca, A., Termini, S.: A definition of a nonprobabilistic entropy in the setting of fuzzy sets theory. Inf. Control **20**(4), 301–312 (1972)
6. Gohain, B., Dutta, P., Gogoi, S., Chutia, R.: Construction and generation of distance and similarity measures for intuitionistic fuzzy sets and various applications. Int. J. Intell. Syst. **36**(12), 7805–7838 (2021)
7. Huang, J., Jin, X., Lee, S.J., Huang, S., Jiang, Q.: An effective similarity/distance measure between intuitionistic fuzzy sets based on the areas of transformed isosceles right triangle and its applications. J. Intell. Fuzzy Syst. **40**(5), 9289–9309 (2021)

8. Hung, W.L., Yang, M.S.: Fuzzy entropy on intuitionistic fuzzy sets. Int. J. Intell. Syst. **21**(4), 443–451 (2006)

9. Jiang, Q., Jin, X., Lee, S.J., Yao, S.: A new similarity/distance measure between intuition-istic fuzzy sets based on the transformed isosceles triangles and its applications to pattern recognition. Expert Syst. Appl. **116**, 439–453 (2019)

10. Jimbo, H.C., Ngongo, I.S., Andjiga, N.G., Suzuki, T., Onana, C.A.: Portfolio optimization under cardinality constraints: a comparative study. Open J. Stat. **7**(4), 731–742 (2017)

11. Joshi, R., Kumar, S.: A novel fuzzy decision-making method using entropy weights-based correlation coefficients under intuitionistic fuzzy environment. Int. J. Fuzzy Syst. **21**(1), 232–242 (2019)

12. Le Thi, H.A., Tran, D.Q.: Solving continuous min max problem for single period portfolio selection with discrete constraints by DCA. Optimization **61**(8), 1025–1038 (2012)

13. Markowitz, H.: Portfolio selection. J. Financ. **7**(1), 77–91 (1952). http://www.jstor.org/stable/2975974

14. Meng, F., Chen, X.: Entropy and similarity measure of Atanassov's intuitionistic fuzzy sets and their application to pattern recognition based on fuzzy measures. Pattern Anal. Appl. **19**(1), 11–20 (2016)

15. Quynh, T.D., Thao, N.X., Thuan, N.Q., Van Dinh, N.: A new similarity measure of IFSs and its applications. In: 2020 12th International Conference on Knowledge and Systems Engineering (KSE), pp. 242–246. IEEE (2020)

16. Shen, F., Xu, J., Xu, Z.: An outranking sorting method for multi-criteria group decision making using intuitionistic fuzzy sets. Inf. Sci. **334**, 338–353 (2016)

17. Singh, S., Sharma, S.: On a generalized entropy and dissimilarity measure in intuitionistic fuzzy environment with applications. Soft. Comput. **25**(11), 7493–7514 (2021). https://doi.org/10.1007/s00500-021-05709-1

18. Song, Y., Wang, X., Quan, W., Huang, W.: A new approach to construct similarity measure for intuitionistic fuzzy sets. Soft. Comput. **23**(6), 1985–1998 (2017). https://doi.org/10.1007/s00500-017-2912-0

19. Szmidt, E., Kacprzyk, J.: Entropy for intuitionistic fuzzy sets. Fuzzy Sets Syst. **118**(3), 467–477 (2001)

20. Thao, N.X., Chou, S.Y.: Novel similarity measures, entropy of intuitionistic fuzzy sets and their application in software quality evaluation. Soft Comput. **26**(4), 2009–2020 (2022). https://doi.org/10.1007/s00500-021-06373-1

21. Thao, N.X., Duong, T.T.T.: Selecting target market by similar measures in interval intuition-istic fuzzy set. Technol. Econ. Dev. Econ. **25**(5), 934–950 (2019)

22. Tran, D.Q., Nguyen, X.T., Nguyen, D.D., Nguyen, Q.T.: A novel entropy of intuitionis-tic fuzzy sets based on similarity and its application in finance. J. Intell. Fuzzy Syst. 1–11 (preprint)

23. Wang, J.Q., Peng, J.J., Zhang, H.Y., Chen, X.H.: Outranking approach for multi-criteria decision-making problems with hesitant interval-valued fuzzy sets. Soft Comput. **23**(2), 419–430 (2019)

24. Yao, H., Li, Z., Li, X., Zeng, Y.: Optimal Sharpe ratio in continuous-time markets with and without a risk-free asset. J. Ind. Manag. Optim. **13**(3), 1273 (2017)

25. Ye, J.: Two effective measures of intuitionistic fuzzy entropy. Computing **87**(1), 55–62 (2010)

26. Zhu, Y.J., Li, D.F.: A new definition and formula of entropy for intuitionistic fuzzy sets. J. Intell. Fuzzy Syst. **30**(6), 3057–3066 (2016)

Edge Computing for Wireless Sensor Network

Thien An Nguyen$^{(\boxtimes)}$ ⓘ, Nhu Hung Huynh ⓘ, and Trong Nhan Le ⓘ

Ho Chi Minh City University of Technology, Ho Chi Minh City, Vietnam
{an.nguyenblue_0610,hung.huynh.tiviluson,trongnhanle}@hcmut.edu.vn

Abstract. Wireless sensor networks (WSNs) are being substantially popular in edge computing these days. In order to establish a reliable and sustainable edge AI system in the network, multiple conditions need to be considered. The performance, accuracy aspects are vital to a successful edge AI node in a WSN as more and more complicated tasks are being solved at the edge of the network. Energy management is also a concern as a great amount of edge devices are being deployed at rural, isolated areas with little to no common energy source. Therefore, energy harvesting is a great solution to the problem. In this paper, we will consider deeply into those categories and provide statistics about the topic.

Keywords: Edge computing · Edge devices · Machine learning ·
Computer vision · Embedded · Wireless sensor network ·
Energy-neutral operation

1 Introduction

Edge computing is a computing paradigm that performs computing at the edge of the network. Thus, the computation may be as near the data as possible rather than concentrating on a far off location like cloud computing. Along with Edge computing, Edge AI is proving itself to be a sustainable and practical way to develop machine learning at the edge. Smart devices utilizing machine learning applications can come in many use cases. In the field of image processing, surveillance cameras are using Edge AI to process long recordings in real time in order to detect and handle suspicious activities. Edge AI can also be used in safety mechanisms in new vehicles to help predict and prevent an accident from happening in real time, which can not be done if only cloud computing was used. IoT overall has seen more and more Edge AI devices being used and AI-powered solutions are believed to be the major technology for years to come. However, there is a wide variety of devices in terms of price, processing speed, form factor, etc. Choosing an appropriate combination for each purpose is a challenge nowadays. In this paper, we will investigate and discuss a few popular options.

Apart from the physical specifications of the devices, the models running on the edge devices are also the keys to a successful implementation. Although

N.-T. Nguyen et al. (Eds.): ICIT 2022, LNDECT 148, pp. 87–97, 2022.
https://doi.org/10.1007/978-3-031-15063-0_8

there are more and more powerful and efficient devices being introduced into the market every year, the sheer size and functionalities of modern machine learning applications tend to overwhelm them. In this paper, we will choose one of the most popular topic which is object detection and discuss different popular object detection models and their effect on different types of edge devices.

Finally, another crucial aspect in IoT is the network connecting the devices. The design of WSNs mainly concentrates on the collecting wirelessly the sensory information and transmission of data from the sensors to the base station. However, their batteries are often a problem when it comes to practicality since they are limited in size and cost. In the experiments in this paper, our devices will come with the energy harvesting ability in order to maintain it's continuity. We will also discuss the power problems that may arise in a system specialized in detecting birds in the forest and our approach in solving them.

The rest of this research contains the following contents. In Sect. 2, related works regarding the contemporary researches about the co-design of software and hardware for the deployment of Edge AI on Edge devices around metrics such as inference time, accuracy, the constraints of computation capability, energy consumption, etc. on Edge devices. In Sect. 3, a brief discussion is given around the reference devices and computer vision models in use, followed by a modified general energy model for Edge devices. Finally, comparison and computation results are presented, alongside with other discussion and comments about the suitability of the devices for specific Edge AI applications.

2   Related Works

There are various factors, when it comes to deploying Edge AI, especially for the field of computer vision [1–3], on Edge devices: on the software aspect, it is the size, the architecture, the required input and the supported operations of the model; on the hardware aspect, some crucial specifications must be taken into account such as the form factor, connection, energy requirement, processing speed, hardware architecture, etc. Together, these factors decide inference speed and accuracy of the model, the generated heat and consumption energy of the system.

The importance of on-the-edge inference of AI shown in the preceding works has motivated more extensive research into the field of Edge AI. Many have worked on the technique related to the trade-off between accuracy and performance to create small models, including pruning [4], quantization [5], development of mobile-specific models [6,7]. Hardware and frameworks are also upgraded to match up with the pace of the counterparts [8,9].

Furthermore, as discussed above, to deploy Edge AI on a WSN for environmental monitoring, energy consumption and energy harvesting should be of the top priorities, because of the highly constrained devices. In [10], a energy model for WSN has been proposed for an Energy Neutral Operation condition. We will incorporate our findings into this model, with some modifications for generalization purpose, to inspect the suitability of each configuration for deployment of Edge AI for a specific purpose.

3   Methodology

3.1   Edge Devices and Machine Learning Models for Comparison

Edge Devices. Our hardware platforms are briefly introduced in this section, which consist of 2 Edge devices and 1 Edge accelerator.

- **Raspberry Pi 4 model B:** Created by the Raspberry Pi Foundation, this 4th version of the Raspberry Pi series is a single-board, embedded computer, suitable for video analysis projects [11] and has proven to be easily accessible yet effective in Edge computing. Many machine learning inference has been sufficiently deployed on it without the need of any extra peripherals [12]. Raspberry Pi 4 runs on Raspberry Pi Desktop Operating System (OS), a configured Debian OS. However, Ubuntu Desktop/Server can also be installed as its OS, which makes Raspberry not only a powerful but also flexible embedded computer.
- **NVIDIA Jetson Nano Developer Kit:** This powerful yet compact embedded computer enables parallel operations of multiple popular neural network applications such as image classification [13], object detection [14], and even speech processing [15].
- **Coral USB Accelerator:** It comes with an Edge Tensorflow Processing Unit (TPU) coprocessor, enabling high-speed machine learning inferencing. It is particularly useful due to its support on a wide range of systems and a simple connection via a USB port. It also supports TensorflowLite models, and offers conversion TensorflowLite model into Edge_TPU TensorflowLite model, which in turn optimizes the utility of TPU on Coral USB Accelerator. Note that this is the only device that needs to connect to a host device in order to accelerate the host's computation capability. In this paper, the Coral USB Accelerator will be used on both of the aforementioned embedded systems.

Table 1. Device specifications of Edge AI devices in use

	GPU	CPU	RAM
Raspberry Pi version 4 Model B	Broadcom VideoCore VI	Quad-core Cortex-A72	4 GB LPDDR4 SDRAM
NVDIA Jetson Nano Developer Kit	NVIDIA Maxwell 128 CUDA cores	Quad ARM A57 MPCore	4 GB LPDDR4 SDRAM at 25.6 GB/s
Coral USB Accelerator	N/A	Quad-core ARM.A53	N/A

Frameworks. There are various frameworks for deploying Machine Learning models onto Edge AI devices [9]. In this paper, we only use **TensorFlow-Lite**

models [16] in order to ensure the fairness in the comparison between the inference time on different hardware setups. TensorFlow-Lite (TFLite) is a compact version of the TensorFlow engine, specifically used for constrained systems. It optimizes the computations for constrained systems like mobile or Edge AI devices by providing prunning, on-training and post-training quantization, computation graph freezing for deployment, etc.

Models. As previously stated, what we are concerned with models that have low inference time and high compatibility and deployability on the Edge AI devices. In this paper, we used 2 well-known models:

- **MobileNet V2** [7]: The architecture of MobileNet V2 is based on its predecessor, MobileNet V1 [17], which used to be a popular breakthrough in computer vision. MobileNet V2 is a 53-layer-deep CNN. The *original* model can classify images into 1000 object categories, from technological devices like keyboard and computer to animals like mice and birds. We also have another *retrained* network that classify images into 900+ bird categories, in order to prove the consistency and adaptability of MobileNet V2.
- **Inception V4** [18]: the `bottleneck` layer used in MobileNet V2 was pioneered by the 1st Inception model [19]. Inception V4 also introduced also applies the technique of `residual` layers to boost the computation as well as to avoid overfitting. This model is originally trained to classify 1000 object categories.

Table 2. Models' specification

Model name	Base validation dataset	Input size	Number of images	Model size
MobileNet V2	ILSVRC2012	$224 \times 224 \times 3$	15000	42.9 MB
	iNaturalist2017/Aves	$224 \times 224 \times 3$	9000	42.9 MB
Inception V4	ILSVRC2012	$224 \times 224 \times 3$	15000	4.1 MB

For a fair comparison, all models in use are *quantized*. The popularity of quantization arises rapidly, which results from, as shown in several studies [3,20,21], the burden on memory and processing speed in all systems can be significantly reduced when more compact data types are used, which is of the essence for constrained systems like our Edge AI devices.

3.2 Energy Model

There is a variety of energy sources for electrical devices available on the market. We can be flexible in choosing the suitable type for our system's purpose. As

Edge devices are being introduced further and further into more distant, isolated areas, the obvious reliable source would be from nature itself.

For long operating autonomous sensors, it's mandatory to make sure that the consumed energy is at most as high as the harvested energy over a protracted period. Trong Nhan Le et al. introduced the term *energy-neutral operation* (ENO) which is a crucial condition to resolve for a long lasting lifetime [10]. In their case, the devices are operating in intervals that the active time is extremely small and the sleep time takes most of the devices' life span. Moreover, when dealing with solar energy, we need to take into account the day time and night time. In a more general term, the formula we should satisfy for day time to achieve ENO is presented below:

$$E_{EI} = P_H T_{EI} - \frac{1}{\eta} \left(\frac{T_{EI}}{T_{Sleep} + T_{Active}} E_{Active} + P_{Sleep} T_{EI} \right) - P_{Leak} T_{EI} \quad (1)$$

where:

E_{EI}: The total harvested energy during daytime.
P_H: The harvesting power.
T_{EI}: Harvesting energy interval.
η: The average DC/DC converter efficiency.
$\dfrac{T_{EI}}{T_{Sleep} + T_{Active}}$: The number of operating periods during T_{EI}
E_{Active}: The consumed energy every time the device is active.
P_{Sleep}: The energy consumed under sleeping mode.
P_{Leak}: The leakage power of the whole system.

As stated in [10], a suitable η should be 0.85 due to certain aspects such as the efficiency range, consumed energy and the impracticality of measuring the exact number. Other specifications from the energy harvesting system we use can be seen in Table 3.

Table 3. Device specifications needed for formulas

Device	Specifications
MONO MSP 40 W Solar panel	P = 40 W
LiFePO4 Deep Cycle Lithium Iron Phosphate Battery	P = 0.35 W

Since the solar panel has P = 40 W, P_H shall be 40 W. We will choose the power leakage to be 8%/month as the battery we are using is of type lithium [22]. Thus, $P_{Leak} = \frac{0.35*8\%}{30} = 0.93$ mW. The energy being consumed while in sleep mode (idle energy) of the Edge devices we are using can be seen in Table 4. Because the energy provided is in the form of solar energy, we can assume on a normal day, the time we can harvest energy from the sun would be 10 h, which means $T_{EI} = 10$ h and $T_{NEI} = 14$ h the combination of them would make a

Table 4. The power specifications of the edge devices in use

Edge device	Power	Idle power
Raspberry pi 4 model B	3.8–5.5 W [23]	3.8–4 W [23]
NVDIA Jetson Nano Developer Kit	5–10 W [24]	1.25 W [9]
Coral USB Accelerator	2.5–4.5 W [25]	3.24 W [9]

full 24 h day. Similar to daytime, at night, we also have a formula to compute the energy being used. The only difference is that at night, there is no sun light which is why there is no $P_H T_{NEI}$.

$$E_{NEI} = \frac{1}{\eta} \left(\frac{T_{NEI}}{T_{Sleep} + T_{Active}} E_{Active} + P_{Sleep} T_{NEI} \right) - P_{Leak} T_{NEI} \qquad (2)$$

It is necessary that the ENO condition after a cycle, which includes T_{EI} and T_{NEI}, be satisfied in order for the achievement of the theoretical infinite lifetime of the WSN node. This requirement is equivalent to the condition of $E_{EI} \geq E_{NEI}$. Therefore, from Eq. 1 and Eq. 2, we can draw out the maximum energy that would be consumed every time the system wakes up as demonstrated in Eq. 3.

$$E_{Active} \leq -(\eta P_{Leak} + P_{Sleep})(T_{Sleep} + T_{Active}) + \frac{\eta T_{EI}(T_{Sleep} + T_{Active})}{T_{EI} + T_{NEI}} \qquad (3)$$

If we tuned the Eq. 3 to be cleaner with 2 apparent variables which are T_{Active} and T_{sleep} as well as add in the fact that $T_{EI} + T_{NEI}$ should be a constant 24 h, the resulting formula would be as such:

$$E_{Active} \leq -\left(\eta P_{Leak} + P_{Sleep} - \frac{\eta P_H T_{EI}}{24} \right) (T_{Sleep} + T_{Active}) \qquad (4)$$

4   Result and Discussion

4.1   Edge Devices and Machine Learning Models for Comparison

Figure 1 illustrate the top 1 and top 5 prediction for each model, sorted by the order of its confidence score. Note that no conclusion should be drawn about the difference of accuracy among any of the models, because each is trained differently and independently. Another point is that although only the data for NVIDIA Jetson Nano, instead for both devices, is shown, because the difference in accuracy between the 2 devices is relatively negligible.

We can see that both the best and the best 5 prediction for all models are acceptably high, which may partially prove their practicality and usefulness for deployment onto Egde AI devices. 3 models follow the same trend: top-1 prediction ranges around 70% to 80% with a whooping increase to about 90% accuracy for top-5 prediction. With further inspection, we can observe that there

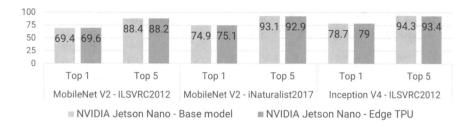

Fig. 1. The best and the best 5 predictions for each model

are usually some similarities among the classes in top-5 predictions. Our result corresponds well with some published data in [26]. After comparing the inference time on different models and configurations, the following conclusions are drawn:

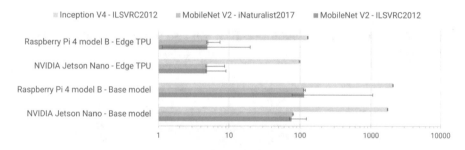

Fig. 2. Average, minimum and maximum inference time for each configuration (with logarithmic scaling)

- Overall, MobileNet V2 model yields acceptable inference speed on embedded devices regardless of the configuration. Inception V4 is not suitable for Edge devices.
- The fast inference speed of MobileNet V2 on edge devices is not hampered by the dataset on which the model was trained.
- Without Coral USB accelerator, Jetson Nano gives a 50%–100% faster inference time compared to Raspberry Pi 4.
- With Coral USB accelerator, the inference time can be reduced by a factor of up to 10x. The speed difference between NVIDIA Jetson Nano and Raspberry Pi 4 ceases to exist.
- The inference time varies very insignificantly due to the fact that the maximum inference time is far from the average inference time.

4.2   Energy Calculation

Raspberry Pi 4 Model B: If we replace the constants with their values in Eq. 4, and graph them, we would achieve 2 graphs for 2 configurations. As the idle power of the Raspberry Pi 4 Model B can vary between 3.8 W and 4 W, we will choose the average which is 3.9 W.

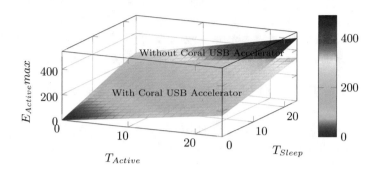

Fig. 3. Maximum active energy plot when using Raspberry Pi version 4 Model B

NVIDIA Jetson Nano Developer Kit: We also provide 2 graphs to display the relationship between T_{Active}, T_{Sleep}, and E_{Active}.

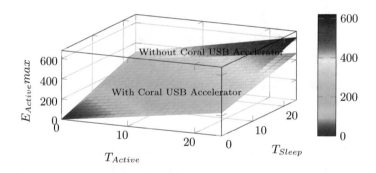

Fig. 4. Maximum active energy plot when using NVIDIA Jetson Nano Development Kit

We can see a trend emerges from the two figures that the combination with Coral USB Accelerator would result in an overall lower max active energy which increases the chance for the system to be not satisfying ENO. It is justifiable as the use of Coral USB Accelerator would result in a higher performance but also more power shall be consumed.

In the case where our system is working non stop (surveillance system, animal detecting system, etc.) meaning $T_{Sleep} = 0$ h and $T_{Active} = 24$ h, we can calculate the maximum energy that can be consumed each day by the system in each configurations as in Table 5:

In this case, we can see that the combination of Raspberry Pi version 4 Model B and Google Coral USB Accelerator doesn't satisfy the ENO while the other

Table 5. Energy consumed and maximum consumed energy while active in a day if work continuously

	Coral USB accelerator	Raspberry Pi version 4 Model B	NVIDIA Jetson Nano Developer Kit
Energy consumed/day	☒	93.6 Wh	30 Wh
	☑	171.36 Wh	107.76 Wh
Max active energy/day	☒	246.38 Wh	309.98 Wh
	☑	168.62 Wh	232.22 Wh

3 configurations do. The Jetson Nano consumes the least power which means it is the most optimal one to use in terms of power in this specific scenario.

4.3   Conclusion

This research analyzed the accuracy of prediction, inference time, energy consumption and energy limit for ENO in Edge AI devices. We hope that the following results may contribute to the users' choice of deploying Edge AI for a specific application, in terms of both frameworks and platforms, despite of our lack of devices and some biases that may have occurred. Additionally, we have inspected the computational optimization of Edge TPU USB. An indisputable point of view is that the combination of software and hardware designs allows low inference time on Edge devices, but there could also be a trade-off between the energy consumption and/or accuracy, which is a research trend that is extensively looked into recently.

For the future work, we intend to broaden the scope of comparison, both on the hardware and software aspects. Energy consumption of actual hardware platforms in operation and the effects of long-term operation will be taken into account to ensure the numerical measurement for the ENO condition.

References

1. Krizhevsky, A., Sutskever, I., Hinton, G.E.: Imagenet classification with deep convolutional neural networks. In: Advances in Neural Information Processing Systems, vol. 25 (2012)
2. Simonyan, K., Zisserman, A.: Very deep convolutional networks for large-scale image recognition. arXiv preprint arXiv:1409.1556 (2014)
3. Vanhoucke, V., Senior, A., Mao, M.Z.: Improving the speed of neural networks on CPUs (2011). https://research.google/pubs/pub37631/
4. Yu, J., Lukefahr, A., Palframan, D., Dasika, G., Das, R., Mahlke, S.: Scalpel: customizing DNN pruning to the underlying hardware parallelism. ACM SIGARCH Comput. Archit. News **45**(2), 548–560 (2017)
5. Courbariaux, M., Bengio, Y., David, J.-P.: Training deep neural networks with low precision multiplications. arXiv preprint arXiv:1412.7024 (2014)

6. Iandola, F.N., Han, S., Moskewicz, M.W., Ashraf, K., Dally, W.J., Keutzer, K.: SqueezeNet: AlexNet-level accuracy with 50x fewer parameters and <0.5 mb model size. arXiv preprint arXiv:1602.07360 (2016)
7. Sandler, M., Howard, A., Zhu, M., Zhmoginov, A., Chen, L.-C.: Mobilenetv2: inverted residuals and linear bottlenecks. In: Proceedings of the IEEE Conference on Computer Vision and Pattern Recognition, pp. 4510–4520 (2018)
8. Pena, D., Forembski, A., Xu, X., Moloney, D.: Benchmarking of CNNs for low-cost, low-power robotics applications. In: RSS 2017 Workshop: New Frontier for Deep Learning in Robotics, pp. 1–5 (2017)
9. Hadidi, R., Cao, J., Xie, Y., Asgari, B., Krishna, T., Kim, H.: Characterizing the deployment of deep neural networks on commercial edge devices. In: 2019 IEEE International Symposium on Workload Characterization (IISWC), pp. 35–48. IEEE (2019)
10. Le, T.N., Pegatoquet, A., Berder, O., Sentieys, O., Carer, A.: Energy-neutral design framework for supercapacitor-based autonomous wireless sensor networks. ACM J. Emerging Technol. Comput. Syst. (JETC) **12**(2), 1–21 (2015)
11. Nikouei, S.Y., Chen, Y., Song, S., Xu, R., Choi, B.-Y., Faughnan, T.R.: Intelligent surveillance as an edge network service: from Harr-cascade, SVM to a lightweight CNN. arXiv preprint arXiv:1805.00331 (2018)
12. Xu, R.: Real-time human objects tracking for smart surveillance at the edge. In: 2018 IEEE International Conference on Communications (ICC), pp. 1–6. IEEE (2018)
13. Mazzia, V., Khaliq, A., Salvetti, F., Chiaberge, M.: Real-time apple detection system using embedded systems with hardware accelerators: an edge AI application. IEEE Access **8**, 9102–9114 (2020)
14. Ullah, S., Kim, D.-H.: Benchmarking Jetson platform for 3d point-cloud and hyperspectral image classification. In: 2020 IEEE International Conference on Big Data and Smart Computing (BigComp), pp. 477–482. IEEE (2020)
15. Gao, C., Rios-Navarro, A., Chen, X., Delbruck, T., Liu, S.-C.: EdgeDRNN: enabling low-latency recurrent neural network edge inference. In: 2020 2nd IEEE International Conference on Artificial Intelligence Circuits and Systems (AICAS), pp. 41–45. IEEE (2020)
16. Tensorflow: Tensorflow lite: for mobile & IoT. https://www.tensorow.org/lite
17. Howard, A.G., et al.: MobileNets: efficient convolutional neural networks for mobile vision applications. arXiv preprint arXiv:1704.04861 (2017)
18. Szegedy, C., Ioffe, S., Vanhoucke, V., Alemi, A.A.: Inception-v4, inception-resnet and the impact of residual connections on learning. In: Thirty-First AAAI Conference on Artificial Intelligence (2017)
19. Szegedy, C., et al.: Going deeper with convolutions. In: Proceedings of the IEEE Conference on Computer Vision and Pattern Recognition, pp. 1–9 (2015)
20. Han, S., Mao, H., Dally, W.J.: Deep compression: compressing deep neural networks with pruning, trained quantization and Huffman coding. arXiv preprint arXiv:1510.00149 (2015)
21. Gong, Y., Liu, L., Yang, M., Bourdev, L.: Compressing deep convolutional networks using vector quantization. arXiv preprint arXiv:1412.6115 (2014)
22. Panatik, K.Z., et al.: Energy harvesting in wireless sensor networks: a survey. In: 2016 IEEE 3rd International Symposium on Telecommunication Technologies (ISTT), pp. 53–58. IEEE (2016)
23. Neukirchen, H.: Power consumption of Raspberry Pi 4 versus Intel J4105 system. https://uni.hi.is/helmut/2021/06/07/power-consumption-of-raspberry-pi-4-versus-intel-j4105-system/

24. Murshed, M.S., Murphy, C., Hou, D., Khan, N., Ananthanarayanan, G., Hussain, F.: Machine learning at the network edge: a survey. ACM Comput. Surv. (CSUR) **54**(8), 1–37 (2021)
25. Coral: USB accelerator datasheet. https://coral.ai/docs/accelerator/datasheet/
26. Coral: Image classification. https://coral.ai/models/image-classification/

Mathematical Expression Detection in Camera Captured Document Images

Bui Hai Phong[1,3]([⊠]) [iD], Thang Manh Hoang[2,4] [iD], and Thi-Lan Le[1,2] [iD]

[1] MICA International Research Institute, Hanoi University of Science
and Technology, Hanoi, Vietnam
{hai-phong.bui,thi-lan.le}@mica.edu.vn
[2] School of Electrical and Electronic Engineering (SEEE),
Hanoi University of Science and Technology, Hanoi, Vietnam
{thang.hoangmanh,lan.lethi1}@hust.edu.vn
[3] Faculty of Information Technology, Hanoi Architectural University, Hanoi, Vietnam
[4] Vietnam-Japan International Institute for Science of Technology, Hanoi University
of Science and Technology, Hanoi, Vietnam

Abstract. The detection of mathematical expressions (MEs) in document images has attracted the attention of many researchers in recent years. Most conventional methods focus on detecting MEs in scanned images. In recent years, mobile devices with integrated digital cameras have been popularly used. Camera captured images that are curved and skewed cause many difficulties in the detection of MEs. Therefore, the detection of MEs in camera captured images needs to be considered. This paper presents a system of two stages to detect MEs in camera captured images. At the first stage, the pre-processing technique is applied to reduce the geometric distortions of camera captured images. The dewarping process is performed using the property of both text-lines and non-text regions in camera captured images. At the second stage, the YOLOv5 network is fine-tuned to detect MEs efficiently. The proposed system has been evaluated on a large dataset of camera captured images. The obtained detection accuracy of 94% of MEs in camera captured images and the performance comparison with various state-of-the-art methods demonstrate the effectiveness of the proposed method.

Keywords: Document image dewarping · Mathematical expression detection · Machine learning · Deep learning · Camera captured images

1 Introduction

The detection of MEs in scientific documents has attracted the attention of many researchers in recent years. The detection of MEs can be considered as the first step of the development of mathematical retrieval systems [1]. In the challenging task, input documents normally exist in three types: handwritten, PDF and image formats. The processing techniques to detect MEs highly relies on input documents. Modern mobile devices are popular and they can be used for

N.-T. Nguyen et al. (Eds.): ICIT 2022, LNDECT 148, pp. 98–109, 2022.
https://doi.org/10.1007/978-3-031-15063-0_9

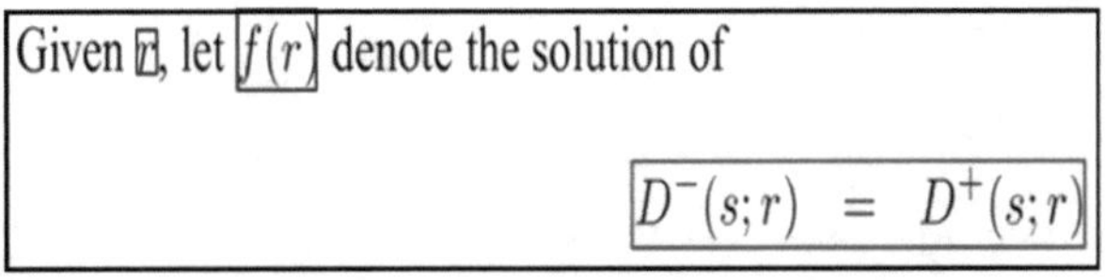

Given a, let $f(r)$ denote the solution of

$$D^-(s;r) \;=\; D^+(s;r)$$

(a) Examples of MEs in scanned images

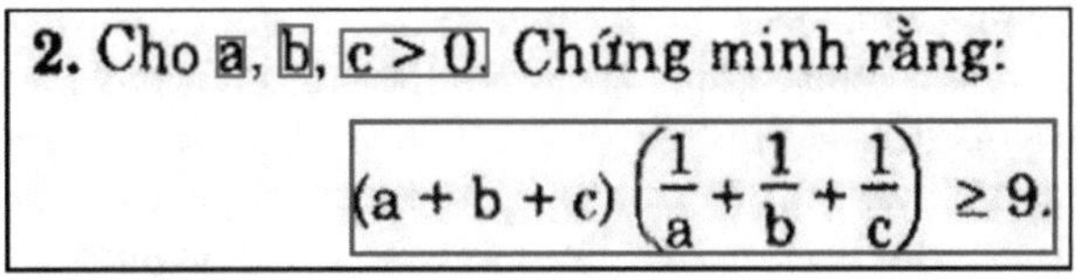

2. Cho a, b, $c > 0$ **Chứng minh rằng:**

$$(a + b + c)\left(\frac{1}{a} + \frac{1}{b} + \frac{1}{c}\right) \geq 9.$$

(b) Examples of MEs in camera captured images

Fig. 1. Demonstration of displayed (red) and inline (blue) MEs in an English scanned (a) and Vietnamese camera captured (b) document image. (Color figure online)

capturing document images. There exists two types MEs in document images: isolated (displayed) and inline (embedded) ones. Isolated MEs typically display in separate lines/regions. Meanwhile, inline MEs can be mixed with other components of pages (e.g. texts, tables and figures) [2]. Figure 1 demonstrates some examples of displayed (red) and inline (blue) MEs in an English scanned (a) and Vietnamese camera captured (b) document image.

Compared to scanned document images, the specific characteristics of camera captured images can be described as following [3]:

- Due to the environment and captured devices, camera captured images may be curved and skew. The characteristics cause many difficulties in the detection of MEs.
- Camera captured images may be blurred and may consist of heterogeneous background that may cause many difficulties in the detection of MEs.
- The variation of lighting environment frequently occurs in camera captured images.

The contributions of the paper are twofold:

(1) The detection of MEs in camera captured document images is more challenging than that in scanned ones. Therefore, we propose a system of two stages to improve the detection accuracy. Firstly, the skew and dewarping techniques are performed in order to enhance the quality of input camera capture images. Then, the YOLOv5 neural network [4] that has shown the high performance for the detection task is fine-tuned to improve the detection accuracy of MEs.
(2) To evaluate the proposed system, we prepare the datasets that consist of a large number of Vietnamese camera capture document images. The images are collected by a large number of end users. Therefore, the datasets are diverse and useful for the evaluation of the detection task.

2   Related Work

2.1   Dewarping Methods of Camera Captured Images

The dewarping techniques aim to enhance the quality of camera captured images. The solution approaches can be performed by using the traditional image processing and the Deep neural networks (DNNs). To rectify curved and skew images, several traditional algorithms were developed based on the build-in information of camera devices [5,6]. Since the text lines can well represent the structure of document images, numerous traditional solutions for the dewarping process are developed based on text lines. The work in [7] extracts the top and bottom text-line information to transform the curved pages to flat ones. In the work of [3], the information of text-lines and non-text regions are extracted to rectify curved images.

Recently, the advanced DNNs [8,9] have been applied for dewarping document images. The DNNs are trained to predict and replace pixels from curved document images to obtain flat ones. The DNNs require a large number of curved document images to achieve high accuracy. Much human effort is required to prepare a large number of document datasets to applying the DNNs.

2.2   Detecting MEs in Document Images

The methods of the ME detection can be classified into two types. Traditional methods (e.g. [2,10,11]) applied the document analysis techniques and the hand-crafted feature extraction to detect MEs. In the last few years, deep neural networks (DNNs) has proven the high performance in the object detection task. Particularly, the AlexNet [12] and ResNet-50 [13] have popularly utilized for the ME detection and recognition [14]. The work proposed by [16] attempted to detect MEs in PDF documents using the deep learning approach. The work combines the Convolutional neural network (CNN) and Recurrent Neural Network (RNN) networks to improve the accuracy of the ME detection. Compared to handcrafted feature extraction methods, the combination of CNNs and RNNs obtains higher accuracy. However, the use of CNNs and RNNs requires much time and resources to train the deep networks.

Recently, several DNNs have been applied and fine-tuned to improve the accuracy of the detection of MEs in heterogeneous documents [15,21]. In the work proposed in [15], a CNN inspired by the U-net is trained on a large and diverse dataset to detect MEs in scanned document images. In the method, document images in the training datasets were divided into various blocks of fixed sizes. Then, the image blocks are utilized to train the CNN. The method has shown high accuracy (95.2%) for the detection of mathematical symbols; however, the additional layout analysis of symbols are necessary to obtain position information of entire MEs. Moreover, the method faced many difficulties in the detection of inline MEs due to the frequent variation of textual font and size styles. The work in [21] applied the Distance Transform of input document images to enhance the differences between MEs and texts. Then, the Faster R-CNN network is optimised to detect MEs accurately. The drawback of the work is the challenge of the estimation of anchor boxes of the Faster R-CNN network.

In recent years, the YOLO neural networks [4] have advanced in the object detection. Three versions of YOLO networks have been released, i.e., YOLOv3, YOLOv4 and YOLOv5. The new versions of the YOLO networks allow to gain high performance in terms of execution time and accuracy in the object detection.

3   Proposed System of ME Detection in Camera Captured Images

Figure 2 presents the flowchart of the proposed method to detect MEs in camera captured images. In the process, input document images are firstly pre-processed using the dewarping technique to improve the quality. Then, the ME detection is performed using the YOLOv5 network. The network is a powerful and efficient framework for the object detection task compared to other DNNs.

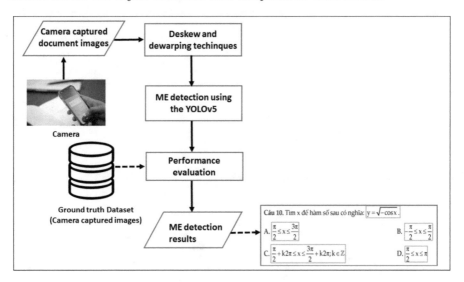

Fig. 2. Proposed system for the detection of MEs in camera captured images.

3.1   Dewarping Input Camera Captured Images

The section presents the dewarping technique for camera captured document images. We apply the technique proposed in [3] to improve the quality of input document images. The dewarping technique allows to obtain the trade-off between the accuracy and the time consuming of the rectification of curved images. In the technique, we extract information of both text-lines and non-text regions to estimate the curved and skew parameters of document images. The text-lines and non-text regions are extracted using the line segment detection algorithm in [17]. Then, the optimization process of the text-line alignment and the horizontally or vertically non-text regions is applied to rectify document images. Figure 3 shows the steps of the dewarping document images. The optimization aims to transform the curved document coordinate (3D) to the flat coordinate (2D). Figure 4 demonstrates the line segmentation and the rectification of skew image using the angle. Figure 5 demonstrates examples of the document images after applying the dewarping technique.

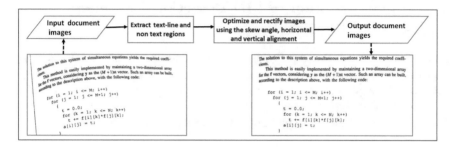

Fig. 3. Dewarping process of camera captured images.

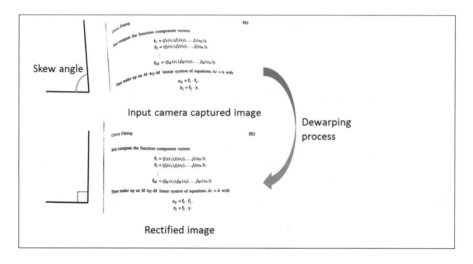

Fig. 4. Dewarping process using the skew angle.

3.2   ME Detection in Camera Captured Images

The ME detection is performed using the YOLOv5 neural network. The network consists of three parts:

(1) The Cross Stage Partial Network (CSPNet) [18] is utilized to efficiently extract features of ME images. In the work, the CSPNet is implemented based on the Darknet-53.

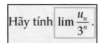

(a) Camera captured image (skew and curved)

Hãy tính lim $\frac{u_n}{3^n}$.

(b) The image after using the dewarping technique

Hãy tính lim $\frac{u_n}{3^n}$.

(c) Detection result

Fig. 5. Example of a camera captured image before (a), after (b) applying the dewarping technique and the detection result (c).

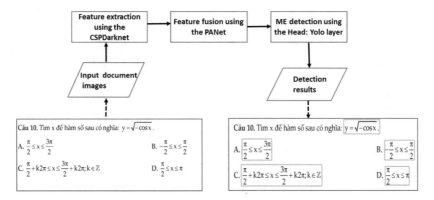

Fig. 6. The detection of MEs using the YOLOv5.

(2) The Path Aggregation Network (PANet) [19] follows the CSPNet for the feature fusion.

(3) The Head: Yolo [4] layer is used for the detection of MEs. The head of YOLOv5 generates various sizes of feature maps to detect multi-scale object detection. Thus, the small and large MEs can be detected correctly using the YOLOv5 network.

Figure 6 demonstrates the structure of YOLOv5 the detection of ME using the network. The training epochs and batch size parameters of the YOLOv5 are selected as 300 and 32, respectively. The learning rate of the network is initially defined at 0.01. Training and testing images are normalized at the size

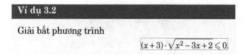

(a) Examples of MEs in camera captured image

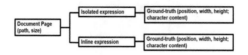

(b) Structure of Json files in the ground truth

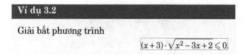

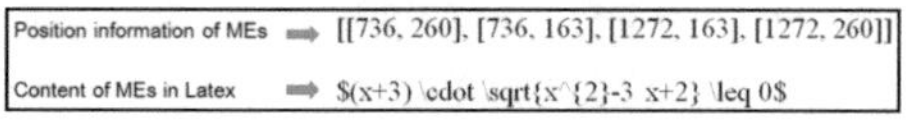

(c) Examples of position information and content of MEs

Fig. 7. Examples of MEs in (a) camera captured image, (b) structure of Json files containing annotation information of MEs in our datasets, (c) position information and content of MEs.

of 640×640. The Stochastic gradient descent (SGD) algorithm [4] was utilized for training the network, in which the momentum is set at 0.93. The network is trained on the Server equipped with Nvidia Tesla V100 (16GB VRAM), CPU 4 cores, and 8GB RAM memory. The implementation environment of the YOLOv5 network is Python 3.7.0 with PyTorch 1.7.

4   Experimental Results

4.1   Datasets

This section provides information of the datasets. Actually, we have collected and prepared the datasets for the performance evaluation. The datasets are collected using camera captured images from high school students in Vietnam. The training, validation and testing datasets consist of 58,322; 10,293 and 7,624 labeled ME images, respectively. The datasets contain 1000 camera captured images that are skew, curved and with complex backgrounds. The document images are captured by mobile devices of thousands of users. The resolution of captured images is 96 dpi. After obtaining camera captured images from users, we have manually annotated the bounding boxes of MEs and the ME content in Latex format.

Groundtruth information is stored in json (JavaScript object notation) files that consist of:

- The path and name of camera captured images.
- Position information of MEs: x and y coordinates of 4 points of bounding boxes.
- Annotation Latex sequences of MEs.

Figure 7 shows examples of the annotation information of the datasets while Table 1 shows statistic information of training, validation and testing sub datasets.

Table 1. Statistic information of training, validation and testing camera captured document images in datasets.

Dataset	Training	Validation	Testing
Number of MEs	58,322	10,293	7,624
Number of camera captured images	31,175	5,503	4,077

4.2   Evaluation Metrics

The Precision and recall metrics [21] are widely used for the ME detection task. Moreover, the Intersection over Union (IoU) metric [21] has been also applied in the work to obtain the in-depth performance evaluation. The IoU metric can be defined as follows:

$$IoU = \frac{area(I_p \cap I_{gt})}{area(I_p \cup I_{gt})} \tag{1}$$

where $I_p \cap I_{gt}$ is the intersection of the predicted (I_p) and ground-truth (I_{gt}) bounding boxes of MEs. Meanwhile, $I_p \cup I_{gt}$ denotes the union of the predicted and ground-truth bounding boxes of MEs. The range of the IoU value is [0,1]. As conventional studies of the ME detection [21,22], a detection is considered to be correct if $IoU \geq 0.5$.

4.3   Performance Evaluation

To emphasize the impact of the dewarping technique in our proposed method, we evaluate the detection accuracy with and without using the dewarping technique. We also compare the performance of the proposed method with other existing detection methods. Table 2 demonstrates the improvement of the detection accuracy of MEs using the dewarping technique. The performance comparison of our and conventional detection methods are shown in Table 3. The proposed method gains the highest detection accuracy because the use of dewarping technique and the efficient detection of the YOLOv5 in comparison with the use of SSD-512, Faster R-CNN and YOLOv3 networks.

The training data is a significant factor that affects the detection accuracy of MEs using DNNs. We evaluate the detection accuracy of MEs depending on various number of MEs in training datasets. Figure 8 demonstrates the impact of the number of MEs in training datasets to the detection precision. It is shown in the figure that the detection precision significantly increases when the number of training datasets increases. Examples of the correct detection of MEs are shown in Fig. 9. In the examples, both inline and displayed MEs are detected correctly.

Table 2. ME detection results using the YOLOv5 with and without applying the dewarping technique ($IoU \geq 0.5$).

Method	Precision	Recall
YOLOv5 without applying the dewarping technique	0.85	0.81
YOLOv5 with applying the dewarping technique	**0.94**	**0.92**

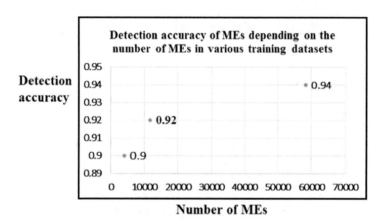

Fig. 8. Comparison of the detection accuracy of MEs in camera captured images with different number of MEs in training dataset.

Table 3. Performance comparison of ME detection with $IoU \geq 0.5$. The highest scores of the proposed method are in bold.

Method	Precision	Recall
Using the SSD-512 network [20]	0.83	0.81
Using the Faster R-CNN network [21]	0.82	0.80
Using the YOLOv3 network [22]	0.81	0.79
Our proposed method	**0.94**	**0.92**

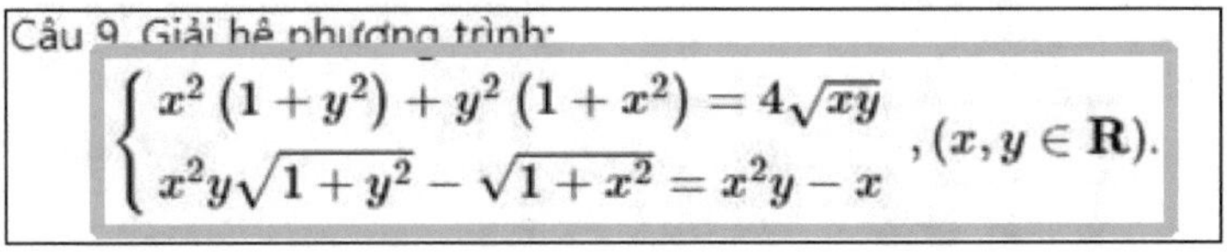

Câu 9. Giải hệ phương trình:

$$\begin{cases} x^2\left(1+y^2\right) + y^2\left(1+x^2\right) = 4\sqrt{xy} \\ x^2 y\sqrt{1+y^2} - \sqrt{1+x^2} = x^2 y - x \end{cases}, (x, y \in \mathbf{R}).$$

(a) Correct detection of isolated MEs.

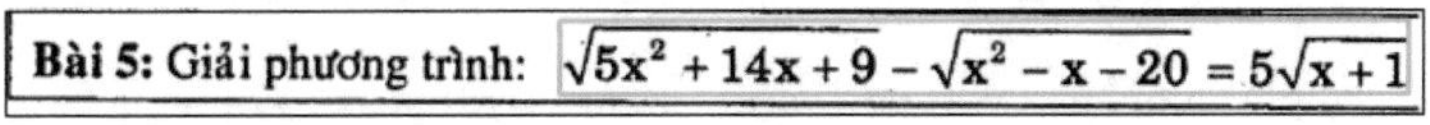

Bài 5: Giải phương trình: $\sqrt{5x^2 + 14x + 9} - \sqrt{x^2 - x - 20} = 5\sqrt{x+1}$

(b) Correct detection of inline MEs.

Fig. 9. Examples of correct ME detection in camera captured images.

Câu 10. Tìm x để hàm số sau có nghĩa: $y = \sqrt{-\cos x}$.

A. $\dfrac{\pi}{2} \le x \le \dfrac{3\pi}{2}$

B. $-\dfrac{\pi}{2} \le x \le \dfrac{\pi}{2}$

C. $\dfrac{\pi}{2} + k2\pi \le x \le \dfrac{3\pi}{2} + k2\pi; k \in \mathbb{Z}$

D. $\dfrac{\pi}{2} \le x \le \pi$

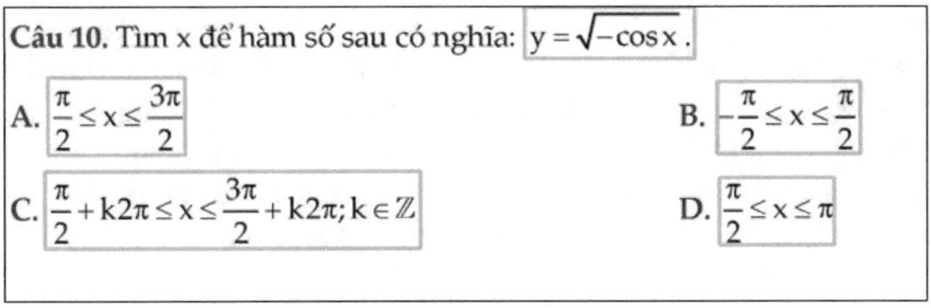

(a) Correct detection of MEs.

Ví dụ 13. Tính : $I = \displaystyle\int_0^{\frac{\pi}{4}} \dfrac{x\sin x + (x+1)\cos x}{x\sin x + \cos x}\,dx$ (CĐ - ĐH khối A – 2011)

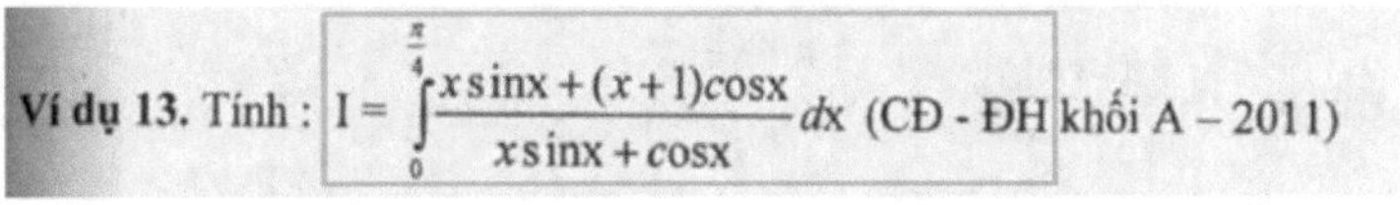

(b) Partial detection of MEs.

Fig. 10. Examples of ME detection in camera captured images.

4.4 Error Analysis and Discussion

Figures 10 and 11 demonstrate examples of some errors of the detection of MEs. Some errors of the ME detection can be summarized as follows:

(i) Camera captured images may consist of complex background that may cause the incorrect detection of MEs.

(ii) Camera captured images may be blurred and there is frequent variation of light that may cause the incorrect detection of MEs.

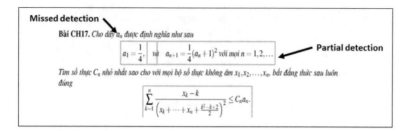

Fig. 11. Examples of missed and partial detection of MEs that are pointed by arrows.

5   Conclusion and Future Work

In the paper, the method for the detection of MEs in camera captured document images has been presented. The use of deskew and dewarping techniques allow to improve the quality of document images. The YOLOv5 is fine-tuned to detect MEs accurately in compared with other detection methods using various DNNs. To evaluate the proposed method, we have collected and prepared large datasets of Vietnamese camera captured document images. The large datasets are useful for developing and testing the detection of MEs in camera captured document images.

In the future, the accuracy of detection will be improved using advanced DNNs. Our datasets of camera captured images can be extended and published for the ME detection and recognition researches.

References

1. Zanibbi, R., Blostein, D.: Recognition and retrieval of mathematical expressions. Int. J. Doc. Anal. Recogn. **15**, 331–357 (2012)
2. Garain, U.: Identification of mathematical expressions in document images. In: Proceedings of the 10th International Conference on Document Analysis and Recognition, pp. 1340–1344. IEEE (2009)
3. Kil, T., et al.: Robust document image dewarping method using text-lines and line segments. In: 14th International Conference on Document Analysis and Recognition (ICDAR) (2017)
4. Redmon, J., et al.: You only look once: unified, real-time object detection. In: 2016 IEEE Conference on Computer Vision and Pattern Recognition (CVPR) (2016). https://doi.org/10.1109/CVPR.2016.91
5. Brown, M.S., et al.: Restoring 2D content from distorted documents. IEEE Trans. Pattern Anal. Mach. Intell. **29**(11), 1904–1916 (2007). https://doi.org/10.1109/TPAMI.2007.1118
6. Meng, G., et al.: Active flattening of curved document images via two structured beams. In: 2014 IEEE Conference on Computer Vision and Pattern Recognition (CVPR) (2014). https://doi.org/10.1109/CVPR.2014.497
7. Fu, B., et al.: A model-based book dewarping method using text line detection. In: International Workshop on Camera-Based Document Analysis and Recognition, pp. 63–70 (2007)

8. Xie, G.-W., Yin, F., Zhang, X.-Y., Liu, C.-L.: Dewarping document image by displacement flow estimation with fully convolutional network. In: Bai, X., Karatzas, D., Lopresti, D. (eds.) DAS 2020. LNCS, vol. 12116, pp. 131–144. Springer, Cham (2020). https://doi.org/10.1007/978-3-030-57058-3_10

9. Das, S., et al.: DewarpNet: single-image document unwarping with stacked 3D and 2D regression networks. In: 2019 IEEE/CVF International Conference on Computer Vision (ICCV) (2019). https://doi.org/10.1109/ICCV.2019.00022

10. Suzuki, M., et al.: INFTY: an integrated OCR system for mathematical documents. In: Proceedings of the 2003 ACM Symposium on Document Engineering. ACM (2003)

11. Jin, J., Han, X., Wang, Q.: Mathematical formulas extraction. In: Proceedings of the Seventh International Conference on Document Analysis and Recognition, vol. 2. IEEE (2003). https://doi.org/10.1109/ICDAR.2003.1227834

12. Krizhevsky, A., et al.: ImageNet classification with deep convolutional neural networks. In: Proceedings of the 25th International Conference on Neural Information Processing Systems, vol. 1. IEEE (2012). https://doi.org/10.1145/3065386

13. He, K., et al.: Deep residual learning for image recognition. In: Proceedings of the 2016 IEEE Conference on Computer Vision and Pattern Recognition. IEEE (2016). https://doi.org/10.1109/CVPR.2016.90

14. Phong, B.H., Hoang, T.M., Le, T.-L.: A hybrid method for mathematical expression detection in scientific document images. IEEE Access **8**, 83663–83684 (2020). https://doi.org/10.1109/ACCESS.2020.2992067

15. Ohyama, W., et al.: Detecting mathematical expressions in scientific document images using a U-Net trained on a diverse dataset. IEEE Access (2019). https://doi.org/10.1109/ACCESS.2019.2945825

16. Gao, L., et al.: A deep learning-based formula detection method for PDF documents. In: Proceedings of the 2017 IEEE Conference on Document Analysis and Recognition. IEEE (2017)

17. Gioi, R.V., et al.: LSD: a line segment detector. Image Process. Online **2**, 35–55 (2012). https://doi.org/10.5201/ipol.2012.gjmr-lsd

18. Wang, C., et al.: CSPNet: a new backbone that can enhance learning capability of CNN (2019). https://arxiv.org/abs/1911.11929

19. Wang, K., et al.: PANet: few-shot image semantic segmentation with prototype alignment (2019). https://arxiv.org/abs/1908.06391

20. Mali, P., et al.: ScanSSD: scanning single shot detector for mathematical formulas in PDF document images (2020). https://arxiv.org/abs/2003.08005

21. Phong, B.H., et al.: An end-to-end framework for the detection of mathematical expressions in scientific document images. Expert Syst. **39**(1), e12800 (2021)

22. Mahdavi, M., et al.: ICDAR 2019 CROHME + TFD: competition on recognition of handwritten mathematical expressions and typeset formula detection. In: 2019 International Conference on Document Analysis and Recognition (ICDAR) (2019). https://doi.org/10.1109/ICDAR.2019.00247

Intelligence Services and Applications

A-DEM: The Adaptive Approximate Approach for the Real Scheduling Problem

Nguyen The Loc[1](✉) ⓘ and Dang Quoc Huu[2] ⓘ

[1] Hanoi National University of Education, Ha Noi, Viet Nam
locnt@hnue.edu.vn
[2] Thuongmai University, Ha Noi, Viet Nam
huudq@tmu.edu.vn

Abstract. This paper proposed a new algorithm to solve the Real-RCPSP problem (Real-RCPSP: Real-Resource Constrained Project Scheduling Problem). The algorithm is developed from the Differential Evolution (DE) algorithm hybrid with the adaptive method, which dynamically changes the crossover probability parameter during the evolution process. That parameter value is calculated from the neighborhood particles. The individuals used to make dynamic crossover probability are found by the star-topology. The new algorithm is called A-DEM. The effectiveness of the new algorithm is verified based on experiments with the iMOPSE dataset, which is the standard data set for this problem. Experimental results show that the proposed algorithm is more effective for this problem.

Keywords: Project scheduling · Evolutionary algorithms · Differential algorithms · Optimization computation

1 Introduction

Scheduling problems have been studied since 1950, which find the best way to assign resources in performing all project tasks. In normal, there are many precedence constraints that have to satisfy while executing the project. The objective of scheduling is to find the solution to minimize the total cost and/or the total time for completing the project. The scheduling problems are applied in many fields, such as scheduling resource coordination in operating systems, scheduling for production lines,… or applications in economics and finance, military… The scheduling problems are proven to be in the NP-Hard, so we can not find the optimal results in the polynomial time.

RCPSP (Resource-Constrained Project Scheduling Problem) [1, 2] is a project scheduling problem with limited renewable resources belonging to the NP-Hard classification. The project has many tasks that need to do based on many priority constraints on execution. This problem has two characteristics:

- Resources are reusable which each resource can only perform one task at a time
- When a task starts, it cannot be interrupted or paused. It has to do until the finish.

© The Author(s), under exclusive license to Springer Nature Switzerland AG 2022
N.-T. Nguyen et al. (Eds.): ICIT 2022, LNDECT 148, pp. 113–123, 2022.
https://doi.org/10.1007/978-3-031-15063-0_10

An extended problem of RCPSP [5] is Multi-Skill RCPSP [6, 8–13] which added skill-levels of the resource, whereby each resource has multi-skills, and each skill has a determined level. The main difference from RCPSP is that MS-RCPSP's resources have many skill types, each having a specific level. Thanks to the new rule of resources, MS-RCPSP extra a new constraint: to perform a task, the resource has to match the skill type and skill level. Whereby, a resource can perform the task if it has the same skill type and the skill level is equal or greater than required by the task. Figure 1 shows an example of a project with four tasks and four renewable resources, which are descriptions of the requirements of tasks and the power of the resources. From two factors, we can build a table presenting the tasks-resources assignment.

✓ Can be assigned		**Tasks**			
✗ Can not be assigned		$S_{2.2}$	$S_{3.1}$	$S_{2.2}$	$S_{1.1}$
		W_1	W_2	W_3	W_4
Resources					
$S_{1.3}, S_{2.2}$	L_1	✓	✗	✓	✓
$S_{2.1}, S_{3.2}$	L_2	✗	✓	✗	✗
$S_{1.2}, S_{2.1}$	L_3	✗	✗	✗	✓
$S_{2.2}, S_{3.3}$	L_4	✓	✓	✓	✗

Fig. 1. The resources-tasks assignment

In Fig. 1, W indicates the task, L demonstrates the resource, $S_{i.j}$ presents that the resource has the skill type is S_i with corresponding skill level is j.

Example: Task W_1 requires the resource with $S_{2.2}$ to execute. That means the resource with S_2 skill type and skill level equal to two or greater can be run W_1. Comparing the resources of the project, easy to identify that resources L_1, L_4 can perform task W_1 and L_2, L_3 can not.

The MS-RCPSP problem has the limitation that a task can be executed in equal time with any suitable resource, regardless of the power of the executing resource. This restriction reduces the practical applicability of the problem, especially in the production line with resources of different capacities. In practice, a resource with a higher skill level often has done tasks better than a smaller one. This paper presents a new problem, expanding from MS-RCPSP problem and having high applicability in practice, called Real-RCPSP. In the Real-RCPSP [3] problem, the duration of a task is not fixed but changes according to the capacity of the resource. Therefore, Real-RCPSP has higher practical applicability, especially in industrial production lines.

2   Related Works

Many scientists have been proposed algorithms to solve the MS-RCPSP problem. The objective of the algorithms is to find the optimal schedule [9, 11] with minimal duration or to find a multi-objective solution (minimizing project execution time and project costs [7, 8, 10, 12]). Myszkowski et al. [11] have published many about the MS-RCPSP problem, using heuristic and metaheuristic methods such as GA, Ant, GR,... Besides, an important contribution of this group is to propose a standard dataset used to evaluate the new algorithms for the MS-RCPSP problem, which is called iMOPSE. Some authors research the multi-objective model of the MS-RCPSP. Their objective is to find the solution by optimizing the cost and time. Ghamginzadeh [7] considers the scheduling of projects with an uncertainty model because of the time-varying factors. The author suggests a method based on the MOICA and NSGA-II algorithms. Hosseinian [8] studied a new variant of the MSRCPSP with transfer-time factor and proposed the MOMAOA algorithm to solve it. Tian [10] et al. studied the MS-RCPSP problem and proposed a method to find a multi-objective solution by limiting the search space with greedy techniques and reorder tasks executions. The authors also propose two mutation operators changing the allocating resources to perform tasks. Zhu [12] et al. use a hybrid algorithm based on GA and heuristic conducted through 5 steps: the decomposition, encoding scheme and an improved repair-based decoding scheme, ten adaptive heuristics, genetic programming, using the Taguchi method of design-of-experiment.

3 Problem Definition

The Real-RCPSP [3] problem is extended from the MS-RCPSP [4, 8, 11] by adding a new constraint that illustrates the task duration depending on the resource's skill level. That means the task execution time is a dynamic value (in the RCPSP problem, this parameter is fixed) controlled by the resource's skill level, which is shorter with the higher skill level. With this new addition, the Real-RCPSP has high applicability in practical, especially in industrial product-line, where products are manufactured depending on the skills of laborers.

To declare the Real-RCPSP, the paper defines the mathematical notations as shown in Table 1.

Using the math symbols shown in Table 1, the real-RCPSP problem can be present as formulations as follow:

$$f(P) \rightarrow min$$

Constraints:

- $S^k \neq \emptyset \quad \forall L_k \in L$ (1)

- $t_j \geq 0 \quad \forall W_j \in W$ (2)

- $E_j \geq 0 \quad \forall W_j \in W$ (3)

- $E_i \leq E_j - t_j \forall W_j \in W, j \neq 1, \ W_i \in C_j$ (4)

- $\forall W_i \in W^k \exists S_q \in S^k g_{S_q} = g_{r^i}$ and $h_{S_q} \geq h_{r^i}$ (5)

Table 1. The notations

Notation	Description
C_i	The precedence of task i
S	Skills of the project's resources
S^i	Skills of the i resource, $S^i \subseteq S$;
S_i	The skill i;
t_j	The time to execute of the j task
L	The set of all project resources
L^k	All the resources which can be performed task k; $L^k \subseteq L$
L_i	The resource i
W	The set of all project tasks
W^k	All tasks can be run by the resource k, $W^k \subseteq W$
W_i	The task i
r^i	The skills of the resource need to perform task i
B_k, E_k	The start time & end time of the task k
A_u, v^t	The value to confirm that the resource v is using to do task u at time t; 1: yes, 0: no;
h_i	The skill level i;
g_i	Type of skill i;
m	The schedule's makespan
P	The feasible schedule
P_{all}	All feasible schedules of the project
$f(P):$	The evaluating function for the P schedule returns makespan
n	Task number
z	Resource number

$$\bullet \quad \forall L_k \in L, \forall q \in m : \sum_{i=1}^{n} A_{i,k}^q \leq 1 \qquad (6)$$

$$\bullet \quad \forall W_j \in W \exists! q \in [0, m], !L_k \in L : A_{j,k}^q = 1; \text{ with } A_{j,k}^q \in \{0; 1\} \qquad (7)$$

$$\bullet \quad t_{ik} \leq t_{il} \text{ với } h_k \leq h_l \; \forall \, (r^k, r^l) \in \{S^k \times S^l\} \qquad (8)$$

4   Proposed Algorithm

The proposed algorithm is built from the DE [13] combined with the adaptive method. A very important coefficient in the DE algorithm used in population evolution is the CR coefficient. Usually, this coefficient is a constant. The adaptive technique will change the CR coefficient after each generation of population evolution to improve the efficiency of the algorithm. In the proposed algorithm, the CR coefficient will be calculated based on a number of neighbors with the best individual. The number of these neighbors is also changed accordingly during the execution of the algorithm.

Star Topology
A star topology [7] is a population architecture in which individuals connect to a central one. Figure 2 is an example of star architecture. In the proposed algorithm, the star topology is used to discover a number of neighborhood individuals of the best particle. These individuals are used to calculate the CR value after each generation.

The Adaptive method
The adaptive method is built based on the star-topology for each evolution generation of the population according to the following steps:

- Find the best particle of the current generation. It is the schedule with the shortest time to complete all tasks of the project
- Calculate total time to execute the project using the best particle (makespan),
- Calculate the number of neighborhoods need to evaluate the CR value and find the neighboring individuals from the best particle
- Calculate the CR value base on the neighboring individuals
- Re-calculate the number of neighboring individuals for the next evolution generation of the population.

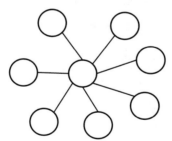

Fig. 2. Star topology

The A-DEM algorithm
Applying adaptive method in CR tuning, the A-DEM algorithm is presented as in Algorithm 1 below.

Algorithm 1. A-DEM
Input: t_{max}: the generation number
Output: the best schedule (P_{best}) and minimum time

```
1 Begin
2 t = 0
3 Load and valid Dataset
4 P_all  ←  initialize(pSize)  // population initialization
with pSize particles
5 {P_best, makespan}  ←  fbest(P_all) // caculate the best par-
ticle
6 μCR = 0.5; w_min = 4; w_max = 10; w = 3
7 while(t<t_max)
8 for (int i = 0; i< pSize; i++)
9     P^i_best = StarAdaptive(P_all,i, w)
10        CR_i = randci (μCR, 0.1)
11        P_i ≠ P_1 ≠ P_2 ← Random from P_all
12        V_i ← P_i + F × (P_best - P_i) + F × (P_1 - P_2)
13        I_rand = randi(1,pSize)
14        for(j=0; j<taskcount;j++)
15            rand_i,j = rand(0,1)
```

$$U_{i,j} = \begin{cases} U_{i,j} \ if \ rand_{i,j} \le CRi \ or \ i = I_{rand} \\ P_{i,j} \ if \ rand_{i,j} > CRi \ or \ i \ne I_{rand} \end{cases}$$

```
17        end for
18        if(f(U_i) ≤ f(P_i))
19            P_i = U_i
20            S_CR = S_CR + {μCR}
21        end if
22     end for
23     c = rand(0,1)
24     μCR = (1 - c) × μCR + c × mean(S_CR)  //Adaptive
factor
25     w = w_max - i/pSize × (w_max - w_min)
26     t ← t+1
27     {P_best, makespan} ← fbest(P_best)
28  end while
29     return {P_best, makespan}
30   end
```

```
f: objective function
fbest: the function to evaluate the best particle
F: Step parameter
μCR: Crossover  Probability
w: number of neighboring particles
randci: Cauchy distribution function
mean: the function to calculate the average value
```

5   Experiment Results

To evaluate the efficiency of the proposed algorithm, we designed the simulator on the Matlab environment executed on the iMOPSE dataset. The experiment results are collected and compared with GA-M algorithm [11]. The simulation is run on a PC with

Intel Core i5-CPU 2.2GHz, 6GB RAM, and using the Windows 10 OS. The dataset used in our simulator is listed in Table 2.

The parameters for the experiment:

- Using the Matlab Software to develop the testing
- Using 15 iMOPSE installers, they are shown in Table 2
- Init a population with 100 particles ($Np = 100$)
- The generations number is 50,000 ($Ng = 50,000$)
- The testing number of each installer data is 35 times

Table 2. iMOPSE benchmark dataset

Name	Tasks	Resources	Precedence relations	Skills
100_5_22_15	100	5	22	15
100_5_46_15	100	5	46	15
100_5_48_9	100	5	48	9
100_5_64_15	100	5	64	15
100_5_64_9	100	5	64	9
100_10_26_15	100	10	26	15
100_10_47_9	100	10	47	9
100_10_48_15	100	10	48	15
100_10_64_9	100	10	64	9
100_10_65_15	100	10	65	15
100_20_22_15	100	20	22	15
100_20_46_15	100	20	46	15
100_20_47_9	100	20	47	9
100_20_65_15	100	20	65	15
100_20_65_9	100	20	65	9

In the iMOPSE dataset, the execution time of tasks is the fixed value with any resource running the task. Therefore, it is not fully suitable for the Real-RCPSP problem. In order to experimentally deploy the new algorithm, the article re-adjusts the task's duration as follows:

- If the resource meets the minimum requirement of the task, the duration is not changed.
- When the skill level of the performance resource is higher than the required one level, the duration will be adjusted down by 3%.
- When the skill level of the implementation resource is higher than two levels, the task execution time will be adjusted down by 5%.

After adjustment, the new dataset is appropriate for Real-RCPSP.

Conduced results

The A-DEM algorithm results with the above settings, as shown in Table 3 below.

Table 3. The experiment results on the iMOPSE dataset

Dataset	GA-M			A-DEM		
	Best	Avg	Std	Best	Avg	Std
100_5_22_15	465	469	3.1	452	454	1.1
100_5_46_15	523	529	5.4	505	506	0.2
100_5_48_9	476	481	4.5	467	471	3.7
100_5_64_15	478	487	8.2	456	461	4.7
100_5_64_9	453	457	3.2	429	430	0.7
100_10_26_15	221	224	2.1	204	208	3.8
100_10_47_9	247	253	5.9	236	238	1.2
100_10_48_15	238	246	7.6	234	237	2.7
100_10_64_9	239	249	9.4	216	222	5.8
100_10_65_15	240	244	3.2	227	230	2.7
100_20_22_15	109	115	5.2	91	95	3.8
100_20_46_15	145	150	4.1	143	147	3.2
100_20_47_9	121	125	3.3	117	119	1.9
100_20_65_15	193	200	6.8	183	185	1.8
100_20_65_9	124	131	6.5	100	101	0.3

In Table 3, GA-M values are the results of the Myszlowski that are public together with the iMOPSE dataset. Experimental results of the A-DEM algorithm corresponding to those datasets get better, which is illustrated through the BEST, AVG, and STD values, specifically as follows:

- With the BEST values (indicating the minimal time to finish the project): the experimental results show that A-DEM is better than GA-M from 1.4% to 19.5%. The detail is shown in Fig. 3.
- With the AVG values (demonstrating the average duration of 35 running times with each instance of the dataset), A-DEM achieved better from 2.0 to 22.9% compared to GA-M. The detail is shown in Fig. 4.
- With the STD values (displaying the algorithm's stability): the GA-M algorithm has a total of STD values of 78.5, and A-DEM is 37.6, which shows that the A-DEM algorithm is more stable than the GA-M algorithm. The detail is shown in Fig. 5.

Experimental results show that the A-DEM algorithm obtains good efficiency for the Real-RCPSP problem. The dynamic change of CR value achieves the results during the evolution of the population. Detail of the effection of the A-DEM is shown in Fig. 3 (for the BEST values), Fig. 4 (for the AVG values) and Fig. 5 (for the STD values).

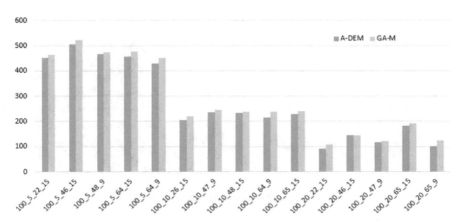

Fig. 3. The BEST values of GA-M in comparison with A-DEM

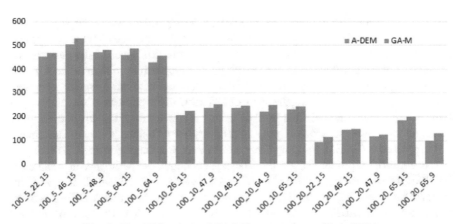

Fig. 4. The AVG values of GA-M in comparison with A-DEM

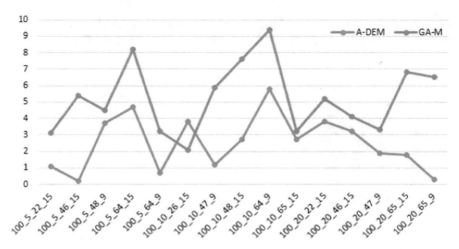

Fig. 5. The STD values of GA-M in comparison with A-DEM

6 Conclusion

The paper presents the Real-RCPSP problem, a new problem with high applicability in practice, especially in the product line. It has added a new constraint on the project execution time that dynamic changing depends on the skill level of the performance resource. A new algorithm is proposed in this paper to solve this problem called A-DEM, a hybrid algorithm between the DE algorithm combined with the Adaptive technique. This technique allows changing the value of CR value in the population's evolution (CR value is important of the DE algorithm) by using the star topology to find out the number of neighbors of the best particle. Applying the Adaptive technique is a significant change that supports the A-DEM, bringing better results than previous algorithms. To verify the effectiveness of the proposed algorithm, the paper conducted experiments on the adjusted iMOPSE dataset, the experimental results were synthesized, evaluated, analyzed, and compared with the GA-M algorithm. Experimental results show that A-DEM has better than GA-M, with iMOPSE dataset, A-DEM is better from 1.4% to 19.5% with the BEST values, and better from 2% to 22.9% with the mean values.

References

1. Christian, A., Demassey, S., Néron, E.: Resource Constrained Project Scheduling: Models, Algorithms, Extensions and Applications. ISBN 978–1–84821–034–9 (2008)
2. Klein, R.: Scheduling of Resource-Constrained Projects. Springer Science & Business Media, vol. 10 (2012)
3. Quoc, H.D., The, L.N., Doan, C.N., Xiong, N.: Effective evolutionary algorithm for solving the real-resource-constrained scheduling problem. J. Adv. Transp. **2020**, 11 (2020). Article ID 8897710
4. Barghi, B., Sikari, S.S.: Meta-heuristic solution with considering setup time for multi-skilled project scheduling problem. Oper. Res. Forum **3**(1) (2022). Springer International Publishing. https://doi.org/10.1007/s43069-021-00117-5

5. De Melo, L.V., de Queiroz, T.A.: Integer linear programming formulations for the RCPSP considering multi-skill, multi-mode, and minimum and maximum time lags. IEEE Latin America Trans. **19**(1), 5–16 (2021)
6. Gnad, D., Hoffmann, J.: Star-topology decoupled state space search. Artif. Intell. **257**, 24–60 (2018)
7. Ghamginzadeh, A., Najafi, A.A., Khalilzadeh, M.: Multi-objective multi-skill resource-constrained project scheduling problem under time uncertainty. Int. J. Fuzzy Syst. **23**(2), 518–534 (2021)
8. Hosseinian, A.H., Baradaran, V.: A multi-objective multi-agent optimization algorithm for the multi-skill resource-constrained project scheduling problem with transfer times. RAIRO-Oper. Res. **55**(4), 2093–2128 (2021)
9. Polo-Mejía, O., et al.: Heuristic and metaheuristic methods for the multi-skill project scheduling problem with partial preemption. Int. Trans. Oper. Res. (2021)
10. Tian, Y., et al.: Multi-Objective multi-skill resource-constrained project scheduling problem with skill switches: model and evolutionary approaches. Comput. Ind. Eng. **167**, 107897 (2022)
11. Myszkowski, P.B., Laszczyk, M., Nikulin, I., Skowroński, M.: iMOPSE: a library for bicriteria optimization in Multi-Skill Resource-Constrained Project Scheduling Problem. Soft. Comput. **23**(10), 3397–3410 (2018). https://doi.org/10.1007/s00500-017-2997-5
12. Zhu, L., et al.: A decomposition-based multi-objective genetic programming hyper-heuristic approach for the multi-skill resource constrained project scheduling problem. Knowl. Based Syst. **225**, 107099 (2021)
13. Zeng, Z., et al.: Enhancing differential evolution with a target vector replacement strategy. Comput. Stand. Interfaces **82**, 103631 (2022)

A Framework to Verify the ABAC Policies in Web Applications

Thanh-Nhan Luong[1]([✉]) [iD], Hong-Anh Le[2], Dinh-Hieu Vo[3],
and Ninh-Thuan Truong[3]

[1] Haiphong University of Medicine and Pharmacy, 72A Nguyen Binh Khiem,
Ngo Quyen, Haiphong, Vietnam
ltnhan@hpmu.edu.vn
[2] Faculty of Information Technology, Hanoi University of Mining and Geology,
Duc Thang, Bac Tu Liem, Hanoi, Vietnam
lehonganh@humg.edu.vn
[3] University of Engineering and Technology, Vietnam National University,
Hanoi, 144 Xuan Thuy, Cau Giay, Hanoi, Vietnam
{hieuvd,thuantn}@vnu.edu.vn

Abstract. Security attacks are increasingly diverse and complex so ensuring security properties has become an important criterion in evaluating the quality of software systems. Besides, attribute-based access control (ABAC) is widely applied to ensure the security properties of software systems, especially large-scale and complex systems. However, deploying this security policy is always likely to contain potential security flaws that are the causes of the failures to guarantee customer security requirements, especially at the programming stage. In this paper, we introduce an approach to check the compliance of implemented ABAC policy with the Spring Security framework of a web application and its specification automatically. The proposal helps developers in detecting security breach situations that lead to disrupting the confidentiality, integrity, and availability of application systems.

Keywords: Checking · ABAC · Security properties · Spring security · Web application

1 Introduction

Attribute-based access control (ABAC) model [6] is a new generation of access control techniques, which is flexible and efficient, especially for large and complex systems. With the mechanism to evaluate the information from the subject's request based on the access policy stored inside the system to make the decision execution or denial of the request, the ABAC model has the ability to limit the access violation to system resources.

However, the implementation of ABAC policy into application systems is diverse and often involves a complex combination of frameworks. For web applications are developed by JavaEE, the MVC architecture and third-party libraries

N.-T. Nguyen et al. (Eds.): ICIT 2022, LNDECT 148, pp. 124–133, 2022.
https://doi.org/10.1007/978-3-031-15063-0_11

like Spring Security [14] are chosen by many programmers. Although the support library does provide mechanisms to ensure the access control policies of software systems, sometimes combining multiple mechanisms, abuse of libraries, or programming complexity can also cause functional conflicts, waste time and effort, and create potential security holes in software systems [13]. Furthermore, the high expressiveness of the ABAC policy formal specification also increases the potential for error in implementation [7]. Meanwhile, systems that implement the ABAC model are often large and can be very complex to manage. As such, the security properties of the application system may not be guaranteed and the conformance between the application and the specification is not achieved. Therefore, testing and verification to ensure the ABAC policies are correctly implemented in web systems are essential in ensuring the quality of software systems.

There have been a number of studies related to the representation, specification, implementation, analysis, and evaluation of ABAC policy of web systems [16]. Meanwhile, studies [1,8] suggest a method to implement the ABAC model into web applications with the Spring Security framework. In the studies [4,11], the authentication and authorization features of Spring Security in real enterprise applications were analyzed and evaluated by the authors. Several methods of verifying ABAC policy have been implemented in studies [9,15]. However, available studies have not performed source code analysis of web applications incorporating Spring Security to test the security, integrity, and availability of the system's access policy. The contributions of this paper include:

- Introduce formal definitions about the confidentiality, integrity, and availability of access policy.
- Propose a framework for verifying the conformance between the web application and its specification through the defined security properties.

The rest of the paper is organized as follows. In the next section, we briefly summarize the ABAC, Spring Security, and our case study. Our proposed approach is detailed in Sect. 3. The next section is about the related work to our study. In the last section, we draw some conclusions and suggestions about future work.

2   Preliminaries

2.1   Attribute-Based Access Control

Access control is an effective method to restrict software systems' resources access violations. However, some traditional models have several limitations when users in affiliated organizations request access to resources under the terms of data-sharing agreements between organizations or access rules based on various attributes. Attribute-based access control (ABAC) is a flexible access control model in which each access decision to resources is based on attributes of the user, attributes of the requested object, and attributes of the environment of

Fig. 1. The ABAC mechanism

request [6]. Figure 1 describes the mechanism of ABAC which has two important components that are PEP (Policy Enforcement Point) and PDP (Policy Decision Point). PEP makes authorization decisions for valid users and enforces resource requests from subjects and PDP assesses the user resource request with the access policy of the system by relying on user attributes, resource attributes, and environmental attributes. From that, it makes a decision to deny or accept the user's request.

The following is a formal definition to describe the ABAC policy model of a system.

Definition 1. *(ABAC model)*
An ABAC model $\mathcal{M}$ is a tuple $\langle \mathbb{S}, \mathbb{RS}, \mathbb{E}, \mathbb{A}_s, \mathbb{A}_{rs}, \mathbb{A}_e, \mathbb{OP}, \mathbb{P} \rangle$, where:

- *$\mathbb{S}$, $\mathbb{RS}$, and $\mathbb{E}$ are subjects, resources, and environments, respectively.*
- *$\mathbb{A}_s$, $\mathbb{A}_{rs}$, $\mathbb{A}_e$ are assigned attributes for the subject $s \in \mathbb{S}$, resource $rs \in \mathbb{RS}$, and environment $e \in \mathbb{E}$.*
- *$\mathbb{OP}$ is the set operations that can be executed on protected resources by subjects $(create, read, update, delete)$.*
- *An access rule R is a Boolean-valued function on the set of attributes of the subject $s \in \mathbb{S}$, operation $op \in \mathbb{OP}$, and the set of attributes of resource $rs \in \mathbb{RS}$ in the set of attributes of the environment $e \in \mathbb{E}$.*

$$R(s, op, rs, e) \leftarrow f(\mathbb{A}_s, op, \mathbb{A}_{rs}, \mathbb{A}_e).$$

- *The ABAC policy $\mathbb{P}$ is the access rule set. It has the form $\mathbb{P} = \{R_1, R_2, ..., R_n\}$ and will be stored in a central repository (e.g. file).*

2.2   Spring Security

Spring Security framework supports for implementing security policies of JavaEE-based enterprise applications. Its architecture includes authentication and authorization sub-systems which can be configured with Java, XML, or

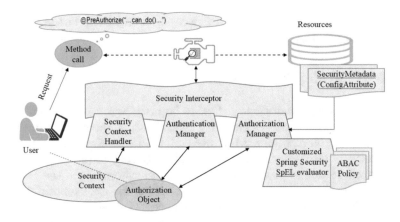

Fig. 2. Spring architecture for implementing ABAC policy in JavaEE

both [2]. Figure 2 shows Spring architecture for implementing ABAC policy in JavaEE. In this architecture, all user's resources requests are interrupted by a security interceptor for authentication and authorization. After the authentication process, if the users are valid subjects in the system, their requests will be evaluated for authorization according to the system's policy. Therefore, only valid users with system policy compliance requests will gain permission to manipulate resources. This architecture helps to limit breaches of access to system resources.

2.3   A Case Study

Our motivational example is the medical record management system of hospitals where the information within medical records is private and sensitive information should be kept confidential. In addition, systems containing medical information always deal with the risks of unauthorized information exploitation [3]. Therefore, medical records are the system's resources and need to be controlled access to ensure that they are only operated by valid users. In our application system, the ABAC policy is described as follows:

- *Patients only read their medical records.*
- *Doctors are only responsible for treating patients in their departments and hospitals. Doctors have the right to read and update the medical records of patients treated by them. In addition, doctors can read their patients' medical records in the same or linked hospitals.*
- *Nurses can read all of the medical records in their department and hospital.*
- *Receptionists have the right to create medical records in their hospitals.*

In this study, each patient has only one identification code and may have many examinations at the same or different hospitals. However, a patient has only one medical record and is treated by a doctor in a hospital at a time. Each doctor only works for one department in a hospital and a doctor can treat many patients. All users login this system must comply with the this access policy.

3   ABAC Policy Verification

ABAC Policy Implementation: Spring Security Framework allows developers to put their access control logic into a centralized component. Spring annotations like *@PreAuthorize, @PostAuthorize* are used to enforce access rules. The ABAC policy is implemented based on SpEL (Spring Expression Language). SpEL is used to define access rules and they are stored in a central repository (e.g.: memory, databases, files,..). The details for each repository format depend on the component implementer. The access rule set of the web application is implemented by JavaScript Object Notation (JSON) as Listing 1.1. Our medical information management system[1] is designed according to MVC architecture [12] and developed by the Java EE Development Kit, NetBeans IDE 8.2, IntelliJ IDEA 2019.1, Maven 3.6.1, and Spring boot.

```
1  [
2      {
3      "name" : "The name of access rule",
4      "description" : "The mean description of access rule",
5      "condition" : "The expression describes the environment
6      condition of access rule"
7      },
8      ...
9  ]
```

Listing 1.1. The implemented ABAC policy file format

ABAC Policy Specification: To check the system's access policy, we will transform the access rules in the specified ABAC policy into the expressions form as shown in the Listing 1.2. All expressions in this format are logical formulas that are used to restrict the subjects, actions, resources, and environment. The occurrence order of the operands in each expression follows the order of predefined attributes of subjects, actions, resources, and the environment.

```
1  <?xml version="1.0" encoding="UTF-8" ?>
2  <Policy>
3      <Rule RuleId = "ruleId" Effect= "Permit">
4          <Subject>
5              Subject_Name
6          </Subject>
7          <Action>
8              Action_Name
```

---

[1] Our web application's source code is available at https://github.com/ltnhan1982/ABACApp.

```
 9|        </Action>
10|        <Resource>
11|            Resource_Name
12|        </Resource>
13|        <Condition>
14|            <Restriction>
15|                ConditionRestrictionExpression
16|            </Restriction>
17|        </Condition>
18|        <Environment>
19|            EnvironmentRestrictionExpression
20|        </Environment>
21|    </Rule>
22|    ...
23|</Policy>
```

Listing 1.2. The expression format of an access rule

As mentioned above, the implementation of the ABAC policy is very complicated because there must be a combination of many components. Therefore, the specified ABAC policy of the system may not be implemented correctly into the web application. The CIA triad (*Confidentiality - Integrity - Availability*) is considered the three most important security properties [5,10]. These security properties are closely related to evaluating the accuracy of implemented access policies in software applications. Because the nonconformity between implementation and specification will lead to potential security breaches and break of these properties. Therefore, this study aims at detecting access breaches that fracture the *confidentiality, integrity* and *availability* of the application system.

Figure 3 shows our verification framework. The input to the process is the initial system policy specification file (**.xml* format) and the application system source code. From the project's code, we will parse it into a set of access rules. And from this analysis result, we will check the C-I-A properties of the application system.

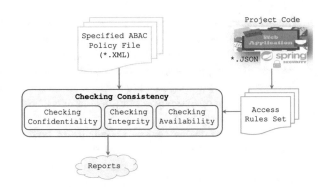

Fig. 3. The ABAC policy verification framework

3.1   Confidentiality of Access Policy

To guard this security property, the access policy of the software systems has to ensure that illegal subjects do not have permission to read systems' resources. This is formalized as *Definition* 2.

Definition 2. *(Confidentiality of access policy)*
Given the ABAC policy $\mathbb{P} = \{R_1, R_2, ..., R_n\}$ *and the set of implemented access rules of the web application* $\mathbb{AR} = \{AR_1, AR_2, ..., AR_m\}$. *The application has confidentiality iff* $((\forall AR_i \in \mathbb{AR}) \wedge (op(AR_i) = "Read"))$ *then* $(\exists R_j \in \mathbb{P} : R_j \Leftrightarrow AR_i)$.

3.2   Integrity of Access Policy

The integrity of a software system is guaranteed based on the principle that no unauthorized user is allowed to perform operations to modify the system's resources. It is clear that if the application system allows unauthorized users to create, update, or delete resources, then the original resources may be changed or lost. These lead to a violation of system integrity.

Definition 3. *(Integrity of access policy)*
Given the ABAC policy $\mathbb{P} = \{R_1, R_2, ..., R_n\}$ *and the set of implemented access rules of the web application* $\mathbb{AR} = \{AR_1, AR_2, ..., AR_m\}$. *The application has integrity iff* $((\forall AR_i \in \mathbb{AR}) \wedge (op(AR_i) \in \{ "Create", "Update", "Delete"\}))$ *then* $(\exists R_j \in \mathbb{P} : R_j \Leftrightarrow AR_i)$.

3.3   Availability of Access Policy

The system's availability ensures that users of the system have enough functions to perform their tasks. A system does not guarantee availability that is a case of inconsistency between implementation and specification (*lack of access rules*).

Definition 4. *(Availability of access policy)*
Given the ABAC policy $\mathbb{P} = \{R_1, R_2, ..., R_n\}$ *and the set of implemented access rules of the web application* $\mathbb{AR} = \{AR_1, AR_2, ..., AR_m\}$. *The application has availability iff* $(\forall R_j \in \mathbb{P})$ *then* $(\exists AR_i \in \mathbb{AR} : AR_i \Leftrightarrow R_j)$.

Assume that the access rules in the specified access policy are correct. Based on the security properties defined above, we will compare each access rule implemented in the application with the specified set of access rules. From there, it is discovered that the security properties are violated and the access rules are not guaranteed. The framework output will include both of these to help developers track down the cause of the inconsistency. Policy implementation is considered consistent if there are no property violations.

The performance of the approach depends primarily on how comparison to judge the equivalence between two access rules and the size of the access policy. Because SpEL is used to define access rules and they are stored in a central repository, policies are defined without being limited to in-memory policies, JSON file policies, or database policies. Therefore, the scope of access rules depends on the expressiveness of the SpEL and the capacity of the memory.

4   Related Work

Many studies have been conducted to specify, analyze, and review the access policies of web applications. In this section, we will briefly discuss some of the proposals related to our study.

In the study [8], the authors showed an approach to implement ABAC policy into web applications by defining subject properties, object properties, and access policy access. The method is illustrated with a web application that provides system members with access to course resources and project files. Meanwhile, Alessandro Armando et al. have developed an extension of the Spring Security framework for ABAC [1] policy specification and enforcement. The study introduced a flexible access control mechanism for APIs. This is an important security mechanism to ensure the enforcement of authorization constraints on resources while calling API functions. In the proposed architecture, all user resource access requests are checked by an interceptor, which evaluates the preconditions of method calls before accessing the data to be protected guard. If the preconditions are met according to the system policy, then the user is granted permission to execute the methods and is denied otherwise.

Constraints are an important factor in access control models. The two main concerns with binding are specification and implementation. Among the types of constraints, the implementation of the Separation of Duty (SoD) constraint is considered the most important in commercial applications. Therefore, the study [7] solved the problem of specification, verification, and implementation of SoD in the ABAC system. The author demonstrated the effect of modifying different components of ABAC on execution. In addition, the study also analyzes and addresses the complexity of ABAC policy enforcement.

The study [4] gave a formal definition of the attribute-based access control model, how to specify the ABAC-based recovery policy using temporal logic in *NuSMV* and how they are implemented, form verification. Recovery verification is implemented by automated model verification techniques, which reduces the required complexity and is a valuable tool for administrators. In addition, Shu *et al.* proposed a method to detect inconsistencies between statistically inconsistent rules in ABAC policy [15]. With rule reduction, the rules are reduced to a compact set of rules that are semantically equivalent through the removal of redundant information between the rules. Then, the binary search technique is applied to discover the inconsistencies between them.

In the studies presented above, some studies have implemented formal definitions of the ABAC model, proposed to implement ABAC in the Spring Security framework, and verified recovery, responsibility division constraints responsibility of ABAC policy. However, the proposed methods have not yet performed the analysis of the ABAC policy implemented in web applications and verified the conformance of the implemented access policy and specification from the application's source code.

5   Conclusion and Future Work

In this paper, we introduced the approach to support checking the compliance of the web application's ABAC policy with its specification. The accuracy of the implemented policy is understood to mean that there is no redundancy or lack of access rules. This is expressed through three security properties that are *confidentiality, integrity* and *availablity*.

The proposed approach is applied to web systems whose attribute-based access policy is implemented against the Spring Security framework. Our proposal can help developers detect some flaws in the system's ABAC policy implementation. We are going to develop a tool to test security properties automatically, and also experiment and evaluate the effectiveness of the proposed method and tool.

In the future, we will take into account how to check the other security properties such as *non-repudiation* and *accountability*. We will also consider the problems related to tracing and storing the activities of users in the system.

References

1. Armando, A., Carbone, R., Chekole, E.G., Ranise, S.: Attribute based access control for APIs in spring security. In: Proceedings of the 19th ACM Symposium on Access Control Models and Technologies, pp. 85–88. ACM (2014)
2. Dikanski, A., Steinegger, R., Abeck, S.: Identification and implementation of authentication and authorization patterns in the spring security framework. In: The Sixth International Conference on Emerging Security Information, Systems and Technologies (SECURWARE 2012) (2012)
3. Gordon, W.J., Fairhall, A., Landman, A.: Threats to information security-public health implications. N. Engl. J. Med. **377**(8), 707–709 (2017)
4. Gouglidis, A., Hu, V.C., Busby, J.S., Hutchison, D.: Verification of resilience policies that assist attribute based access control. In: Proceedings of the 2nd ACM Workshop on Attribute-Based Access Control, pp. 43–52 (2017)
5. Greene, S.: Security Policies and Procedures: Principles and Practices. Prentice Hall Security Series. Prentice-Hall, Inc. (2005)
6. Hu, C.T., et al.: Guide to attribute based access control (ABAC) definition and considerations [includes updates as of 25 February 2019]. Technical report, NIST (2019)
7. Jha, S., Sural, S., Atluri, V., Vaidya, J.: Specification and verification of separation of duty constraints in attribute-based access control. IEEE Trans. Inf. Forensics Secur. **13**(4), 897–911 (2018)
8. Kauser, S., Rahman, A., Khan, A.M., Ahmad, T.: Attribute-based access control in web applications. In: Malik, H., Srivastava, S., Sood, Y.R., Ahmad, A. (eds.) Applications of Artificial Intelligence Techniques in Engineering. AISC, vol. 698, pp. 385–393. Springer, Singapore (2019). https://doi.org/10.1007/978-981-13-1819-1_36
9. Martin, E., Hwang, J., Xie, T., Hu, V.: Assessing quality of policy properties in verification of access control policies. In: 2008 Annual Computer Security Applications Conference (ACSAC), pp. 163–172. IEEE (2008)

10. Mead, N.R., Allen, J.H., Barnum, S., Ellison, R.J., McGraw, G.R.: Software Security Engineering: A Guide for Project Managers. Addison-Wesley Professional, Boston (2004)
11. Meng, N., Nagy, S., Yao, D., Zhuang, W., Arango-Argoty, G.: Secure coding practices in Java: challenges and vulnerabilities. In: Proceedings of the 2018 IEEE/ACM 40th International Conference on Software Engineering (ICSE), pp. 372–383. IEEE (2018)
12. Principe, M., Yoon, D.: A web application using MVC framework. In: Proceedings of the International Conference on e-Learning, e-Business, Enterprise Information Systems, and e-Government (EEE), p. 10. IEEE (2015)
13. Rashid, F.Y.: Library misuse exposes leading Java platforms to attack (2017)
14. Scarioni, C.: Pro Spring Security. Apress, New York (2013)
15. Shu, C., Yang, E.Y., Arenas, A.E.: Detecting conflicts in ABAC policies with rule-reduction and binary-search techniques. In: 2009 IEEE International Symposium on Policies for Distributed Systems and Networks, pp. 182–185. IEEE (2009)
16. Xu, D., Zhang, Y.: Specification and analysis of attribute-based access control policies: an overview. In: 2014 IEEE Eighth International Conference on Software Security and Reliability-Companion, pp. 41–49. IEEE (2014)

Aspect-Based Sentiment Analysis with Deep Learning: A Multidomain and Multitask Approach

Trang Uyen Tran[1](✉) [iD], Ha Thanh Thi Hoang[2] [iD], Phuong Hoai Dang[3] [iD],
and Michel Riveill[4] [iD]

[1] Vietnam-Korea University of Information and Communication Technology, The Danang University, Danang, Vietnam
tutrang@vku.udn.vn, trang.tranuyen@gmail.com
[2] University of Economics, The Danang University, Danang, Vietnam
ha.htt@due.edu.vn
[3] University of Science and Technology, The Danang University, Danang, Vietnam
dhphuong@dut.udn.vn
[4] Université Côte d'Azur, CNRS, Inria, Nice, France
michel.riveill@univ-cotedazur.fr

Abstract. Sentiment analysis aids in obtaining the opinion of the users towards a particular product, service or policy. Focusing on classifying the sentiment that corresponds to each aspect of the entity in the document will help to identify the sentiment more clearly. This is also the mission of aspect-based sentiment analysis (ABSA). The vast majority of prior studies in ABSA have implemented single-task execution models on single-domain datasets. This is inconvenient when it is necessary to perform the full range of tasks in ABSA and on domain-independent datasets. In this paper, we offer to operate the advanced arrangement of deep learning techniques for multidomain and multitask approach in ABSA. The main tasks in ABSA: aspect extraction, category identification, sentiment classification and domain classification are all finished by an integration framework of Convolutional Neural Network (CNN), Bidirectional Independent Long Short Term Memory (BiIndyLSTM) and Attention mechanism. In addition, we use a POS tag layer combined with GloVe in word embedding layer to get the morphological attributes of each token word from review sentences. Through the experimenting process in the Laptop_Restaurant_Hotel multidomain dataset, we found that our proposed model has achieved high precision in multitasking ABSA. With this approach, we hope our proposed model will lay the foundation for ensuring flexibility and multiutility compared to previous opinion analysis models.

Keywords: Aspect-based sentiment analysis · Deep learning · Multitask-ABSA · Multidomain-ABSA

1   Introduction

Nowadays, along with the development of social networks and user communities, the trend of getting opinions from online communities has also increased markedly. Specifically, each of us before making a decision to use a product or service needs to get opinions from the majority of users who have experienced it before. On that basis, we can choose the service or product that best suits our needs. It is for this reason that the urgent need to create automatic opinion mining systems is gradually more attracted in study. The sentiment analysis [1] focuses on researching the ability to automatically extract and classify users' sentiments on services, products, policies…experienced. Aspect-based sentiment analysis is the best sentiment analysis scope that helps users get close to the point of view correctly expressed on each aspect of the entity in the comment. This is also the direction of research that we are focusing on. Additionally, our offered ABSA model is also multitask on domain independent comments ensuring usability and ease of use.

In this study, we suggest to utilize the technology of deep learning for our multitask multidomain ABSA approach. We use the incorporated architecture of the CNN, the BiIndyLSTM and Attention for the proposed model. In fact, these two techniques can complement each other in order to extract features more accurately, thereby increasing the performance of the model. Local attributes can be efficiently extricated by diverse-magnitude convolution kernel procedure of CNN. Furthermore, with the proficiency to maintain continuing memorial; manipulate the time-throughout backpropagation of gradient to settle the problem of vanishing or exploding gradient; and maybe enhance the model's depth with multiple layers of IndyLSTM without affecting the gradient-decay accross layers IndyLSTM is entirely competent of controlling long sequences. Moreover, the additional use of the Attention mechanism makes our model able to focus more on the main keywords in opinion text increasing the accuracy of the classifier. We also apply GloVe [2] word embedding tool for the domain of Restaurant and the multidomain of Laptop-Restaurant-Hotel. In which, the Restaurant domain belongs to the dataset of SemEval 2016 and the multidomain belongs to the SemEval 2016 and our self-construction dataset. The experimentation show that our suggested multitask multidomain approach used for ABSA proves to be more effectual in terms of model usability and performance than former state-of-the-art methods that only performed single task and on single domain.

We establish the extant parts of our manuscript in this manner: The correlated ABSA researches are expressed in Sect. 2; Sect. 3 displays our CNN_BiIndyLSTM_Attention joined design for the model of recommended multitask_multidomain ABSA; The experimental results of our proposed method in the domain of Restaurant and the multidomain of Laptop-Restaurant-Hotel and our argument for established consequences are explained in Sect. 4; Sect. 5 at last, emphases on our assumption and upcoming study trend.

2   Related Works

Current sentiment analysis studies have concentrated on one of the two approaches: lexicon-based and machine learning-based approach.

- The first approach with a group of statistical measurements computes the correlation level of words from the sentiment lexicon and the document to determined the opinion polarity of the document.
- The second approach uses classified techniques with supervised or unsupervised learning to find out opinion polarity of the document.

In the lexicon-based approach, [3–5] based on a group of already defined opinion words in the dictionary and the phrases' relationship to specify sentiment polarity for adjectives in document. With PMI and LSA, authors in [5] estimated phrases' average semantic orientation in text to obtain the sentiment. Similarly, two seed words "good" and "bad" were also used in [6] to estimate the shortest distance between them and words to forecast the sentiment of the words in text. Also with this approach, [7] verified the opinion polarity on each aspect of the document based on set of natural language processing rules. [8] extricated opinion target word by using a appropriate rule set resulted from the combination of greedy algorithm and local search algorithm. [9] used a new intense opinion vocabulary list with an unsupervised learning method to predict the most negative and positive opinions.

As for the latter approach, which is attracting a lot of recent research has been giving positive results in the field of sentiment analysis. [10] extracted aspects and sentiments in text by applying an joined model of Skip-CRF and Tree-CRF. Also with this approach, authors in [11] used SVM to classify opinion target word and the sentiment exhibited in comments' aspects. [12] suggested a classifier with logistic regression and LSTM to obtain sentiment on aspect of the opinion sentence. A model of LSTM combined with attention was used in [13] for classifying sentiment polarity in the document. [14] classified sentiment based on gathering the vital level of context terms with a model of deep memory network and layers of neural attention. Also with this method, authors in [15] got opinion phrases in ABSA exactly. [16, 17] proposed two stacked models: BiGRU-CRF and BiIndyLSTM-CRF for extracting aspect in SA. With the advantage of bidirectional association architecture for both GRU and BiIndyLSTM in feature extrication, the two models achieved results in fairly high accuracy. [18] used an integrated CNN-BiGRU model for extracting aspect and sentiment. The combination of these two deep learning techniques in machine learning-based approach has also yielded a desirable result. [19] used CNN parameterized filters and CNN parameterized gated mechanism to accomplish the definite features for aspect. This suggestion helped to improve the accurateness of their model.

In this paper, uponed on deep learning techniques we recommend a multidomain multitask approach for ABSA. Our offered model has achieved enhanced correctness in four measures: Precision, Recall, F1, Accuracy assessed based on the earlier studies that performed only one or two main ABSA tasks in single domain.

3   Methodology

Recurrent neural network (RNN) [20] one of deep learning techniques, has competence to model sequences of random length owing to repeat of a recurrent unit along tokens in the subsequence. Nevertheless, vanishing and exploding gradients [21] are the biggest problems that an RNN proves ineffective when working with long-term dependencies. Long short-term memory (LSTM) is deep learning techniques studied to solve the stated limitations of RNN. However, LSTM's activation function in the form of a hyperbolic tangent or a sigmoid functions may initiate gradient decay across layers. Therefore, modeling and training a deep LSTM is actually unachievable. For this case, we suggest using the variant IndyLSTM_an improved model of IndRNN [22] as the solution for our proposed model. This proposal is founded on the leads of IndyLSTM over RNN and LSTM.

In this paper, we apply an incorporated CNN_BiIndyLSTM_Attention framework for multitask aspect based sentiment analysis. Our model has been tested on domain Restaurant to have a basis of comparison with previous studies and at the same time has also tested on multidomain Laptop_Restaurant_Hotel to consider the efficiency achieved. It is probable that CNN has good ability to extract local features round indicated words in input sentence while IndyLSTM can work well with long-term dependencies. As a result, both techniques complement each other when integrated in the same model. Moreover, the use of bidirectional IndyLSTM can support the learning ability at long distances from both forward and backward of the current token. Besides, adding Attention to BiIndyLSTM also shows to be an efficient method through catching the suitable noteworthy level values of words in opinion document. The purpose of this joined methodology is to improve the accurateness and assist to complete key tasks of multitask multidomain aspect based sentiment analysis: aspect, category, opinion polarity extraction and domain classification in comments of the multidomain dataset.

3.1 Convolutional Neural Network (CNN)

Recently, CNNs have illustrated optimistic outcomes when applied to NLP tasks [23, 24]. CNNs allow to extract high-level local features around each word in the comment. This helps to increase the accuracy of the model in near-range feature mining. Specifically in this model, an input sentence of length S (w_1, w_2, . . . , w_S). after being vectorized over word embeding tool and the added POS tag, will be provided to CNNs to extract word features.

3.2 Independently Long Short-Term Memory (IndyLSTM)

LSTMs are architectures controls memory access by gates. LSTM supports RNN's problem solving in gradient vanishing and exploding issue. By dividing the memory state vector into two halves: the first half is the memory cells for memory storage and the second half is the working memory, LSTM has demonstrated outstanding performance for learning long dependencies. At each input state, the gate mechanism helps to decide how many inputs should be kept in memory cells and how much of the cell content should be forgotten. IndyLSTM is an improved version of LSTM with the gradient decay solving ability when applying hyperbolic tangent and sigmoid functions as activation functions in multiple LSTM in a deep network. Based on the independent neurons' rules on each layer, IndRNNs and LSTMs, IndyLSTM was composed and represented as follows:

$$\begin{aligned}
f_t &= \sigma_g(W_f x_t + u_f \circ h_{t-1} + b_f) \\
i_t &= \sigma_g(W_i x_t + u_i \circ h_{t-1} + b_i) \\
o_t &= \sigma_g(W_o x_t + u_o \circ h_{t-1} + b_o) \\
c_t &= f_t \circ c_{t-1} + i_t \circ \sigma_c(W_c x_t + u_c \circ h_{t-1} + b_c)
\end{aligned} \tag{1}$$

$x_t \in \mathbb{R}^M$: input state at time step t.
$h_t \in \mathbb{R}^N$: hidden state at time step t.
$W \in \mathbb{R}^{N \times M}$: weight of current input.
$U \in \mathbb{R}^{N \times N}$: weight of recurrent input.
$b \in \mathbb{R}^N$: bias.
σ: activation function.

Neuron in a layer can realize its private state h and c. It can't see all states. Neuron associations happen only when overlapping layers of IndyLSTM. Particularly, the output of all neurons in any one layer will be processed by each neuron in the subsequent layer.

3.3 Bidirectional Independently Long Short-Term Memory (Bi-indyLSTM)

A major problem with unidirectional IndyLSTM is that it enables to learn representations from earlier time steps. As a result, it keep only past information based on the only inputs it has seen are from former time steps. However, in case we desire to learn representations from future time steps to better understand the context and eliminate the ambiguity incurred by learning one way. Bidirectional IndyLSTM (Bi-IndyLSTM) can solve this issue.

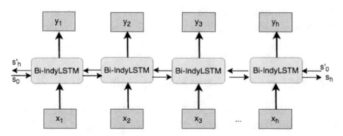

Fig. 1. Bidirectional IndyLSTM model

Figure 1 presents the illustration of the basic Bi-IndyLSTM structure with two components. Bi-IndyLSTM will run inputs in two ways, one from the past to the future and one from the future to the past. Thanks to this advantage, it can preserve information from both past and future and maybe show better results as it can understand context better.

3.4   Attention Mechanism

Attention mechanism actually helps the model to concentrate on important parts of the data by creating an alignment model a calculates the alignment score a_{ij} to reweight the hidden states h_j. Adding Attention layer in LSTM or Bi-LSTM can enhance the performance of the model and also assists in making prediction in an accurate sequence. In general, Attention proves to be very effective for NLP models on big data.

3.5   Proposed Multitask Multidomain Aspect Based Sentiment Analysis Model

After the data preprocessing step, we encode all properties correlated to semantic and sentence structure by the word embedding layer. In this paper, we use GloVe with 300-dimension feature vector. Besides, we utilize an additional layer of POS tagger to verify parts of speech of each word. Afterward, the above GloVe and POS tagger feature vectors are linked to each other and fed to the CNN layer for extracting local features around each token. The architecture that puts BiIndyLSTM next after CNN supports the ability to learn long dependencies from both backward and forward sides to the recent token. The supplemental use of the Attention layer in the next position makes our proposed model able to focus more on the main keywords that bring a lot of information in the text. Using this mechanism helps to increase the model's accuracy in feature extraction. The fully connected layer with the softmax activation function at the last position acts as a classifier to determine the main sentiment analysis tasks to be extracted and classified the domain. Finally, we obtain the classification and extraction progression: domain, category, aspect and sentiment concurrently. Our suggested multitask multidomain ABSA framework has achieved remarkable exactness on aspect extraction tasks on both single domain and multidomain compared to earlier models that experimented only on single domain of the same dataset. Moreover, our model has accomplished high results on remain main tasks of ABSA. Our offered model is shown in Fig. 2.

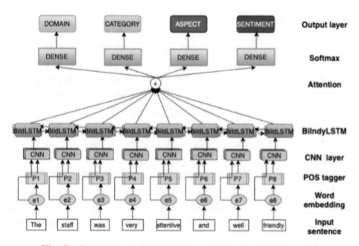

Fig. 2. Our suggested multitask multidomain ABSA model

4   Experiments and Results

4.1   Dataset

Our suggested multitask multidomain aspect based sentiment analysis model has been tested on multidomain dataset that include three domains: Restaurant, Laptop and Hotel mixed arbitrarily:

– Hotel domain is the dataset we built ourselves
– Restaurant and Laptop domains are two domains of the benchmark dataset[1]_the SemEval 2016;

Our suggested multitask multidomain aspect based sentiment analysis model has been tested on multidomain dataset that include three domains: Restaurant, laptop and hotel mixed arbitrarily: Restaurant and laptop domains are two domains of the benchmark dataset[2]_the semeval 2016; hotel domain is the dataset we built ourselves (Table 1).

Table 1. Laptop, restaurant, hotel domain and laptop_restaurant_hotel multidoman

Domain	Training sentence	Testing sentence
Laptop (the SemEval 2016 dataset)	2500	808
Restaurant (the SemEval 2016 dataset)	2000	676
Hotel (our dataset)	1454	506
Mixed laptop_restaurant_hotel (multidomain dataset)	5954	1990

4.2   Experimental Results

The model testing process is divided into two stages: experimenting on Restaurant domain; experimenting on Laptop_Restaurant_Hotel multidomain. We use *Accuracy, F1, Precision* and *Recall* to assess the effectiveness of the proposed model. The feature vectors obtained from GloVe and POS layer are applied to the next layer_CNN. The parameters of our proposed model is presented in Table 2.

All the main tasks in ABSA: aspect, category, sentiment polarity classification along with domain identification of our proposed model are assessed through comparison with three baseline models MIN [25], THA & STN [26], BiDTreeCRF [27] that experimented on the same SemEval 2016 dataset. The difference here is that these models are all experimental on Restaurant single domain and only finished the aspect extraction task. Meanwhile, our model is tested on both Restaurant single domain and Laptop_Restaurant_Hotel multidomain and has achieved quite satisfactory results for multitask.

---

[1] http://alt.qcri.org/semeval2016/task5/.

[2] http://alt.qcri.org/semeval2016/task5/.

Table 2. Parameters for our suggested model

Parameters	Value
Kernel size	3
Optimizer	Adam
Learning rate	0.001
Convolution filter channel	50
Dropout rate	0.4
Window size	128
Epoch number	120

In Table 3 below, we only have comparison the aspect extraction task of the suggested model with previous models tested on single domain Restaurant. Besides, we also test our proposed model on multidomain Laptop_Restaurant_Hotel. The model's performance was evaluated through the F1 measure and gave outstanding results.

Table 3. F1 aspect extraction results of state-of-art and our suggested model on restaurant and mixed laptop_restaurant_hotel domain

Models	F1_score	
	Restaurant domain	Mixed laptop_restaurant_hotel domain
MIN	73.44	–
THA&STN	73.61	–
BiDTree-CRF	74.49	–
Our proposed model	93.25	82.3

Table 3 shows the aspect extraction results of previous models and our suggested model in *F1_score* on single domain and on mixed domain. It can be found that the accuracies of the three previous models for this task are relatively good in the range of over 73%. However, our model achieves superior results on the same Restaurant domain with the same evaluation measure. Specifically, our model has exceeded over 19.8% compared to MIN, over 19.6% compared to THA&STN and over 18.7% compared to BiDTree-CRF. Moreover, when experimenting on a Laptop_Restaurant_Hotel multidomain dataset, our proposed model also achieves high accuracy on the same aspect extraction task with 82.3%. In particularly, when considering the F1 measure, the suggested model on the Laptop_Restaurant_Hotel multidomain is 8.86% superior to MIN, 8.69% superior to THA&STN and 7.81% superior to BiDTree-CRF tested on Restaurant domain only.

In Fig. 3, we can be realized visibly that the suggested model accomplishes considerable accurateness in *F1_score* on single domain and multidomain over previous baseline models.

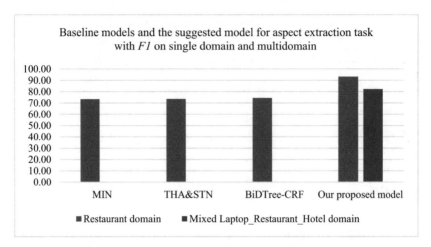

Fig. 3. Performance of models on single and multidomain

Table 4. Evaluating the performance of our proposed model with multitasking on multidomain dataset through measures of precision, recall, F1 and accuracy

Our proposed model	Laptop_restaurant_hotel multidomain			
	Precision	Recall	F1	Accuracy
Domain classification	*99.5*	*99.5*	*99.4*	*99.4*
Category extraction	*88.5*	*80.2*	*85.6*	*86.2*
Aspect extraction	*86.2*	*78.8*	*82.3*	*83.5*
Sentiment classification	*94.6*	*93.2*	*94.3*	*94.1*

Our proposed model for aspect, category, domain and sentiment classification main tasks on Laptop_Restaurant_Hotel multidomain from Table 4 and Fig. 4 displays effective results.

- 86.2%, 78.8%, 82.3% and 83.5% on Precision, Recall, F1 and Accuracy respectively for aspect extraction.
- 88.5%, 80.2%, 85.6% and 86.2% on Precision, Recall, F1 and Accuracy respectively for category classification.
- Over 99% on all Precision, Recall, F1 and Accuracy for domain classification.
- 94.6%, 93.2%, 94.3% and 94.1% on Precision, Recall, F1 and Accuracy respectively for sentiment classification.

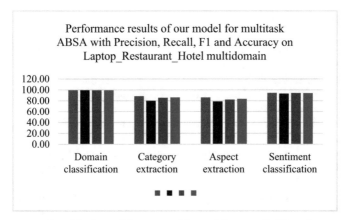

Fig. 4. Our proposed multitask model results with precision, recall, F1 and accuracy score on laptop_restaurant_hotel multidomain

From the figures illustrated in Fig. 4, our recommended model shows remarkable consequences in multitask ABSA on multidomain dataset but former models did not implement these tasks. Moreover, as can be seen clearly from Table 3 and Table 4, our recommended model experimented on both Restaurant domain and Laptop_Restaurant_Hotel multidomain also displays greater results in the subtask of aspect extraction.

5   Conclusion

We have offered a new sentiment analysis approach_multitask multidomain ABSA. Our proposed multitask model with the application of deep learning techniques has performed most of the main tasks in ABSA such as aspect, category, sentiment extraction as well as domain classification simultaneously on a multidomain dataset. This dataset is a random mix of three domains: the Restaurant and the Laptop domain of the SemEval 2016 benchmark dataset and our self-constructed Hotel domain. We propose to apply multidomain dataset in this study in order to increase the amount of data to a large enough level to ensure that the proposed deep learning model can work well with accurate results. In addition, experimenting the model on a multidomain dataset creates a ground to support the design of a less cubersome and complex model based on the idea that only one model needs to be used for all reviews from multiple domains while still ensuring the performance of this model. Our offered model has attained noteworthy improvement in performance on single Restaurant domain and on Restaurant_Laptop_Hotel multidomain over the former baseline models tested only on Restaurant domain of the benchmark dataset.

In the future, we intend to build the multilingual dataset and test our proposed model in this newly built dataset to enhance the usability of the model. Additionally, we wanted to survey substitute variant models of RNNs for our approach to find more effectual performance. Along with the improvement of techniques based on RNN for sentiment analysis, our future research direction can achieve high performance and meet model flexibility on multidomain and multilingual dataset.

References

1. Liu, B.: Sentiment Analysis: Mining Sentiments, Opinions and Emotions, 2nd edn. Cambridge University Press, Chicago (2020)
2. Pennington, J., Socher, R., Manning, C.: GloVe: global vectors for word representation. In: EMNLP'14 Proceedings of the 2014 Conference on Empirical Methods in Natural Language Processing, pp. 1532–1543 (2014)
3. Hu, M., Liu, B.: Mining opinion features in customer reviews. In: AAAI'04 Proceedings of the 19th National Conference on Artificial Intelligence, pp. 755–760 (2004)
4. Hu, M., Liu, B.: Mining and summarizing customer reviews. In: KDD'04 Proceedings of the 10th ACM SIGKDD International Conference on Knowledge Discovery and Data Mining, pp. 168–177 (2004)
5. Turney, P.D., Littman, M.L.: Measuring praise and criticism: inference of semantic orientation from association. In: ACM Transactions on Information Systems, vol. 21, issue 4, pp. 315–346 (2003)
6. Kamps, J., Marx, M., Mokken, R., Rijke, M.: Using WordNet to measure semantic orientation of adjectives. In: Proceedings of LREC-04, 4th International Conference on Language Resources and Evaluation, vol. 4, pp. 1115–1118 (2004)
7. Ding, X., Liu, B., Yu, P.S.: A holistic lexicon-based approach to opinion mining. In: Proceedings of the 2008 International Conference on Web Search and Data Mining, pp. 231–240 (2008)
8. Liu, Q., Gao, Z., Liu, B., Zhang, Y.: Automated rule selection for opinion target extraction. Knowl. Based Syst. **104**, 74–88 (2016)
9. Almatarneh, S., Gamallo, P.: A lexicon based method to search for extreme opinions. PLoS One **13**(5), e0197816 (2018)
10. Li, F., et al.: Structure-aware review mining and summarization. In: COLING'10 Organizing Committee, pp. 653–661 (2010)
11. Kiritchenko, S., Zhu, X., Cherry, C., Mohammad, S.: NRC-Canada-2014: detecting aspects and sentiment in customer reviews. In: the 8th International Workshop on Semantic Evaluation (SemEval 2014), pp. 437–442 (2014)
12. Saeidi, M., Bouchard, G., Liakata, M., Riedel, S.: SentiHood: targeted aspect based sentiment analysis dataset for urban neighbourhoods. In: COLING 2016 arXiv:1601.03771v1 [cs.CL] (2016)
13. Han, H., Li, X., Zhi, S., Wang, H.: Multi-attention network for aspect sentiment analysis. In: Proceedings of the 2019 8th International Conference on Software and Computer Applications, pp. 22–26 (2019)
14. Tang, D., Qin, B., Liu, T.: Aspect Level Sentiment Classification with Deep Memory Network. In: arXiv: 1605.08900v2 [cs.CL] (2016)
15. Liu, N., Shen, B., Zhang, Z., Zhang, Z., Mi, K.: Attention-based Sentiment Reasoner for aspect-based sentiment analysis. Human-centric Comput. Inform. Sci. **9**(1), 1–17 (2019)
16. Trang, T.U., Ha, H.T.T., Hiep, H.X.: Aspect Extraction with Bidirectional GRU and CRF. In: The 2019 IEEE-RIVF International Conference on Computing and Communication Technologies (RIVF) pp. 60–64 (2019)
17. Tran, T.U., Hoang, H.-T., Huynh, H.X.: Bidirectional independently long short-term memory and conditional random field integrated model for aspect extraction in sentiment analysis. In: Satapathy, S.C., Bhateja, V., Nguyen, B.L., Nguyen, N.G., Le, D.-N. (eds.) Frontiers in Intelligent Computing: Theory and Applications. AISC, vol. 1014, pp. 131–140. Springer, Singapore (2020). https://doi.org/10.1007/978-981-13-9920-6_14
18. Trang, T.U., Ha, H.T.T.: Deep learning in aspect-based sentiment analysis. In: the 10th Conference on Information Technology and Its Application (2021)

19. Huang, B., Carley, K.: Parameterized convolutional neural networks for aspect level sentiment classification. In: arXiv: 1909.06276v1 [cs.CL] (2019)
20. Cho, K., et al.: Learning phrase representations using RNN Encoder-Decoder for statistical machine translation. In: arXiv:1406.1078v3 [cs.CL], 3 Sep 2014
21. Bengio, Y., Simard, P., Frasconi, P.: Learning long-term dependencies with gradient descent is difficult. IEEE Trans. Neural Networks 5(2), 157–166 (1994)
22. Li, S., Li, W., Cook, C., Zhu, C., Gao, Y.: Independently recurrent neural network (IndRNN): building a longer and deeper RNN. In: arXiv: 1803.04831v3 [cs.CV], 22 May 2018
23. Poria, S., Cambria, E., Gelbukh, A.: Aspect extraction for opinion mining with a deep convolutional neural network. Knowl. Based Syst. **108**, 42–49 (2016)
24. Kim, Y.: Convolutional neural networks for sentence classification. In: EMNLP 2014, pp. 1746–1751 (2014)
25. Li, X., Lam, W.: Deep multi-task learning for aspect term extraction with memory interaction. In: EMNLP 2017, pp. 2886–2892 (2017)
26. Li, X., Bin, L., Li, P., Lam, W., Yang, Z.: Aspect term extraction with history attention and selective transformation. In: arXiv:1805.00760v1 [cs.CL] (2018)
27. Luo, H., Li, T., Liu, B.: Improving aspect term extraction with bidirectional dependency tree representation. In: IEEE/ACM Transactions on Audio, Speech and Language Processing, vol. 27, pp. 1201–1212 (2019)

Combined Local and Global Features for Action Recognition from Motion Sensors

Trung Hieu Le[1,2] , Quoc Tuan Nguyen[1] , Thanh-Hai Tran[1(✉)] ,
and Cuong Pham[3]

[1] School of Electrical and Electronic Engineering,
Hanoi University of Science and Technology, Hanoi, Vietnam
`hai.tranthithanh1@hust.edu.vn`
[2] Dai Nam University, Hanoi, Vietnam
[3] Posts and Telecommunications Institute of Technology, Hanoi, Vietnam

Abstract. Hand gestures are becoming one of the most convenient means of communication between human and machine. In this paper, we propose a method for hand gesture recognition using wearable motion sensors. The method composes of three main steps. The first step removes noise using smoothing filters. Then features extracted from each sliding window will be inputted into a machine learning model for recognition. While deep learning models have achieved impressive results on many tasks, some hand-crafted features-based techniques are still preferable due to their lightweight in terms of memory requirement and computational times. Mostly, they are easier to be deployed on edge or low-resource computers. We follow the hand-crafted features approach which extracts statistical features from the given signal. Extracting features from the whole signal can characterize the global information but not local ones. We then propose to split the original signal into several non-overlapped segments and compute the same features on each of the local segments. The final feature vector is the combination of both local and global features. We have evaluated the proposed method on three datasets (CMDFALL, C-HMAD, and DaLiAc) and show that the proposed method achieves better performance than using the global features. This shows the potential to apply such a technique in practical applications.

Keywords: Gesture recognition · Classification · Accelerometer sensor · Global features · Local features

1 Introduction

Smart infrastructure has become a fascinating area of research in recent years. It integrates sensing, control, and Internet of Things (IoT) technologies to provide a safer, more efficient, and convenient living space for residents [11]. While energy management and IoT networks are some of the research aspects, another growing trend is human-machine interaction, specifically the control of smart devices.

© The Author(s), under exclusive license to Springer Nature Switzerland AG 2022
N.-T. Nguyen et al. (Eds.): ICIT 2022, LNDECT 148, pp. 146–155, 2022.
https://doi.org/10.1007/978-3-031-15063-0_13

Conventional methods to interact with these devices, such as using handheld controls or smartphones, were not really effective.

To interact with these devices, a natural and intuitive interface is needed to interpret human intention. Hand gesture recognition is an appropriate tool, as we naturally use our hands to convey information while communicating with others [9]. In the previous research, hand gesture recognition has been applied to a variety of smart home applications, including controlling devices such as lights and TVs, interacting with computer games, and robot control. In the literature, many works have been proposed for hand gesture recognition from visual or motion sensors. Although the visual data is informatively rich, it is limited by the field of camera view and sometimes violates user privacy. Besides, their processing is commonly time-consuming. Data obtained by motion sensors (e.g. accelerometer or gyroscope) is more flexible due to its mobility. In addition, it only requires light-weight processing then more feasible to be deployed on edge or low-resource computers. Then hand gesture recognition from wearable motion sensors is more advantageous.

In this paper, we aim at developing a technique for hand gesture recognition from motion wearable sensors for human-machine interaction applications. Important requirements for such an application are accuracy, memory, and computational time. Deep learning techniques are usually unsuitable for the two last cues because of their high complexity. Hand-crafted features-based techniques are preferable when their accuracy is competitive in case the number of gestures is not so high [5]. Previously, hand-crafted features-based methods basically followed the main steps: data collection, pre-processing, data segmentation (e.g. window sliding), feature extraction, and finally classification. However, direct feature extraction from sliding windows will not show the characteristic of the gestures in detail. We propose to equally divide each gesture sample into several segments. Feature extraction will be conducted on each of those segments, called local features. Global features extracted from the whole gesture sample will be combined with local features to produce the final representation of the gesture.

In summary, our main contributions are two-fold. First, we propose a framework that equally divides the data stream into several segments then extracts local and global statistical features from the segments and the whole data. These features are then combined and inputted into a machine learning algorithm (e.g. Support Vector Machine, K-Nearest Neighbor, Random Forest) for gesture classification. Secondly, we evaluate the proposed method on several public datasets and deeply analyze its performance in terms of machine learning algorithm as well as the role of local features for fine-grained characterization of hand gestures. Different aspects such as the window size will be reported.

2   Related Works

Akin et al. [2] presented an overview of studies using accelerometer wearable sensors. Due to specific applications, accelerometers can be mounted at one or several locations of the human body (thighs, hips, chest, wrists, arms, ankles, soles of shoes, or at all major joints of the body). The activities measured from the accelerometer can take place indoors or outdoors, from daily life activities (cooking, teeth brushing, walking, and climbing stairs) to sports activities (running, playing ball/sports, climbing). For example, in a study assisting the elderly with Stylianos Paraschiakos [12] accelerometers were mounted at various locations. In the early days, the methods mainly extracted hand-crafted features in the temporal or frequency domain to represent activities/gestures. In Kern's work [8], mean and variance were calculated over windows with a width of 50 samples (corresponding to 0.5 s), then the Bayesian classifier was used to identify the activities involving the legs (standing, sitting, walking, going up and downstairs) and the activities involving the hands (such as shaking hands, writing on the board, typing on the keyboard). Bao et al. [3] performed the recognition of 20 natural gestures obtained from five accelerometers mounted on the body. Features such as mean value, energy, entropy in the frequency domain, correlation features calculated in a window of 512 samples size (corresponding to 6.7 s) are passed through different classifiers such as Bayes, decision trees. Bayat's contribution [4] with accelerometer data obtained from the phone served to identify 6 types of activities (running, slow walking, brisk walking, aerobic exercise, climbing stairs, descending stairs). The authors used filters of low and high-frequency components to represent the operation, then many classifiers such as SVM, Random Forest, Neural Network are applied. In [9], the authors have proposed a method for hand gesture recognition from wearable accelerometer sensors.

In recent times, when deep neural networks have been strongly developed and obtained good results in many problems. Deep learning techniques have also been widely applied to the problem of recognizing human activities based on the accelerometer. In the study of Alsheikh and his colleagues [1], the authors converted the acceleration signal to the frequency domain (spectrogram) before being fed into the Deep Belief Network (DBN). The results show that using a deep neural network gives better results than using traditional methods on 03 different databases. Shakya et al. [14] conducted a comparative study on traditional architectures and deep learning networks (CNNs) and RNNs (recurrent neural networks). Experiments showed that CNN gives better results than RNN and traditional methods. Tran et al. [15] proposed a method to detect falls over time from the accelerometer. In another work, Pham et al. [13] presented a real-time fall detection and activity recognition system with low implementation cost and easy deployment using use the Wii Remote worn on the body. Continuous 3-D data streams are segmented into sliding windows and then preprocessed to remove noisy signals. Features including Mean, Standard deviation, Energy, Entropy, Correlation between acceleration axes extracted from the sliding window are used to train the activity recognition model. The trained model is then used to detect falls and identify 13 activities. Experiments on 12 subjects were performed to closely evaluate the performance with recognition accuracy rates as high as 95%.

Most of the existing hand-crafted features-based methods mainly extracted statistical features from the whole gesture sample or a sliding window. These methods can characterize the global features but not the fine-grained features inside each gesture. This paper deals with this issue by proposing local features in addition to global features then improving the performance of the recognition.

3   Method for Action Recognition from Motion Sensors

3.1   Proposed Framework

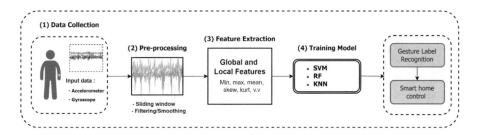

Fig. 1. Proposed framework for hand gesture recognition from wearable sensors

Our proposed framework for hand gesture recognition is illustrated in Fig. 1. It composes of three steps:

- Pre-processing: At this step, we apply a smoothing filter to reduce the noise in the original signal, then we apply the window sliding technique to take a data sample from the signal data stream. Next, we equally divide the data sample into N segments.
- Feature extraction: We compute statistical features for each of those segments (namely local features) and a global feature for the entire data sample. These features are then concatenated to produce the final feature vector.
- The extracted feature vector is fed into a classifier to output the gesture label. In this paper, we will investigate different classifiers such as SVM, K-NN, and Random Forest.

In the following, we will describe in detail each step of our framework.

3.2   Pre-processing

Data Segmentation: The raw data will be loaded into the framework through the data loader. The data loader will be responsible for removing redundant or missing data gestures (noise filtering) and then assigning data labels. The data loader outputs gesture data from the accelerometer and gyroscope. Let (a_x, a_y, a_z) and (g_x, g_y, g_z) denote the data from accelerometer and gyroscope sensors respectively.

We use sliding window technique with 50% overlap for all experimented datasets. The signal in each window is considered to contain a specific gesture. In DaLiAc dataset [10], we selected the windows size as 5 s, as used in [11]. In C-MHAD [16] and CMDFALL [15] datasets, we experiment with different window sizes to choose the best one.

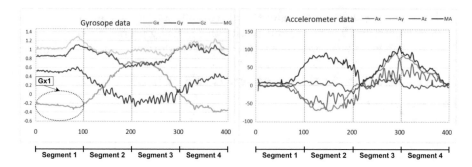

Fig. 2. An example of a segmentation of the motion data.

Noise Removing: The data needs to be smoothed before feature extraction to maximize the performance of the method. In this paper, we utilize the following smoothing function: $x_i = \frac{(x_{i-2}+x_{i-1}+x_i+x_{i+1}+x_{i+2})}{5}$ where x_i is the data signal at the time i. This formula is applied for all channels of accelerometer and gyroscope data and also the magnitudes of these two signals. Once the signal is smoothed, it is equally divided into N segments, as shown in Fig. 2. When $N = 1$, our method extracts only the global features.

3.3   Feature Extraction

In this paper, the features extracted from the sliding window forms the "global features", and the ones extracted from each segment of the sliding window are called as "local features". We calculate time-domain features to describe an overall perspective of the gestures. In addition, we calculate time-domain features for each segment to provide more specific details of gestures in a subtle period of movement.

In [13], six generic features were computed for each sliding window in each of the three accelerometers and gyroscope axes. In this paper, we extracted eight features as global features for each sliding window and local features for each sub-time frame from every axis. Besides the three-axis (x, y, z) signals, its magnitude is also computed for feature extraction. The eight global features include Mean, Max, Min, Energy, Skewness, Kurtosis, Standard deviation, Entropy. Suppose we have a time series of M samples $\{s_1, s_2, ..., s_M\}$, following statistical values will be computed:

– F_1(Mean):

$$Mean(s) = \frac{\sum_{i=1}^{M} s_i}{M} \tag{1}$$

– F_2(Max):

$$Max(s) = \max_{i=1}^{M}\{s_i\} \tag{2}$$

- F_3(Min):

$$Min(s) = \min_{i=1}^{M}\{s_i\} \quad (3)$$

- F_4(Standard Deviation):

$$SD(s) = \sqrt{\frac{1}{M}\sum_{i=1}^{M}(s_i^2) - Mean(s)^2} \quad (4)$$

- F_5(Skewness):

$$Skewness(s) = \frac{\sum_{i=1}^{M}(s_i - Mean(s))^3/M}{SD(s)^3} \quad (5)$$

- F_6(Kurtosis):

$$Kurtosis(s) = \frac{\sum_{i=1}^{M}(s_i - Mean(s))^4/M}{SD(s)^4} \quad (6)$$

- F_7(Energy):

$$Energy(s) = \frac{\sum_{i=1}^{M}s_i^2}{M} \quad (7)$$

- F_8(Entropy):

$$Entropy(s) = -\sum_{i=1}^{M}p(s_i)log(p(s_i)) \quad (8)$$

Eight features will be extracted from both acceleration and gyroscope data to input to a separated classifier.

3.4 Classification

In machine learning algorithms, data points are sometimes measured with different units or have two components (of the data vector) that differ too much from each other. At this point, we need to normalize the data before performing the next steps. One of the commonly used methods is MinMax Scaler. Accordingly, a feature vector s will be normalized by:

$$s(Scaled) = \frac{s - s_{min}}{s_{max} - s_{min}} \quad (9)$$

where s_{min} and s_{max} are the minimal and maximal elements of the vector s. After normalization, data (features and labels) are put into classification models. The effectiveness of the method will be evaluated based on the accuracy and confusion matrix of the method using global and local features compared with using only global features. We also change parameters such as the number of segments, the number of window sizes, the value of K in the K-NN classifier method, etc. to evaluate in detail the effectiveness of our method.

4   Experiment

4.1   Datasets

We use three available datasets to evaluate our method, including CMDFALL [15], C-MHAD [16] and DaLiAc [10].

- **CMDFALL** [15]: This is a rather large dataset collected from 50 people wearing 2 sensors at wrist and waist position. The data set includes 9 normal activities (such as walking, lying on the bed, sitting down in a chair, etc.) and 11 abnormal movements (such as falling on the back, falling side, staggering, slipping ...). The sampling frequency of the data set 50 Hz.
- **C-MHAD** [16]: The continuous dataset contains 5 actions of interest of the smart TV gestures performed by 12 subjects (10 males and 2 females). 10 continuous streams of video and inertial data, each lasting for 2 min, were captured for each subject. The inertial signals in this dataset consist of 3-axis acceleration signals and 3-axis angular velocity signals which were captured by the commercially available Shimmer3 wearable inertial sensor at a frequency 50 Hz on a laptop via a Bluetooth link.
- **DaLiAc** [10]: In the DaLiAc dataset, 4 inertial measurement units each consisting of a triaxial accelerometer and a triaxial gyroscope were used. The sensors were placed on the right hip, chest, right wrist, and left ankle. The sampling rate was 204.8 Hz. The dataset includes 19 subjects (8 female and 11 male). Each subject had to perform 13 daily life activities.

4.2   Experimental Results

In the experiment, we will compare the use of only global features ($N = 1$) with different combination of global features and local features ($N = 2, 4, 8$). We also analyze the performance of classification algorithms.

- **CMDFALL dataset:** With this dataset, we evaluate our method for two groups of activities S_2 (6 activities) and S_3 (20 activities) [15]. Table 1 shows the results obtained by using global features ($N = 1$) and combined global and local features ($N = 2, 4, 8$). We observe that with the use of SVM classifier, combined global and local features achieve the highest accuracy of 67.09% with $N = 4$ on the group S_2. The same conclusion is made with the group S_3 where the combination of global features and local features helps to improve the accuracy from 48.91% ($N = 1$) to 54.94% ($N = 4$). With KNN classifier, on the group S_2, the accuracy increases slightly from 58.40% ($N = 1$) to 58.99% ($N = 2$). However, on the group S_3, global features get the highest accuracy of 44.96%. With RF classifier, the accuracy on both groups S_2 and S_3 is better with the combined global and local features compared to the global feature only. This shows that the local features help to improve the recognition accuracy.

  We also compare our proposed method with the method introduced in [15]. Figure 3 shows F1 score on two groups S_2 and S_3 of CMDFALL dataset produced by [15] method and our method using combined global and local features (N = 4, SVM classifier). The results show that the F1 score of our method both increased compared to the results of the previous study. On the group S_2, the F1 score increased only slightly, while it increases significantly on the group S_3 by 14.3% (Table 1).

Table 1. Comparison of recognition accuracy (%) on the group S_2 and the group S_3 of CMDFALL dataset

Group	S_2				S_3			
Classifier method	N = 1	N = 2	N = 4	N = 8	N = 1	N = 2	N = 4	N = 8
SVM	61.96	66.40	**67.09**	59.98	48.91	54.74	**54.94**	49.80
KNN	58.40	**58.99**	56.82	53.06	**44.96**	43.92	43.38	36.07
RF	63.93	64.03	**65.12**	61.36	50.89	53.95	**54.25**	53.36

Table 2. Comparison of recognition accuracy (%) on C-MHAD dataset

Classifier method	N = 1	N = 2	N = 4	N = 8
SVM	78.53	84.12	**84.53**	83.85
KNN	76.52	81.66	**83.53**	79.93
RF	82.34	86.37	**88.21**	86.47

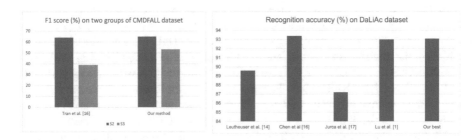

Fig. 3. Left: Comparison of F1 score (%) on two groups S_2, S_3 of CMDFALL dataset. Right: Comparison of accuracy (%) on DaLiAc dataset

- **C-HMAD dataset:** Similar to the CMDFALL dataset, the accuracy of our proposed method on the C-MHAD dataset is significantly improved (Table 2). At N = 2 and N = 8, there is a slight increase with SVM and RF. Especially at N = 4, the accuracy increases to more than 5% for SVM and about 7% for KNN. It shows that the recognition is improved with additional use of local features.

Table 3. Recognition accuracy (%) on DaLiAc dataset

Classifier method	N = 1	N = 2	N = 4	N = 8
SVM	**90.65**	90.08	91.09	89.02
KNN	**91.92**	86.65	84.66	81.09
RF	**93.11**	91.45	90.58	89.61

– **DaLiAc dataset:** Table 3 shows the results obtained by our methods with different values of N (N = 1, 2, 4, 8). We observe that on this dataset, the local features do not improve the recognition rate. The reason is that this dataset contains daily life activities which usually do not have a clear starting and ending time. In addition, there is no significant difference between segments inside the activity. As the result, the global features may be more suitable. The global features (N = 1) produce the highest result for all classification algorithms. RF achieves the best accuracy of 93.11 %, the following are KNN with an accuracy of 91.22% and SVM with an accuracy of 90.65%. We compare the performance of some existing works on DaLiAc dataset [6,7,10] and [11]. Figure 3 shows the comparison result.

5   Conclusion

In this paper, we have presented our framework for action/gesture recognition from motion sensors. Our framework consisted of three steps: pre-processing, feature extraction, and classification. For feature extraction, we have extracted both local and global statistical features. We showed that the combination of local features with the global ones helps to improve the recognition accuracy on two datasets CMDFALL and C-HMAD where the stating and ending times of each gesture/action are clear. For the dataset DaLiAc, the global features achieved the highest scores but the performance of combined global and local features are still comparable. We also evaluated different classification algorithms such as SVM, RF, KNN and showed that the combination of local features and global features are still good for different classifiers. The proposed methods achieved competitive or higher results compared to some existing methods. This result shows the potential for device control applications in smart homes and offices.

Acknowledgment. This material is based upon work supported by the Air Force Office of Scientific Research under award number FA2386-20-1-4053.

References

1. Alsheikh, M.A., Selim, A., Niyato, D., Doyle, L., Lin, S., Tan, H.P.: Deep activity recognition models with triaxial accelerometers. In: Workshops at the Thirtieth AAAI Conference on Artificial Intelligence (2016)
2. Avci, A., Bosch, S., Marin-Perianu, M., Marin-Perianu, R., Havinga, P.: Activity recognition using inertial sensing for healthcare, wellbeing and sports applications: a survey. In: 23th International Conference on Architecture of Computing Systems 2010, pp. 1–10 (2010)
3. Bao, L., Intille, S.S.: Activity recognition from user-annotated acceleration data. In: Ferscha, A., Mattern, F. (eds.) Pervasive 2004. LNCS, vol. 3001, pp. 1–17. Springer, Heidelberg (2004). https://doi.org/10.1007/978-3-540-24646-6_1
4. Bayat, A., Pomplun, M., Tran, D.A.: A study on human activity recognition using accelerometer data from smartphones. Procedia Comput. Sci. **34**, 450–457 (2014)
5. Chen, F., Deng, J., Pang, Z., Baghaei Nejad, M., Yang, H., Yang, G.: Finger angle-based hand gesture recognition for smart infrastructure using wearable wrist-worn camera. Appl. Sci. **8**(3), 369 (2018)
6. Chen, Y., Guo, M., Wang, Z.: An improved algorithm for human activity recognition using wearable sensors. In: International Conference on Advanced Computational Intelligence (ICACI), pp. 248–252 (2016)
7. Jurca, R., Cioara, T., Anghel, I., Antal, M., Pop, C., Moldovan, D.: Activities of daily living classification using recurrent neural networks. In: RoEduNet Conference: Networking in Education and Research (RoEduNet), pp. 1–4 (2018)
8. Kern, N., Schiele, B., Schmidt, A.: Multi-sensor activity context detection for wearable computing. In: Aarts, E., Collier, R.W., van Loenen, E., de Ruyter, B. (eds.) EUSAI 2003. LNCS, vol. 2875, pp. 220–232. Springer, Heidelberg (2003). https://doi.org/10.1007/978-3-540-39863-9_17
9. Le, T.H., Tran, T.H., Pham, C.: The Internet-of-Things based hand gestures using wearable sensors for human machine interaction. In: MAPR, pp. 1–6 (2019)
10. Leutheuser, H., Schuldhaus, D., Eskofier, B.M.: Hierarchical, multi-sensor based classification of daily life activities: comparison with state-of-the-art algorithms using a benchmark dataset. PLoS One **8**(10), e75196 (2013)
11. Lu, J., Zheng, X., Sheng, M., Jin, J., Yu, S.: Efficient human activity recognition using a single wearable sensor. IEEE Internet Things J. **7**(11), 11137–11146 (2020)
12. Paraschiakos, S., et al.: Activity recognition using wearable sensors for tracking the elderly. User Model. User-Adap. Inter. **30**(3), 567–605 (2020). https://doi.org/10.1007/s11257-020-09268-2
13. Pham, C., Diep, N.N., Phuong, T.M.: A wearable sensor based approach to real-time fall detection and fine-grained activity recognition. J. Mob. Multimed. 015–026 (2013)
14. Shakya, S.R., Zhang, C., Zhou, Z.: Comparative study of machine learning and deep learning architecture for human activity recognition using accelerometer data. Int. J. Mach. Learn. Comput. **8**(6), 577–582 (2018)
15. Tran, T.H., et al.: A multi-modal multi-view dataset for human fall analysis and preliminary investigation on modality. In: ICPR, pp. 1947–1952 (2018)
16. Wei, H., Chopada, P., Kehtarnavaz, N.: C-MHAD: continuous multimodal human action dataset of simultaneous video and inertial sensing. Sensors **20**(10), 2905 (2020)

Combining Artificial Intelligence with the Internet of Things in Microalgae Cultivation

Nga Le-Thi-Thu[(✉)], Trieu Nguyen-Van-Duong, and Tinh Doan-Van

Vietnam - Korea University of Information and Communication Technology, Danang, Vietnam
{lttnga,nvdtrieu.17it1,dvtinh.17it3}@vku.udn.vn

Abstract. Microalgae is the lower organism that gets that attention because they not only have specific mechanisms but also grow and develop extremely fast. Microalgae's current research and applications have created a revolution in agriculture, the food industry, the pharmaceutical industry, new materials, and environmental protection. The rapid development of the microalgae industry in the world is due to many advantages of the organism compared to higher plants such as short life cycle, high yield, high efficiency of use, and high yield. A series of technologies applied to the microalgae treatment process has been completed to help reduce costs and improve biomass quality. This paper presents the application of the Internet of Things (IoT) in microalgae cultivation and builds a Machine Learning (ML) model based on the collected data from the IoT system to predict the biomass of microalgae.

Keywords: Microalgae · Cultivation · Biomass of microalgae · Internet of Things · Artificial intelligence · Machine learning

1 Introduction

Vietnam is a coastal country, the sea area is three times larger than the land area, containing many rich resources and resources. Therefore Vietnam has the potential to develop the marine economy followed by the development of the seafood processing industry. Due to the technological characteristics of the seafood processing industry, it has discharged the environment, a large amount of wastewater with solid wastes and emissions, polluting water sources and polluting the environment, affecting the environment, the health of the surrounding community. The pollution problem of seafood processing companies is the top concern of environmental managers. The study of wastewater treatment for the seafood processing industry, as well as other industries, is an urgent requirement not only for environmentalists but also for everyone we to overcome the pollution problem of aquatic product processing wastewater, microalgae in water treatment is considered a candidate and decisive factor. In addition to overcoming the pollution problem of aquatic wastewater, the cultivation and harvesting of microalgae also create economically beneficial products such as biofuels, animal feeds, biological materials, etc.

N.-T. Nguyen et al. (Eds.): ICIT 2022, LNDECT 148, pp. 156–166, 2022.
https://doi.org/10.1007/978-3-031-15063-0_14

Microalgae is the lower organism that gets that attention because they not only have specific mechanisms but also grow and develop extremely fast. Microalgae's current research and applications have created a revolution in agriculture, the food industry, the pharmaceutical industry, new materials, and environmental protection. The rapid development of the microalgae industry in the world is due to many advantages of the organism compared to higher plants such as short life cycle, high yield, high efficiency of use, and high yield.

IoT combines various information sensing devices with the internet to form a ubiquitous huge network, which can realize the interconnection of people, machines, and things at any time and any place. Among them, the sensor is an indispensable element and a branch of the Internet of things. To give a comprehensive review of how IoT has applied to microalgae biorefinery, the development, application, and integration of various types of sensors used in microalgae biorefinery are therefore necessary. ML is rapidly becoming an integral feature of IoT devices. The typical data model of traditional data analysis is usually static, which has limitations in dealing with rapidly changing and unstructured data. When it comes to the IoT, it is usually necessary to determine the correlation between dozens of sensor inputs and external factors that rapidly generate millions of data points. In this regard, predictive analysis through machine learning is very valuable for many IoT applications. ML is a study of how computers simulate or realize human learning behavior to acquire new knowledge or skills and reorganize the existing knowledge structure to improve its performance. Researchers have innovatively applied ML including deep learning to the field of microalgae biorefinery [1].

Many methods currently exist to monitor the viability of microalgae cells in microalgae cultivation, and the most common method is the involvement of machine performance to help reduce costs and improve biomass quality. Although machines can analyze lots of cells within short hours, it is essential to develop appropriate protocols for machine operation and analysis. Therefore, to automate this process, the researchers introduced ML processing (such as flow cytometry reading) into the corresponding measurement process [2–4]. This paper presents the application of IoT in microalgae cultivation and builds a ML model based on the collected data from the IoT system to predict the biomass of microalgae.

The rest of the paper is organized as follows: Sect. 2 describes the type of microalgae used for cultivation, factors affecting the growth cycle of microalgae, the cultivation environment, and the application of IoT and ML in microalgae cultivation. IoT system for microalgae cultivation and machine learning models to predict microalgae biomass are presented in Sect. 3. Section 4 shows some experimental results. Finally, Sect. 5 gives some conclusions and possible extended works.

2   Microalgae Cultivation Technique

This section describes the type of microalgae used for cultivation, factors affecting the growth cycle of microalgae, cultivation environment.

2.1 Source of Microalgae Seed

The microalgae used in our research and cultivation process is Chlorella Vulgaris. It has a fast, strong growth rate, does not need clean water, especially capable of growing in wastewater, the life cycle lasts about 15–20 days in natural environmental conditions or when there is sufficient nutrition and lasts through 4 stages of development, going from lag phase (adaptation phase), growth phase (log), equilibrium phase (stable phase) and finally decline phase (death phase). For microalgae biomass culture, usually harvested at the end of the growing phase, and at the beginning of the equilibrium phase. The algal varieties were preserved, activated, and propagated in Antoine medium. Specific environmental components are in Table 1 and Table 2 [6].

Table 1. Components of the antoine

Chemical	Concentration (mg/l)	Chemical	Concentration (mg/l)
NH_4Cl	720	KH_2PO_4	350
$CaCl_2.2H_2O$	25	$NaHCO_3$	840
$MgSO_4.7H_2O$	140		
Add 0.5 ml of trace solution for a total volume of 1 L			

Table 2. Components of trace solutions

Chemical	Concentration (g/l)	Chemical	Concentration (g/l)
$ZnSO_4 \cdot 7H_2O$	22	H_3BO_3	11,4
$MnCl_2 \cdot 4H_2O$	5,06	$FeSO_4 \cdot 7H_2O$	4,99
$CoCl_2 \cdot 6H_2O$	1,61	$(NH_4)_6Mo_7O_{24} \cdot 4H_2O$	1,1
$CuSO_4 \cdot 5H_2O$	1,57	EDTA	50

Use the preservation method by spreading on agar plates to preserve the seed source. The variety was then activated and propagated under light conditions of 4000 lx for 7–10 days in Antoine medium before inoculation at 10% concentration in wastewater culture medium. About the benefits and applications of microalgae: treatment of polluted water sources, biofuels (diesel), bioactive substances, prevention and treatment of cancer diseases, soil improvement, etc.

2.2 Factors Affecting the Growth Cycle of Microalgae

Light is essential for the growth and development of microalgae. Because microalgae are photoautotrophs. Microalgae absorb sunlight and CO_2 in the air to convert them into organic compounds for the body and form microalgae biomass, through the process of photosynthesis. While light is essential for microalgae growth, darkness is also required for microalgae respiration. The ratio of light time to dark time is called the photoperiod. The photoperiod is similar to the circadian rhythm mechanism of life. The balance between light and dark is necessary for photosynthetic cells and carbon metabolism. Each microalgae growth process will require different temperatures for maximum growth rate °C. Vulgaris grows well in the environment with a temperature of 25–32 °C, and can withstand the temperature of 37 °C, however, at this temperature limit, cells are easily deformed, grow poorly, have a shortened life cycle and easy to die. The optimum temperature for the cultivation of C. Vulgaris ranges from 28–30 °C [5].

The purpose of using aeration is to mix the microalgae solution and create a homogenization in the solution. CO_2 is essential for microalgae to carry out photosynthesis. The right amount of CO_2 will help microalgae grow and develop hundreds of times faster than without CO_2. Studies have shown that the addition of CO_2 to the medium by about 1% by volume or a combination of aeration and CO_2 results in photosynthetic efficiency many times higher than vigorous aeration or no aeration [4]. Turbidity of water is an ever-changing factor, not only due to the lighting regime, temperature, and nutrient content, but also due to the opposite effect on the growth state of the algae population. As algae grow stronger, the turbidity of the water in the environment is changed and becomes a decisive factor for the growth and development of microalgae.

2.3   Cultivation Environment

Previous research has shown that the microalgae Chlorella Vulgaris has the ability to adapt well to the environment of organic-rich wastewater, specifically the wastewater taken from the by-product wastewater tank of the biogas digester at the livestock farm [5]. In this study, we continue to select another source of organic-rich wastewater from a seafood processing plant as a medium for cultivating this microalgae strain. Characteristics of wastewater from seafood processing factories contain a large amount of organic waste, have an unpleasant odor, and cause environmental pollution due to its unsecured natural self-cleaning ability, leading to eye damage by ecosystem. However, with its rich nutrient content, especially ammonia, it is the main source of protein for microalgae to synthesize cellular proteins through photosynthesis; Phosphorus, magnesium and potassium are also nutrients that affect microalgae growth. If we take advantage of this source of nutrients to grow microalgae, we can save production costs because it can be combined with environmental treatment.

3   Microalgae Cultivation System and Machine Learning Model

This section proposes the IoT system for microalgae cultivation and a ML model to predict microalgae biomass.

3.1   Microalgae Cultivation System

The microalgae cultivation IoT system will include 4 main objects: a microalgae aquarium, control circuit, database, and mobile/web application. Basically, the operating process of the system will be as follows: first, the sensors from the control circuit will send information about the environment of the aquarium to the database server, this information will be displayed in real-time on the applications. Users can review the variables on their own mobile devices as well as can send control feedback to the server and the control circuit will perform the control of the devices according to the user's wishes. The following Fig. 1 will describe in detail so that we can have an intuitive view of each module as well as how that module works.

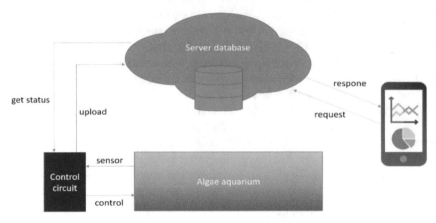

Fig. 1. Microalgae cultivation system

The factors affecting the growth of microalgae when growing experimentally were determined as light, temperature, water turbidity and aeration rate, so the control circuit will have sensors and devices:

- Light sensor is used to measure the light intensity received by the microalgae.
- Temperature sensor to measure the temperature of the aquarium (microalgae solution).
- Turbidity sensor to measure turbidity in microalgae solution.
- Aeration motor to maintain a stable level of aeration.
- Motor controls the curtain to control the curtain to open and close automatically if the light intensity is too high.

If only the above sensors are not enough, to read and control the sensors, the brain of the circuit is Arduino UNO R3, or to connect and send data to the server, the ESP8266 module is indispensable, NodeMCU. There are also some other components such as wires, screws, 8mm shafts, ball bearings,…

3.2   Machine Learning Model

After completing the model and control circuit, we carried out experimental cultivation. The microalgae used is called Chlorella Vulgaris. The environment for farming is aquatic wastewater. During the cultivation process, we collected data and extracted culture samples to measure the biomass of microalgae by UV/Vis machine. These data will be used to build a ML model to predict the biomass of microalgae.

Building a multivariable linear regression model with the dependent variable Y being the biomass of microalgae and the remaining variables being the independent variables X such as light intensity, temperature, turbidity, and airflow. From a mathematical perspective, it is possible to estimate the dependent variable Y according to the following formula:

$$y = \theta_0 + \theta_1 x_1 + \theta_2 x_2 + \theta_3 x_3 + \theta_4 x_4 \tag{1}$$

In which, y is the estimated value (or predicted value) of Y x_1, x_2, x_2, x_4 are the independent variables of light intensity, temperature, turbidity, and airflow. θ_0 is a constant, θ_1 to θ_4 are the coefficients of the relationship between x_1, x_2, x_2, x_4 and y.

If we simply use linear regression (1), we will not be able to build a model that fits the data because the independent and dependent variables in the scatterplots are not close to the straight. It is necessary to use polynomial regression to solve the data fitting problem. Formula (1) is reduced to the following general form.

$$y = \theta_0 + (x_1 + x_2 + x_3 + x_4)^n \tag{2}$$

Expanding (2) we get:

$$y = \theta_0 + \theta_1 x_1 + \theta_2 x_2 + \theta_3 x_3 + \theta_4 x_4 + \theta_5 x_1^2 + \theta_6 x_1 x_2 + \\ \dots + \theta_{k-1} x_3 x_4^{m-1} + \theta_k x_4^m \tag{3}$$

Our task is to find the $i(i = 1..k)$ where k is the number of elements expanding from the exponent n of the polynomial (2). Our cost function is an error squared function defined by the following formula:

$$J(\theta_i) = \frac{1}{2m} \sum_{i=1}^{m} (y^{(i)} - Y^{(i)})^2 \tag{4}$$

In which m is the number of samples. Formula (4) represents the average deviation (distance) between the actual data points and the predicted result after parameter estimation. The goal is to find θ_i so that the function $J(\theta_i)$ is minimized. There are two methods used to find θ_i the gradient descent algorithm and normal equation. Some of the methods that we used to evaluate our regression model are R^2 score, *explained variance score*, and *mean squared error* (MSE).

R^2 Score

R^2 score represents the proportion of variance (of y) that has been explained by the independent variables in the model. It provides an indication of the goodness of fit and therefore a measure of how well-unseen samples are likely to be predicted by the model, though the proportion of explained variance.

As such variance is dataset dependent, R^2 may not be meaningfully compared across different datasets. The best possible score is 1.0 and it can be negative (because the model can be arbitrarily worse). A constant model that always predicts the expected value of y, disregarding the input features, would get a R^2 score of 0.0.

If $y^{(i)}$ is the predicted value of the i-th sample and $Y^{(i)}$ is the corresponding true value for total samples, the estimated R^2 is defined as:

$$R^2(Y, y) = 1 - \frac{\sum_{i=1}^{m}(Y^{(i)} - y^{(i)})^2}{\sum_{i=1}^{m}(Y^{(i)} - \bar{y})^2} \quad \text{where} \quad \bar{y} = \frac{1}{m}\sum_{i=1}^{m} y^{(i)} \tag{5}$$

Explained Variance Score

If y is the estimated target output, Y the corresponding (correct) target output, and *Var* is Variance, the square of the standard deviation, then the explained variance is estimated as follows:

$$explained\_variance(Y, y) = 1 - \frac{Var\{Y - y\}}{Var\{y\}} \tag{6}$$

Mean Squared Error (MSE)

This function computes mean square error, a risk metric corresponding to the expected value of the squared (quadratic) error or loss. If $y^{(i)}$ is the predicted value of the i-th sample, and $y^{(i)}$ is the corresponding true value, then the mean squared error (MSE) estimated over $m_{samples}$ is defined as:

$$MSE(Y, y) = \frac{1}{m_{samples}} \sum_{i=1}^{m_{samples}} (Y^{(i)} - y^{(i)})^2 \tag{7}$$

Formula (7) is different from formula (4) in that it does not divide by *2m* and only divides by *m*, essentially the meaning of these two formulas is the same, but in formula (4) divided by *2m* is to make it easier to calculate the derivative.

4   Experimental Results

This section presents achieved results when implementing experiments. To serve to monitor the growth and development of microalgae and collect data, we built a system of microalgae cultivation models including hardware and software. Which hardware includes aquariums, and electronic components. And the software is the microcontroller code and the applications. The system runs with time 24/7, data is updated regularly, and sampling time is 1 h/time. Figure 2 below shows the cultivation system in practice.

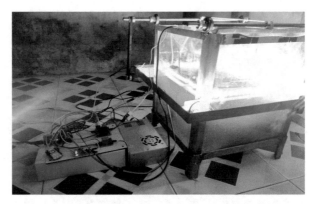

Fig. 2. Experimental microalgae cultivation system

We have carried out test planting on 3 samples and obtained the results (Fig. 3, Fig. 4 and Fig. 5).

- The first sample: It was raised within 14 days (from 11/12/2021 to 11/26/2021).
- The second sample: We raised microalgae for 12 days from 01/12/2021 to 12/12/2021.
- The third sample: In this sample, we cultivated microalgae for 12 days from 12/12/2021 to 24/12/2021.

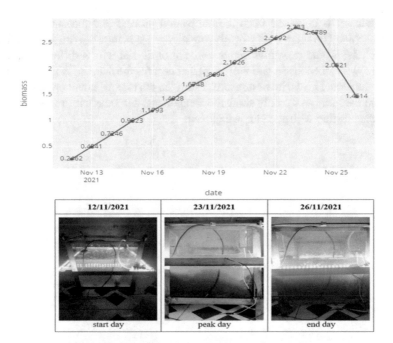

Fig. 3. Results of microalgae culture of the first sample

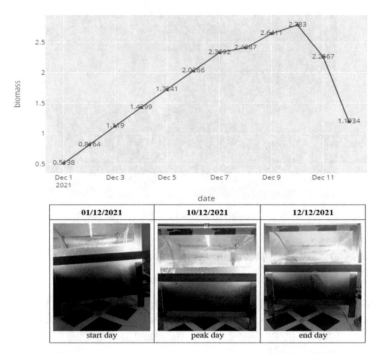

Fig. 4. Results of microalgae culture of the second sample

The data set is collected from the cultivation process and measures biomass by UV/Vis machine The ML model is a polynomial regression model. We use the evaluation method R^2, *MSE*, and *explainvariancescore* for each trial with n different exponents. From there, we will choose n and the model that best fits the data. In Table 3 we find that for $n = 3$ the model gives the best results. From the 4th exponent and up, the evaluation metrics are the same so we only show the 5 exponents and otherwise to reduce the cost of time when testing with each large exponent.

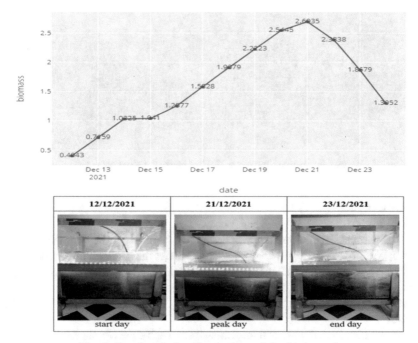

Fig. 5. Results of microalgae culture of the third sample

Table 3. Test the polynomial regression model with different exponents n and choose the most optimal model.

n	R^2	Explained_variance	MSE
1	0.9214	0.9221	0.0346
2	0.9444	0.9451	0.0245
3	0.9515	0.9518	0.0213
4	0.9513	0.9517	0.0214
5	0.9513	0.9517	0.0214

Figure 6 shows a comparison chart between actual biomass value and predicted biomass value. The red line is the predicted value, the blue line is the actual value. The difference between the two lines is not too big. The machine learning model gives relatively good results.

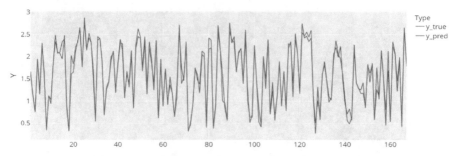

Fig. 6. Actual biomass value and predicted biomass value

5 Conclusions

The paper proposes an automatically operated IoT system to grow microalgae, the culture data will be used to build a ML model to predict the biomass of microalgae. The experimental results we obtained: Successfully building a microalgae farming system applied outdoors, incorporating IoT to automate and collect data from the farming process, using the collected data to propose a machine learning model to predict microalgae biomass.

Currently, the system has been implemented and achieved certain results, which can be applied in practice, however, it is still necessary to continue researching and developing the following points: Continue to study to shorten the growth cycle of microalgae and study the stable growth of microalgae, and collect more culture data and peak microalgae growth day to improve machine learning model that can predict harvest day.

References

1. Wang, K., et al.: How does the Internet of Things (IoT) help in microalgae biorefinery? Biotechnol. Adv. **54**, 107819 (2022)
2. Teng, S.Y., Yew, G.Y., Sukačová, K., Show, P.L., Máša, V., Chang, J.-S.: Microalgae with artificial intelligence: a digitalized perspective on genetics, systems and products. Biotechnol. Adv. **44**, 107631 (2020)
3. Nayak, M., Dhanarajan, G., Dineshkumar, R., Sen, R.: Artificial intelligence driven process optimization for cleaner production of biomass with co-valorization of was tewater and flue gas in an algal biorefinery. J. Cleaner Prod. **201**, 1092–1100 (2018)
4. Ochiai, H., Teranishi, K., Toda, K., Sugimoto, K., Ota, S.: Application of machine learning-driven label-free flow cytometry to analyze T cell products. Cytotherapy **22**, S132–S133 (2020)
5. Singh, S.P., Singh, P.: Effect of temperature and light on the growth of algae species: a review. Renew. Sustain. Energ. Rev. **50**, 431–444 (2015)
6. Zhu, L., Li, Z.: Microalgae Chlorella vulgaris biomass harvesting by natural flocculant: effects on biomass sedimentation, spent medium recycling and lipid extraction. Biotechnol. Biofuels **11**, 183 (2018)

Computer-Aided Provisional Diagnosis System Using Machine Learning

Vu-Thu-Nguyet Pham, Quang-Chung Nguyen, Van-To-Thanh Nguyen,
and Quang-Vu Nguyen[✉] ⓘ

The University of Danang, Vietnam – Korea University of Information and Communication
Technology, Da Nang, Vietnam
{pvtnguyet.19it1,nqchung.19it1,nvtthanh.19it1,nqvu}@vku.udn.vn

Abstract. The volume of fresh information from scientific researches is expanding at a quicker rate due to the rapid growth of technology. Because there is so
much data, doctors have a difficult time diagnosing the disease, which can lead
to confusion. Every three years, the volume of medical information doubles. It
is estimated that a doctor needs to read 29 h every day to remain up to date on
all medical material. Furthermore, big data sources such as data from electronic
health records (EHRs), "omic" data – genomics data, metabolic data, proteomics
data, as well as sociodemographic and lifestyle data, are data sources that would
be useless without extensive analysis. Artificial Intelligence (AI) technology is the
only way to obtain access to and utilise huge amounts of information in the medical
profession. Besides that, an accurate and timely examination of any health-related
problem is critical for sickness prevention and treatment. This study proposed an
AI-based system that can generally predict diseases based on patients' symptoms.
We have designed our system using many Machine Learning (ML) algorithms
such as Naïve Bayes, Random Forests, and Decision Trees.

Keywords: Diseases prediction · Machine learning · Naïve Bayes · Decision
trees · Random forests · Medical informatics

1 Introduction

The number of internet users has been increasing at an exponential rate throughout the
years. Finney Rutten LJ et al. [1] discovered that 74.9% of individuals in the United
States went online first during their most recent search for health information in 2017.
People typically post health-related questions (such as inquiring what type of ailment
they may be suffering from) on numerous healthcare forums. Another group of people
replies to those messages, gives advice, and sometimes forecasts potential illnesses.
However, these projections are not always accurate, and users cannot always expect to
receive a response to their messages. Furthermore, some posts are fabricated or made
up, which may lead the patient astray.

AI is presently being used in a variety of fields, most notably medicine, where the
government is spending heavily on digital transformation technology for the healthcare

N.-T. Nguyen et al. (Eds.): ICIT 2022, LNDECT 148, pp. 167–174, 2022.
https://doi.org/10.1007/978-3-031-15063-0_15

industry. The current pressing issue is sickness diagnosis in the context of expanding disease data and the emergence of new diseases, which makes diagnosis more difficult. Furthermore, having a big amount of medical data may require clinicians to wait longer to make a diagnosis. As a result, using AI technology to enhance health information processing through the collection, storage, and effective retrieval is critical (in due time and place).

In this paper, we build an online illness detection tool that employs a variety of machine learning algorithms. There are five sections in this study. Section 2, which follows the introduction, will discuss about some clinical diagnosis researches utilizing computers in recent years. A unique model was proposed in Sect. 3. The experiment technique is depicted in Sect. 4. The last section is the article's conclusion.

2 Related Works

Deep learning and machine learning technologies are actively used to produce intelligent applications in a variety of industries [2, 3]. A disease prediction system is a critical component in the development of healthcare infrastructure. In light of this, the current study suggests a predictive model, as well as a recommender system, in order to provide a more tactical and resilient healthcare environment. A detailed assessment of the present e-healthcare system reveals that there are few illness-specific solutions available. A diversified disease prediction system that takes into account all major diseases is also still absent.

Several studies have been conducted to predict diseases based on symptoms displayed by an individual using machine learning algorithms. M. Chen et al. [7] employed machine learning methods to forecast chronic illness outbreaks in disease-prone populations. They used data from real-life hospital data obtained in central China between 2013 and 2015 to test updated prediction algorithms. To tackle the challenge of missing data, they employed a latent component model to rebuild it. They also tested a localised chronic illness of cerebral infarction. Monto et al. [4] created a statistical model to predict whether or not a patient had influenza. Their study included 3744 unvaccinated adults and adolescent influenza patients who had a fever as well as at least two other flu symptoms. Based on this data, their model scored 79%. Sreevalli et al. [5] predicted the disease using the random forest algorithm. The system has a low time consumption and a low cost for disease prediction. The algorithm is 84.2% accurate. Langbehn et al. [6] created many tools to detect Alzheimer's disease. For the purpose of training the ML algorithm, data from 29 adults were used. These are all highly accurate algorithmic approaches. However, its diagnostic scale only stops specific diseases, not a wide range of diseases, some studies are algorithmic and do not provide end-user interfaces. The lack of a graphical interface provides an impediment for users. Following that, the research concentrated on a wide variety of disorders in order to suit the demands of a big number of individuals. We also provides a user-friendly online application.

3 Proposed Methodology

A thorough and timely investigation of any health-related issue is crucial for disease prevention and treatment. Furthermore, as the number of patients and ailments rises year after year, the medical system in many countries becomes overcrowded and, as a result, expensive. The majority of disorders require a doctor's visit to be treated. With adequate data, disease prediction by an algorithm might be reasonably straightforward and low-cost. The ability to predict sickness based on symptoms is an essential aspect of treatment. In our investigation, we sought to accurately predict a disease based on the symptoms of the patient.

The suggested method is made up of two parts: a web application and a machine learning model. The web application allows the user to engage with the system, whilst the prediction system anticipates the disease based on the symptoms presented.

3.1 Web Application

A web application is created for users to interact with the platform. It is divided into two key sections: Front-end and Back-end.

The front-end provides a simple GUI using HTML and JavaScript, whilst the back-end of the website is designed utilizing the Flask engine and SQLite as the database.

3.2 Disease Prediction System

The prediction system is built using the ensemble approach, which utilizes Naïve Bayes, Decision Trees, and Random Forests. The Python-based Flask framework was used to develop the back-end. Our database is stored using SQLite. The module is made up of a user-friendly graphical user interface (GUI) and a back-end server with integrated machine learning algorithms. The next section goes through the entire system in detail.

4 Experiments

4.1 Dataset

The first dataset for this research came from a study by Xiaoyan Wang et al. [11]. They presented a knowledge database of disease-symptom associations derived from summarized records of patients admitted to New York-Presbyterian Hospital (NYPH) in 2004. We constructed an excel sheet using this open-source dataset to identify many ailments and their related symptoms. The data was then preprocessed, yielding a list of around 150 disorders with approximately 1000 different symptoms.

A second disease/symptoms database is compiled from a Kaggle source as well. It has 132 unique symptoms as well as 41 distinct diseases.

These datasets are used to create the models' training and testing sets. The cleaned datasets include symptoms as column names, with rows denoting the ailments. Symptoms of an illness are marked by 1 in the column, whereas the remaining elements are designated by 0 in the column. We divided our data into three sets: 60% for training, 20% for cross-validation, and 20% for testing.

4.2 Classification Strategies

Classification is a two-stage procedure in machine learning that includes two phases: learning and prediction. During the learning step, the model is built using the available training data. In the prediction step, the model is utilized to forecast the given testing data. In this paper, we used three different kinds of classification techniques in order to categorize the data and detect the disease. These models include: Random Forests, Decision Trees, Naïve Bayes. Using these classifiers, our prediction system employs an ensemble method. In order to improve the prediction methodologies, a voting mechanism is used to assign values to different classifiers, which act as weights for our ensemble algorithm. The higher the weight, the more the classifier's output will be prioritized.

Random Forests
Random forest [8, 9] is an algorithm for classification, regression, and other problems that works by constructing a huge number of decision trees. In the creation of the model, the random forests method uses decision trees as the primary building component. The random forests approach is based on the premise that obtaining output from a single decision tree may not be totally accurate. Because there are so many factors involved in arriving at the desired result, it isn't really the closest thing to the true solution. As a result, a random forest generates many decision trees and analyzes them all at the same time. Random forest also has advantages over decision trees. Random forest selects random smaller data and creates trees based on them, so avoiding the disadvantage of overfitting in decision trees. As a result, the numerous decision trees strategy utilized in the random forest produces better results than a single decision tree. In classification tasks, the outcome of a random forest is the one which selected by the majority of trees.

Decision Trees
The decision tree algorithm [10] is a very successful and adaptable categorization approach. The goal of using a decision tree is to create a training model which can be used to predict the target by learning fundamental decision - making rules from training data. Because of its versatility, it is utilized for categorization in exceedingly complicated issues. It is also capable of dealing with higher-dimensional challenges. It is made up of three parts: the root, the nodes, and the leaves. To anticipate a class label for a record in a decision tree, we start at the bottom of the tree and work our way up. The values of the root attribute are compared to the values of the record's attribute. Based on the comparison, we advance to the next node by following the branch relating to that value.

Naïve Bayes
The Naïve Bayes algorithm is a member of the Bayes Classification family. It is based on Bayes' theorem, which is widely used in probability analysis in mathematics and computer science. The Naïve Bayes algorithm treats each attribute as an independent property that contributes to the final classification. As a result, Naïve Bayes has been praised for its accuracy and dependability in a variety of studies. The Bayes' theorem, which is employed in the Naïve Bayes algorithm, is expressed by Eq. (1) and Eq. (2).

$$P(A|B) = \frac{P(B|A) \cdot P(A)}{P(B)} \tag{1}$$

$$P(x|c) = P(x_1|c) \cdot P(x_2|c) \cdots P(x_n|c) \tag{2}$$

4.3 Evaluation Metrics

The experimental findings are discussed in this part. Equation (3), Eq. (4), Eq. (5), and Eq. (6) are the benchmark performance assessment matrices that were employed.

$$Accuracy = \frac{TP+TN}{TP+TN+FP+FN} \tag{3}$$

$$Precision = \frac{TP}{TP+FP} \tag{4}$$

$$Recall = \frac{TP}{TP+FN} \tag{5}$$

$$F1 - score = \frac{2*(precision*recall)}{(precision+recall)} \tag{6}$$

4.4 Results

For the available datasets, we examined illness prediction using several machine learning models. We were able to get at least 83.55% accuracy for all three models. Random forest is the most accurate model, with 98.03% and 93.33% accuracy on the Kaggle and NYPH datasets, respectively. Our findings are summarized in Table 1.

Table 1. Training results of different ML models

Dataset	Model	Accuracy (%)	F1-score (%)	Precision (%)	Recall (%)
Kaggle	Naïve bayes	95.07	93.66	92.96	95.07
	Decision tree	83.55	84.54	88.43	83.55
	Random forest	98.03	97.4	97.05	98.03
NYPH	Naive bayes	91.67	91.11	90.83	91.67
	Decision tree	91.67	91.11	90.83	91.67
	Random forest	93.33	92.78	92.5	93.33

We compared our technique to the other methodologies, as indicated in Table 2. For illness prediction, many studies have employed some traditional models such as SVM, KNN, Gaussian Naïve Bayes, RUSBoost, etc. For the prediction of over 150 illnesses and over 1000 discrete symptoms, we employed three different models with two independent datasets. With the Random Forest model, we attained the maximum accuracy of 98.03%, which is deemed high when compared to most of the other approaches presented.

Table 2. Comparison of different models

Authors	Models used	Maximum accuracy (%)
Mir et al. [12]	Naive Bayes, SVM, Random forest and simple CART	79.13
Vijayarani et al. [13]	SVM	79.66
Mohan et al. [14]	HRFLM	88.4
Sriram et al. [15]	Random forest	90.26
Rinkal Keniya et al. [16]	Coarse, Medium, Fine and Weighted KNN; Gaussian Naive Bayes; Kernel Naive Bayes; Coarse, Medium and Fine Decision Tree; SubSpace KNN; RUSBoost algorithm	93.5
Our proposed method	Naïve Bayes	98.3
	Decision trees	
	Random forest	

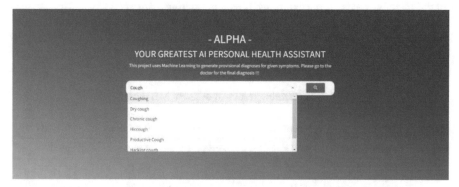

Fig. 1. Symptoms searching interface

Fig. 2. Disease prediction results interface

5   Conclusion

A disease prediction system can assist to enhance the whole healthcare industry of a country. It can help with not just "capturing a disorder", but also organizing medical data, improving research operations, and regulating illicit practices. The authors of this research proposed an innovative way for constructing a general sickness prediction system. The framework reported in this paper is comprised of three distinct algorithms: Naive Bayes, Decision Trees, and Random Forest. The experiment is carried out using two datasets, one from the Kaggle website and the other from a Colombia University study, both of which reflect the related symptoms of numerous illnesses. As shown in Fig. 1 and Fig. 2, a web-based application with a graphical user interface has also been developed to aid users in engaging with and using the sickness prediction system. According to the experimental data, the Random Forest model has achieved the highest accuracy of 98.03%.

In the near future, the project will expand to keep up with new ailments and the advancement of healthcare technology. With the help of multiple medical professionals and hospitals, a larger dataset and more accurate model may be obtained. The system cannot be used as a substitute or as a quick fix for diagnosing. It may, however, augment the physicians' experience and help them in reaching a decision. The doctor always has the option in determining whether or not to use the algorithm's diagnosis. With enough self-learning and a large database of medical records to mine from, this may be used to build sophisticated medical assistance software that can be of enormous use to all physicians, particularly new practitioners and students. In the long term, it will benefit the medical community by assisting them in obtaining appropriate diagnoses and providing chances for information exchange, allowing for faster research and the saving of many lives.

References

1. Finney Rutten, L.J., Blake, K.D., Greenberg-Worisek, A.J., Allen, S.V., Moser, R.P., Hesse, B.W.: Online health information seeking among US adults: measuring progress toward a healthy people 2020 objective. Public Health Rep. **134**(6), 617–625 (2019)
2. Agarwal, S., et al.: Unleashing the power of disruptive and emerging technologies amid COVID-19: a detailed review. arXiv preprint arXiv:2005.11507 (2020)
3. Punn, N.S., Agarwal, S.: Multi-modality encoded fusion with 3D inception U-net and decoder model for brain tumor segmentation. Multimed. Tools Appl. **80**(20), 30305–30320 (2020)
4. Monto, A.S., Gravenstein, S., Elliott, M., Colopy, M., Schweinle, J.: Clinical signs and symptoms predicting influenza infection. Arch. Intern. Med. **160**(21), 3243–3247 (2000)
5. Delshi, H.D.R., Sreevalli, P., Keerthana, A., Prathyusha, D., Asia, M.: Prediction of diseases using random forest classification algorithm. Zeichen **6**, 19–26 (2020)
6. Langbehn, D.R., Brinkman, R.R., Falush, D., Paulsen, J.S., Hayden, M.R.: A new model for prediction of the age of onset and penetrance for Huntington's disease based on CAG length. Clin. Genet. **65**(4), 267–277 (2004)
7. Chen, M., Hao, Y., Hwang, K., Wang, L., Wang, L.: Disease prediction by machine learning over big data from healthcare communities. IEEE Access **5**, 8869–8879 (2017)
8. Ho, T.K.: Random decision forests. In: Proceedings of 3rd international conference on document analysis and recognition, vol. 1, pp. 278–282. IEEE (1995)
9. Breiman, L.: Random Forests. Mach. Learn. **45**, 5–32 (2001)
10. Rokach, L., Maimon, O.: Decision trees. In: Maimon, O., Rokach, L. (eds.) Data mining and Knowledge Discovery Handbook, pp. 165–192. Springer-Verlag, New York (2005). https://doi.org/10.1007/b107408
11. Wang, X., Chused, A., Elhadad, N., Friedman, C., Markatou, M:. Automated knowledge acquisition from clinical narrative reports. In: AMIA Annual Symposium Proceedings, vol. 2008, p. 783. American Medical Informatics Association (2008)
12. Mir, A., Dhage, S.N.: Diabetes disease prediction using machine learning on big data of healthcare. In: 2018 Fourth International Conference on Computing Communication Control and Automation (ICCUBEA), pp. 1–6. IEEE (2018)
13. Vijayarani, S., Dhayanand, S.: Liver disease prediction using SVM and Naïve Bayes algorithms. Int. J. Sci. Eng. Technol. Res. (IJSETR) **4**(4), 816–820 (2015)
14. Mohan, S., Thirumalai, C., Srivastava, G.: Effective heart disease prediction using hybrid machine learning techniques. IEEE Access **7**, 81542–81554 (2019)
15. Sriram, T.V., Rao, M.V., Narayana, G.S., Kaladhar, D.S.V.G.K., Vital, T.P.R.: Intelligent Parkinson disease prediction using machine learning algorithms. Int. J. Eng. Innov. Technol **3**, 212–215 (2013)
16. Keniya, R., et al.: Disease prediction from various symptoms using machine learning. SSRN Electron J. (2020)

Design of Yaw Controller for a Small Unmanned Helicopter Based on Fuzzy Logic Controller and PID-Based Adaptation

Tri-Quang Le[✉] [iD]

Hung Yen University of Technology and Education, Dan Tien, Khoai Chau, Hung Yen, Vietnam
quangletri@gmail.com

Abstract. This paper aims to design a controller for the movement of the yaw channel of a small unmanned helicopter. The developed controller is a combination fuzzy logic controller and PI-based adaptation, which was developed based on the proportional-integral controller in considering the relationship between the error amplitude and tendency. This relation will determine which adaptive control rule is applied. The software-in-the-loop (SIL) is then used to validate the performance of the designed controller in the ideal and turbulence environment. The received results are much better than the fuzzy logic controller.

Keywords: Helicopter · PID · Fuzzy logic

1 Introduction

A helicopter is widely applied in civil and military, especially in smart agriculture, recuse and transportation, etc.[1]. Hence, it has been investigated and attracted many researchers worldwide for several decades. The helicopter features are inherently nonlinear and complex due to torques of the primary and tail rotor, which affect position and attitude. It led to the designed controller for helicopters being extremely challenging. Yaw angle is precarious. So any slight change in the main rotor torque or external effect wind gust, etc., all affect significant yaw motion [2]. So keeping the yaw motion stable in a turbulence environment is a pretty exciting and challenging topic.

It has many synthesized control algorithms for controlling an unmanned aerial vehicle [3, 4]. It can be categorized into four groups, including linear control methods, model-based controller methods, nonlinear control methods, and learning-based control methods, some remarkable control algorithms as a proportional–integral–derivative controller (PID), the linear–quadratic regulator (LQR), sliding mode control (SMC), model predictive controller, Fuzzy logic controller (FLC), reinforcement learning-based controllers (RL). As mentioned above, helicopter dynamics are inherently nonlinear; Many unmeasured parameters; Also strongly coupled between the dynamic states [5]. Therefore, determining the accuracy of the helicopter model for all of the operation state is a challenging task. It takes a lot of time and budget to implement. So, the performance of the linear and model-based control methods cannot guarantee the vast range operation

N.-T. Nguyen et al. (Eds.): ICIT 2022, LNDECT 148, pp. 175–183, 2022.
https://doi.org/10.1007/978-3-031-15063-0_16

of the helicopter. This problem is often overcome by combining control algorithms. And it is called an adaptive or hybrid controller. For example, the authors used fuzzy logic to adjust the value of control gains (Kp, Ki, Kd) to enhance the PID controller's performance in a more extensive range[6]. Besides, Reinforcement learning-based/ learning-based adaptive control algorithms are applied to eliminate error and improve system quality. Hence overall system qualify achieves an expected performance [7–10].

FLC belongs to the class of nonlinear control methods. The nature of fuzzy control is to translate the pilot knowledge into the control rule, and linguistic variables are used to present the states. This method has many advantages, such as without the required model, strong robustness, and response to a wide range of operations. It is popularly applied to control the unmanned aerial vehicle (UAV) [11–14]. But, the FLC cannot wholly eliminate the static error because of a lack of integral links [3]. Therefore, FLC is often combined with others, such as a hybrid fuzzy–PID controller [6, 14], backstepping, and sliding mode control algorithm [15].

In this paper, FLC is considered the central controller applied to control the yaw movement of the helicopter. And enhance the quality of the developed control system, such as system response static error, the author focuses on analyzing the relationship between yaw error and the trend of error change to predict the direction of system error. It is the basis of the adaptive control rules. The essence, adaptive rules are constants or PI controllers with different control gains. Besides, we also adopted the software-in-the-loop (SIL) environment, which is developed based on MATLAB&Simulink and X-Plane simulator [16], for validating the designed controller under ideal and turbulence conditions.

The rest of the paper has presented the contents: Sect. 2 introduces FLC and PID controller. Section 3 represents the proposed method. Section 4 describes the controller structure, simulation results, and discussions. Finally, the conclusions and future work is mentioned in Sect. 5.

2 Proposed Method

The helicopter tail rotor and primary rotor dynamics have a coupling relationship. The thrust created by the tail rotor maintains the yaw rate at zero and against the helicopter rotates around the z-axis. The Eq. (1) presented yaw angle in the form of the Newton-Euler [2]

$$\dot{\Psi} = \frac{q \sin \phi + r \cos \phi}{\cos \theta} \tag{1}$$

where ϕ, θ, and ψ mean the roll, pitch, and yaw angles, respectively. Symbols of r and q denote the yaw and pitch rate, respectively. Since the helicopter operates at the trim conditions as hovering or cruise mode. Pitch and roll angle can be approximated to zero. Therefore, Eq. (1) is written as

$$\dot{\Psi} \approx r \tag{2}$$

Equation (2) has shown that the yaw motion is so sensitive that any slight change in the control signal or external noise also significantly changes the yaw channel. Otherwise, collective input also affects the yaw angle [11, 17]. It is to keep yaw channel balance. The pedal must increase the proportion to the collective. So, in this research, the author proposed the yaw controller with three inputs, including yaw error, collective and yaw rate, and one output as pedal, shown in . 1. It is a combination of two fuzzy logic controllers and PID-Based adaptation, Fig. 2.

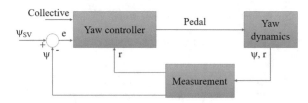

Fig. 1. Diagram of the yaw controller

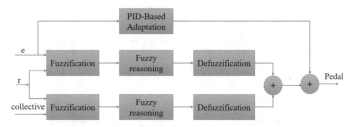

Fig. 2. Diagram of the proposed controller

Where collective and pedal mean altitude and yaw channel control inputs, respectively. ψ_{sv} and ψ denote the desired and current yaw angle. r means the yaw rates. e presents the derivation of the process yaw comparison with target your.

2.1 Fuzzy Logic Controller

The linguistic variables of FLC were defined: NBB (very negative significant), NB (great negative), NM (medium), NS (slight negative), Z (zero), PS (minor positive), PM (medium positive), PB (predominantly positive), PBB (very positive significant). The membership function of the fuzzy controllers are illustrated in Figs. 3, 4, 5, 6, 7 and 8.

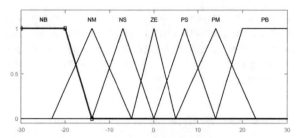

Fig. 3. Membership function of yaw error of FCL1

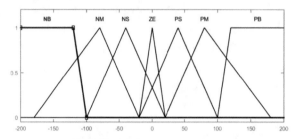

Fig. 4. Membership function of yaw rate of FCL1

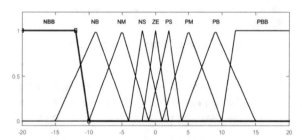

Fig. 5. Membership function of the ped1 of FCL1

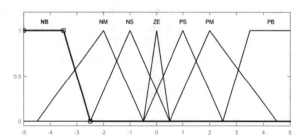

Fig. 6. Membership function of yaw rate of FCL2

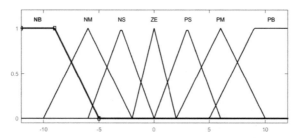

Fig. 7. Membership function of collective of FCL2

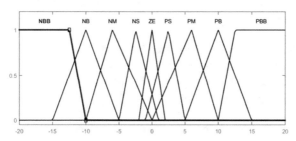

Fig. 8. Membership function of ped2 of FCL2

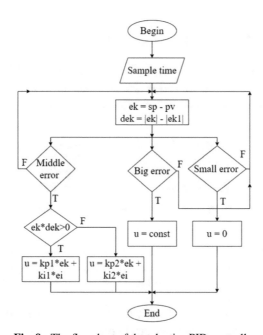

Fig. 9. The flowchart of the adaptive PID controller

2.2 PID-Based Adaptation

Defined:

$$e_k = \psi_{SV} - \psi_k \qquad (3)$$

$$de_k = |e_k| - |e_{k-1}| \qquad (4)$$

where e_k denotes the yaw error at k-time; ψ_k presents the measured yaw angle at k-time; de_k is the change of the error at k-time. We can predict the system performance by considering the error and the error change at k-time. The concept of PID-based adaptation is explained as the following rules:

if $(e_k < 0)\&\&(e_k * de_k < 0)$ then error tends bigger
if $(e_k < 0)\&\&(e_k * de_k > 0)$ then error tends smaller
if $(e_k > 0)\&\&(e_k * de_k > 0)$ then error tends bigger
if $(e_k > 0)\&\&(e_k * de_k < 0)$ then error tends smaller

It means that depending on the error trend. The designed controller will add a strong control signal, weak control signal, or nothing. The adaptive algorithm is presented in Fig. 9

3 Software-in-the-Loop and Simulation Results

The simulation environment is adopted by using Simulink and X-Plane via user datagram protocol (UDP) for communication [16]. X-Plane provides flight parameters, and Simulink is utilized to implement the controller. This process is shown in Fig. 10.

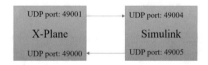

Fig. 10. Illustration of communication between Simulink and X-Plane

This study only mentions solutions for the control of yaw motion. However, pitch, roll, and height controllers were also developed for flight control. The entire software-in-the-loop (SIL) system is presented in Fig. 11. And simulation results are represented in Figs. 12, 13, 14 and 15.

The developed system was tested under the ideal environment and a light turbulent. The weather parameters for the turbulent condition were set up in the X-Plane following: the wind speed is 10 knots (approximately 5.14 m/s), the shear speed is 2 knots (about 1.02 m/s), and the turbulence is level 3. Both FLC and the proposed method well matched the desired value for the ideal testing, as shown in Fig. 12. However, the current process gave a minor error and a rapid response, illustrated in Fig. 13. The peak value appearing

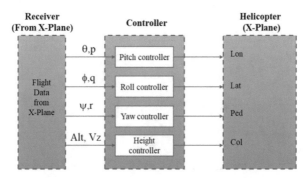

Fig. 11. Diagram of the developed SIL system

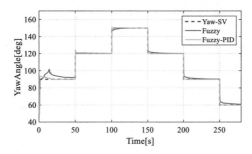

Fig. 12. Simulation results of the FLC and FLC-PID in the ideal condition

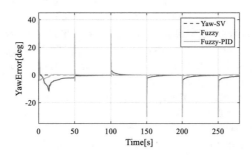

Fig. 13. The error of yaw angle in the ideal condition

in Figs. 13 and 15 is the more slow down response of the designed system than the desired values changes.

Similarly, for turbulent testing, the designed system was affected by the bad weather condition, leading to simulation results appearing oscillation. However, the current solution is still a well-kept yaw angle agreement with setpoints paths. The FCL had a prominent instability, Figs. 14 and 15. Significantly, while the helicopter was taking up, the current controller tried to keep the yaw angle at a balanced state. And FCL could not perform these tasks well, Fig. 14.

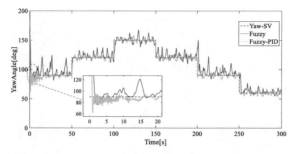

Fig. 14. Simulation results of the FLC and FLC-PID in the turbulent condition

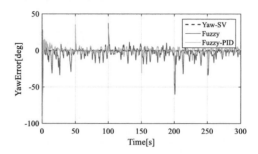

Fig. 15. The error of yaw angle in the turbulent condition

Although FCL is well-known as a high-performance solution for dealing the nonlinear system, it depends on the expert experience for defining the rule table to reach the required quality. It is even more difficult since the control object is the helicopter, which features changes dramatically in a wide range (taking off, landing, cruise, and hovering mode). On the contrary, combining the traditional fuzzy logic controller with the PID-based adaptation has improved the system's performance. The reasons are, firstly, the PI controller helps improve the system response. Secondly, in this study, the error and change of error were used to analyze the trend of the system, then using expert experience will choose a control rule corresponding to the system state. So that, system quality is improved in an all working range. However, the role of the fuzzy controller is critical. It keeps the system stable.

4   Conclusions

The article has presented the solution of controller design for a small unmanned helicopter by combining fuzzy logic controller and PID-based adaptation. The adaptive control rules were based on the error's information and the error's change. The achieved results have an essential significance for enhancing the quality of the flight control system. Moreover, fuzzy logic and PID controller are entirely feasible implementations on the embedded system. So proposed method will be applied to the actual scenario in the future.

References

1. Lee, J.-W., Xuan-Mung, N., Nguyen, N.P., Hong, S.K.: Adaptive altitude flight control of quadcopter under ground effect and time-varying load: theory and experiments. J. Vib. Control 10775463211050169 (2021)
2. Ding, L., Ma, R., Wu, H., Feng, C., Li, Q.: Yaw control of an unmanned aerial vehicle helicopter using linear active disturbance rejection control. Proc. Inst. Mech. Eng., I: J. Syst. Control Eng. **231**(6), 427–435 (2017)
3. Li, L., Sun, L., Jin, J.: Survey of advances in control algorithms of quadrotor unmanned aerial vehicle. In: 2015 IEEE 16th International Conference on Communication Technology (ICCT), pp. 107–111. IEEE (2015)
4. Nguyen, N.P., Mung, N.X., Thanh, H.L.N.N., Huynh, T.T., Lam, N.T., Hong, S.K.: Adaptive sliding mode control for attitude and altitude system of a quadcopter UAV via neural network. IEEE Access **9**, 40076–40085 (2021)
5. Mettler, B., Tischler, M.B., Kanade, T.: System identification of small-size unmanned helicopter dynamics. In: Annual Forum Proceedings-American Helicopter Society, vol. 2, pp. 1706–1717. Citeseer (1999)
6. Shim, T., Kim, Y., Bang, H.: Hybrid fuzzy-PID control and modeling of coaxial rotor helicopter. In: 2016 International Conference on Unmanned Aircraft Systems (ICUAS), pp. 689–694. IEEE (2016)
7. Xian, B., Zhang, X., Zhang, H., Gu, X.: Robust adaptive control for a small unmanned helicopter using reinforcement learning. In: IEEE Transactions on Neural Networks and Learning Systems (2021)
8. De Paula, M., Acosta, G.G.: Trajectory tracking algorithm for autonomous vehicles using adaptive reinforcement learning. In: OCEANS 2015-MTS/IEEE Washington, pp. 1–8. IEEE (2015)
9. Ouyang, Y., Dong, L., Xue, L., Sun, C.: Adaptive control based on neural networks for an uncertain 2-DOF helicopter system with input deadzone and output constraints. IEEE/CAA J. Autom. Sin. **6**(3), 807–815 (2019)
10. Chun, T.Y., Park, J.B., Choi, Y.H.: Reinforcement Q-learning based on multirate generalized policy iteration and its application to a 2-DOF helicopter. Int. J. Control Autom. Syst. **16**(1), 377–386 (2018)
11. Khizer, A.N., Yaping, D., Ali, S.A., Xiangyang, X.: Stable hovering flight for a small unmanned helicopter using fuzzy control. In: Mathematical Problems in Engineering, vol. 2014 (2014)
12. Aras, M.S.M., et al.: Evaluation and analysis of pid and fuzzy control for auv-yaw control. J. Theor. Appl. Inf. Technol. **95**(17) (2017)
13. Limnaios, G., Tsourveloudis, N.: Fuzzy logic controller for a mini coaxial indoor helicopter. J. Intell. Rob. Syst. **65**(1), 187–201 (2012)
14. Leal, I.S., Abeykoon, C., Perera, Y.S.: Design, simulation, analysis and optimization of pid and fuzzy based control systems for a quadcopter. Electronics **10**(18), 2218 (2021)
15. Liang, X., Wan, L., Blake, J.I., Shenoi, R.A., Townsend, N.: Path following of an underactuated AUV based on fuzzy backstepping sliding mode control. Int. J. Adv. Rob. Syst. **13**(3), 122 (2016)
16. Bittar, A., Figuereido, H.V., Guimaraes, P.A., Mendes, A.C.: Guidance software-in-the-loop simulation using x-plane and simulink for uavs. In: 2014 International Conference on Unmanned Aircraft Systems (ICUAS), pp. 993–1002. IEEE (2014)
17. Le, T.-Q., Lai, Y.-C., Yeh, C.-L.: Adaptive tracking control based on neural approximation for the yaw motion of a small-scale unmanned helicopter. Int. J. Adv. Rob. Syst. **16**(1), 1729881419828277 (2019)

Discovery the Quality of UML Models Through an Exploration of Dataset in Practice

Khanh-Hoang Doan[1,2]([✉]) [iD], The-Loc Nguyen[1] [iD], and Mai-Huong Tran[1] [iD]

[1] Hanoi University of Mining and Geology, Hanoi, Vietnam
{doankhanhhoang,nguyentheloc,tranmaihuong}@humg.edu.vn
[2] University of Bremen, Bremen, Germany

Abstract. Within Model-Driven Engineering (MDE) software development, models are considered as central artifacts. Thus, determining and correcting models quality issues occurring in the design phase is essential to ensure the success of the outcome product. Design smell is a kind of quality issue that indicates the weaknesses or drawbacks in the design (models) that might negatively affect the software development and maintainability. In this paper, we utilize an intelligent technique to uncover the typical design smells and quality issues of UML (Unified Modeling Language) models. The answer is achieved through an evaluation and analysis of a large dataset of models in practice.

Keywords: UML models · Model quality · Design smell · Intelligent software engineering

1 Introduction

Model-Driven Engineering (MDE) focuses on the abstract representation of the system being developed. Therefore, software models are highlighted as center artifacts within the development process [15]. Other software elements, such as code, documentation, and test cases, can be obtained from models through mode transformations [17]. As models are the central artifacts, one of the major concerns in MDE is the model quality because fault or weakness in software models can propagate to other software elements and lead to defects or malfunctions in the final software product. This has been specified in several studies in the literature [2,3]. As a result, ensuring the quality of software models in the design phase has become one of the main research topics in MDE software engineering [2,3].

Theoretically, design smells can be considered as early indicators of model quality. Design smells are poor design decisions that violate the design principle and, as the result, negatively affect the evolution and maintainability of the software system. Bad design decisions (design smells) might reduce the evolution and maintainability of the system under construction. Initially, bad smells were applied to source code [18], however, their concepts can also be applicable to the context of modeling because design smells might have the same negative impacts on the software development quality as code smells. As a consequence,

N.-T. Nguyen et al. (Eds.): ICIT 2022, LNDECT 148, pp. 184–193, 2022.
https://doi.org/10.1007/978-3-031-15063-0_17

design smells detection and refactoring can significantly enhance the software development process. Indeed, a number of approaches have been presented in the literature regarding this topic [1, 2, 4, 11].

In this paper, we evaluate and analyze a large dataset of UML models collected from practice to discover and draw common characteristics and quality of UML models. Our work is considered to have a connection to Intelligent Software Engineering (ISE). According to Xie [21], one of two aspects of ISE is the utilization of artificial intelligence (AI) or intelligent technique to software engineering. Moreover, the authors in [13] have indicated that "exploration of data to discover knowledge using analysis technique" can be considered as an "*intelligent technique*" applicable to software engineering.

The rest of the paper is structured as follows. In Sect. 2, we introduce and explain the evaluation dataset and tool support. The main contribution of this paper is presented in Sect. 3. In this section, we analyze the design smells on models in the dataset to uncover the quality of models in practice and answer the research questions. The contribution ends with concluding remarks in Sect. 5

2   Evaluation Dataset and Tool Support

2.1   Dataset

Our aim is to analyse the data for a common design errors or weaknesses of UML models in practice through a set of design smells. Therefore, the dataset should contain a significantly high number of models. The more models, the more reliable the analysis result. Moreover, the size of the analyzed models should fluctuate from small ones to very large ones.

To achieve a dataset satisfying the above requirements, the models using in this experiment are collected from two sources. The first source is a series of models gathered from systematically selected GitHub repositories. How this dataset is collected and organized is presented in [12]. In this evaluation, we only consider the structure of the models, therefore, only .ecore files are taken into account. From 16502 .ecore files in this data set, 1118 models are selected based on the following reasons.

- There are many duplicate files (files with the same names) in the original data set. To prevent bias when analyzing the data, the duplicate files are removed. We only include unique files.
- We conduct the evaluation on the tool USE, therefore, the model definition stored in the Ecore models must be transformed into USE specifications. Due to e.g., syntax errors or unsupported features of Ecore models in USE, some original models are unable to transform to USE specifications.
- The models which are too small, i.e., models which have fewer than 3 classes, are excluded.

The second source is the Atlan Ecore zoo[1] dataset of 305 Ecore models. This dataset has been chosen because it contains a set of modes that MDE

---

[1] https://web.imt-atlantique.fr/x-info/atlanmod/index.php?title=Zoos.

practitioners built in practice. Some models in this dataset are also excluded due to the same criteria as in the first data source.

As a result, we have accumulated a dataset of 1338 publicly available models with 28884 classes in total used for the evaluation[2]. Figure 1 illustrates the statistical information of the dataset used in the experiment. The size of models in this repository varies from small ones with four class to large ones with more than 100 classes. The largest model has 332 classes, and 21.6 is the average number of classes of models in the data set. We believe that by analysing metrics and the design issues on a larger number of models in this dataset, one can learn about the typical level of object-oriented mechanisms as well as the popularity of the issues occurring in the models used in practice.

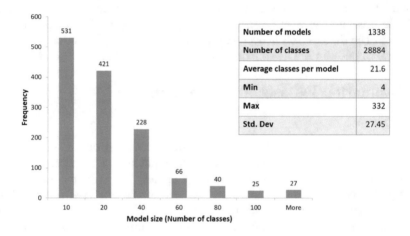

Fig. 1. The size of models using in the evaluation.

As can be seen in Fig. 1, more than **70%** (952 of 1338) are small models with less than 20 classes. The number of models decreases when their size increases. However, the variability of the model size is acceptable with nearly 150 models which contain more than 60 classes and 52 models are considered to be big models (including more than 100 classes).

2.2 Tool Support

Tool USE [9,10] has been utilized for our work in this paper. USE is an information system modeling tool, which originally only supports two-level modeling mechanisms. USE specifications are based on Unified Modeling Language (UML) and OCL (Object Constraint Language). The main functionalities of this tool are

---

[2] The dataset is available at: https://github.com/doankh/USE_Dataset.

information system visualization, validation, and verification. We have extended
the tool USE to support our work in this paper. Figure 2 illustrates an overview
of this extension regarding the evaluation.

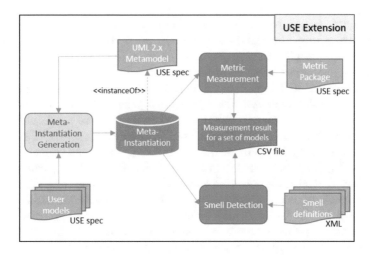

Fig. 2. Extension of tool USE.

Some new features, which mainly relate to model quality assurance, i.e.,
metric measurement and smell detection, are now available on the extended
version[3]. We firstly extend tool USE from a two-level to three-level modeling
system by adding UML 2.4 metamodel (the OMG superstructure) to the top
of the modeling hierarchy. The new functions of metric measurement and smell
detection will utilize this metamodel and metamodel instantiation, i.e., a UML
metamodel instantiation that corresponds to the UML user model, for metrics
and design smells definition. To perform the evaluation, a set of predefined smells
must be evaluated on a large number of class models. Therefore, we also extend
the tool USE to support working with a set of input models and collecting
measurement results from the input dataset.

The result is recorded in a CSV (Comma-separated values) file, and it can
be utilized for further analysis.

3   Design Smells Evaluation

To assess the quality of the UML class models in practice regarding design smells,
we have defined a library of design smells utilizing the metamodelling approach
presented in [5,6]. This library covers typical flaws or issues that modelers tend
to commit. In this evaluation we assess smells of four categories, i.e., essential

---

[3] Download the tool at: https://github.com/doankh/my_use.

property, best practice, metrics, naming convention. Some of them are collected from [1,11,20]. We exclude the trivial smells and syntactic smells because they can mostly be checked by modeling tools (tool USE as well) while compiling the user models. The list of design smells used in the evaluation is presented in Table 1.

Firstly, we run the tool USE to perform a batch evaluation of all predefined smells on every model in the dataset. The result is recorded in a CSV file. The evaluation result will be then analyzed to discover the quality of models used in practice. Table 2 represents a summary of the result.

As can be seen, 3069 issues are found in 1338 analyzed models. The minimum number of issues detected in a single model is 0, while the maximum is 12. Issues are discovered in 1092 models (79%) and on average there are 2.29 issues in a user model. Within the smell library used in the evaluation, the best practice smells are most likely to occur, while naming convention smells have the lowest probability. Figure 3 illustrates the number of flawed models regarding the number of detected smells. Models with less than three issues are most popular. The most common number of the detected smells is 2 with 348 models. From this peak, the number of faulty models significantly decreases corresponding to the increase in the number of the found issues. Only one model has 12 issues, the maximum number of issues founded in a model in the dataset.

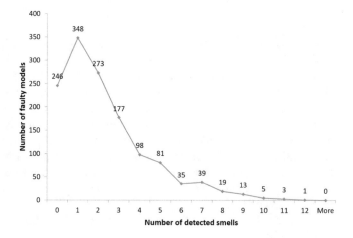

Fig. 3. Flawed models regarding number of detected smells.

With regard to the number of faulty models corresponding to each smell, Fig. 4 presents a bar chart which depicts how many analyzed models violate each design smell in Table 1.

The chart shows that the smell BP1 is discovered in the largest number of models with 558 models (40% of the analyzed models). This smell inspects the existing of a class A, which is contained in two other classes, e.g., B and C. The stronger version of the BP1 smell is the EP6 smell, which also fails in 155

Table 1. Predefined design smells using in the evaluation.

ID	Description	Severity
	Best Practice	
BP01	A class is contained in two classes.	Low
BP02	A root class is not an abstract class. A root class has subclasses and does not participate in any association.	Low
BP03	An attribute is repeated (with the same definition) among all specific classes of a hierarchy.	Medium
BP04	There exists an abstract class which is a general class of only one specified class.	Medium
BP05	There exist redundant generalization. A generalization is redundant if it could be implied from other generalizations in the model.	Medium
BP06	A class that associates to a descendant class.	Low
	Essential Property	
EP01	A non-abstract class is a superclass of an abstract class.	Critical
EP02	There exist composition cycles in the design.	Critical
EP03	A superclass is contained in one of its subclasses and the multiplicity value in the composition end is 1 (this is finitely unsatisfiable).	Critical
EP04	A reflexive association that has two member ends $x1..x2$ and $y1..y2$, where $y1>x2$.	Critical
EP05	There exist isolated classes, i.e., classes which donot have any relation (association or inheritance hierarchy) to other classes, in the design.	Medium
EP06	A class is contained in two classes, and the cardinality in the composition end of one of them is 1.	Medium
	Metrics	
ME01	A class has to many attributes (10-max by default). This smell regards to the NOA (Number of Attributes) metric.	Low
ME02	A class participates in more than ten association. This smell regards to the NAS (Number of Associations) metric.	Medium
ME03	There exists an inheritance hierarchy which is too deep (by default, 5 levels is maximum) ÃćâĆňâĂİJ This smell regards to the DIT (Depth of Inheritance) metric.	Medium
ME04	A class is overloaded with direct children (10-max by default) – This smell regards to the NOC (Number of Children) metric.	Medium
ME05	There exists a class that is too big, i.e., has more than 60 attributes and operations (God class).	Medium
	Naming Convention	
NA01	The first letter of class name is not a capital letter (PascalCase convention)	Medium
NA02	An attribute name is named by a combination of its feature class and its ordinary name (e.g., an attribute personID in class Person).	Low
NA03	A name of a class is a Java keyword.	Low
NA04	A name of a class is a C++ keyword.	Low
NA05	A name of an attribute or operation does not start with a lowercase letter (camelCase convention).	Medium

Table 2. Summary of the evaluation result.

	BP	ES	ME	NA	Total
Detected smells	1577	513	572	407	3069
Faulty models	910	378	390	340	1092
Min	0	0	0	0	0
Max	6	4	4	4	12
Average	1.18	0.38	0.43	0.3	2.29

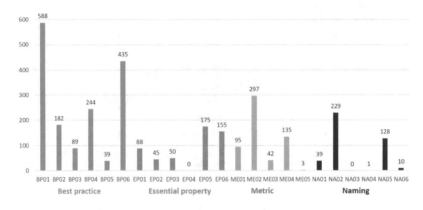

Fig. 4. Number of faulty models corresponds to each smell.

models. This smell specifies a class contained in not only two other classes but also the cardinality in the composition end of one of them being greater than 1.

The high percentage of the faulty models also can be seen in the evaluation of the BP06 and ME02 smells, with 435 (31%) and 297 (21%) non-fulfilled models, respectively. The BP06 smell checks the classes that associate to their descendant class and the ME2 smell inspects the existence of classes which participate in too many associations, i.e., ten associations. Interestingly, there are two smells, i.e., EP04 and NA03, which are not violated in any analyzed model. EP04 checks the presence of a reflective association which has a lower bound at one end being greater than the upper bound of the other end. Creating a valid system state of a model containing this kind of reflective association is impossible. Zero faulty models regarding the NA03 smell indicates that there is no class of any analyzed model named as a Java keyword. This is understandable since the analyzed models are collected from the EMF community. A very few numbers of flawed models are also found in the evaluation of the ME5 and NA04 smells, with only 3 and 1 models, respectively.

We also conducted some analysis to answer the question of how the model size (with respect to the number of classes) affects the number of non-fulfilled models. The result is presented with four charts in Fig. 5. In these charts, the vertical axis represents the average amount of faulty models, and the horizontal axis is the model size measured by the number of classes.

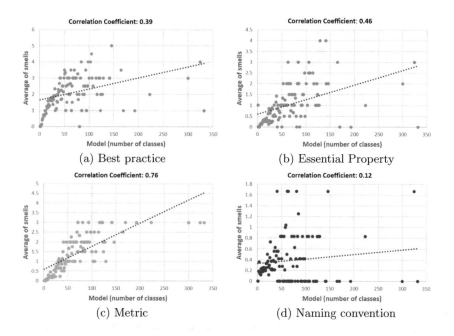

Fig. 5. Correlation between number of faulty models and model size.

The chart Fig. 5c indicates a quite strong relationship between model size and the number of faulty models regarding metric smells, with a correlation coefficient of 0.76. This means the number of non-fulfilled models regarding the metric issues seems to increase when the size of the model grows. This trend is understandable because the growth of the number of classes usually escalates the complexity of the model. The correlation coefficient of 0.39 and 0.46 corresponding to the best practice and essential property categories show that there is a positive correlation between the number of faulty models and the model size, but it is not strong and likely insignificant. The increase of the model size seems to be irrelevant to the number of flawed models regarding the naming convention smells since the correlation coefficient value is 0.12.

4   Related Work

Some intelligent techniques, e.g., statistical analysis, metric analysis and machine learning techniques, have been utilized to explore software project quality. In particular, [8] and [7] use a data distribution analysis method to classify the studied metrics into three ranges: Good, Regular, Bad. For example, [8] recommends the DIT metric can be 2/2–4/4 as Good/Regular/Bad categories. In [16] metric thresholds are calculated by a statistical method using a logistic regression model. These approaches calculated and extracted metric thresholds at the source code level.

On the other hand, some authors has introduced other works on exploring the quality of models in practice. An empirical analysis of some basic size metrics, e.g., number of classes, has been presented in [19]. The authors analyze data over publicly available Ecore metamodels from different repositories to discover what the metamodels look like in real life. In the same direction, [14] presents an approach to uncover the structural characteristics of metamodels. In this work, some metrics regarding the usage of inheritance have been calculated on repositories containing more than 450 metamodels and then analyzed using statistical tools.

5   Conclusion

This paper represents an evaluation of UML models quality in practice regarding design smells. The contribution of this evaluation is to answer the following research questions: (1) *How do the models in practice look?* (2) *What is the quality of the models in practice?*. To answer these research questions, we use a modeling tool, i.e. the tool USE, to record the data regarding the design smells of all models in the dataset collected from practice. This dataset contains 1338 models varying from small models to very large models. The data are subsequently analyzed, and the results are visually illustrated as tables and charts.

Acknowledgement. This research is funded by the project CT.2019.01.07 of the Ministry of Education and Training, Vietnam.

References

1. Aguilera, D., Gómez, C., Olivé, A.: A method for the definition and treatment of conceptual schema quality issues. In: Atzeni, P., Cheung, D., Ram, S. (eds.) ER 2012. LNCS, vol. 7532, pp. 501–514. Springer, Heidelberg (2012). https://doi.org/10.1007/978-3-642-34002-4_39
2. Basciani, F., Rocco, J.D., Ruscio, D.D., Iovino, L., Pierantonio, A.: A customizable approach for the automated quality assessment of modelling artifacts. In: 10th International Conference on the Quality of Information and Communications Technology, QUATIC 2016, Lisbon, Portugal, 6–9 September 2016, pp. 88–93 (2016)
3. Bertoa, M.F., Vallecillo, A.: Quality attributes for software metamodels (2010)
4. Bettini, L., Ruscio, D.D., Iovino, L., Pierantonio, A.: Quality-driven detection and resolution of metamodel smells. IEEE Access **7**, 16364–16376 (2019)
5. Doan, K.H., Gogolla, M.: Extending a UML and OCL tool for meta-modeling: applications towards model quality assessment. In: Schaefer, I., Karagiannis, D., Vogelsang, A., Mindez, D., Seidl, C. (eds.) Modellierung 2018, pp. 205–220. GI, LNI 280 (2018)

6. Doan, K., Gogolla, M.: Quality improvement for UML and OCL models through bad smell and metrics definition. In: 22nd ACM/IEEE International Conference on Model Driven Engineering Languages and Systems Companion, MODELS Companion 2019, Munich, Germany, 15–20 September 2019, pp. 774–778 (2019)

7. Ferreira, K.A.M., da Silva Bigonha, M.A., da Silva Bigonha, R., Mendes, L.F.O., Almeida, H.C.: Identifying thresholds for object-oriented software metrics. J. Syst. Softw. **85**(2), 244–257 (2012)

8. Filó, T.G.S., da Silva Bigonha, M.A.: A catalogue of thresholds for object-oriented software metrics. In: Proceedings of the Advances and Trends in Software Engineering, SOFTENG 2015, Barcelona, Spain, 19–24 April 2015 (2015)

9. Gogolla, M., Büttner, F., Richters, M.: USE: a UML-based specification environment for validating UML and OCL. Sci. Comput. Program. **69**(1–3), 27–34 (2007)

10. Gogolla, M., Hilken, F., Doan, K.: Achieving model quality through model validation, verification and exploration. Comput. Lang. Syst. Struct. **54**, 474–511 (2018)

11. López-Fernández, J.J., Guerra, E., De Lara, J.: Assessing the quality of metamodels. In: Proceedings of the 11th Workshop MoDeVVa@MODELS 2014, pp. 3–12 (2014)

12. Noten, J., Mengerink, J., Serebrenik, A.: A data set of OCL expressions on GitHub. In: Proceedings of the 14th International Conference on Mining Software Repositories, MSR 2017, Buenos Aires, Argentina, 20–28 May 2017, pp. 531–534 (2017)

13. Perkusich, M., et al.: Intelligent software engineering in the context of agile software development: a systematic literature review. Inf. Softw. Technol. **119**, 106241 (2020)

14. Rocco, J.D., Ruscio, D.D., Iovino, L., Pierantonio, A.: Mining metrics for understanding metamodel characteristics. In: Proceedings of the 6th International Workshop on Modeling in Software Engineering, pp. 55–60. ACM (2014)

15. Selic, B.: The pragmatics of model-driven development. IEEE Softw. **20**(5), 19–25 (2003)

16. Shatnawi, R.: A quantitative investigation of the acceptable risk levels of object-oriented metrics in open-source systems. IEEE Trans. Softw. Eng. **36**(2), 216–225 (2010)

17. Stahl, T., Völter, M., Bettin, J., Haase, A., Helsen, S.: Model-Driven Software Development - Technology, Engineering, Management. Pitman (2006)

18. Suryanarayana, G., Samarthyam, G., Sharma, T.: Refactoring for Software Design Smells: Managing Technical Debt, 1st edn. Morgan Kaufmann Publishers Inc., San Francisco (2014)

19. Williams, J.R., et al.: What do metamodels really look like? In: Proceedings of the 3rd International Workshop on Experiences and Empirical Studies in Software Modeling, vol. 1078, pp. 55–60 (2013)

20. Wüst, J.: SD Metrics. https://www.sdmetrics.com/index.html. Accessed 02 May 2022

21. Xie, T.: Intelligent software engineering: synergy between AI and software engineering. In: Proceedings of the 11th Innovations in Software Engineering Conference, ISEC 2018, Hyderabad, India, 09–11 February 2018, p. 1:1. ACM (2018)

Evaluation of Feature Extraction Methods for Bee Audio Classification

Thi-Thu-Hong Phan[1](✉) (ID), Huu-Du Nguyen[2], and Doan-Dong Nguyen[3]

[1] Department of Information Technology, FPT University, Danang, Vietnam
hongptt11@fe.edu.vn
[2] School of Applied Mathematics and Informatics,
Hanoi University of Sciences and Technology, Hanoi, Vietnam
[3] Faculty of Applied Sciences, International School, Vietnam National University,
Hanoi, Vietnam

Abstract. In recent years, machine learning (ML) methods have been widely used as a powerful tool for monitoring beehives based on bee sound data. In the use of ML algorithms, it is necessary to extract important features from the raw audio. In this study, we investigate the performance of many ML algorithms using five different feature extraction methods. We also compare the results of our experiment with the results from a previous study in the literature. The obtained results show that by choosing the right method of extracting important features, the performance of the ML methods on classifying bee sounds can be improved significantly.

Keywords: Feature extraction · ML algorithms · Audio classification · Chroma · MFCC · Spectral contrast · STFT

1 Introduction

Bee buzzing sounds are an important data source for monitoring the beehives as they contain much useful information related to the colony's health and behavior. In the literature, it has been documented that the bees buzz specific sounds depending on the special situations they are exposed to such as swarming, missing queen, or feeling airborne toxicants [3,8]. In practice, many experienced beekeepers listen to their hives to identify the state of their honey bee colonies. To support the apiarists in improving the efficiency of monitoring their beehives, there are increasing studies aiming to develop systems and methods to analyze and classify the bee sound data. Among these studies, the machine learning (ML) based methods are perhaps the most robust technique. For example, the ML methods have been applied to automatically recognize different states in a beehive using audio as input in [16]. Cejrowski et al. [5] suggested a method to detect swarming in honey bee colonies using the SVM algorithm. A deep neural network IoT-based acoustic swarm classification is proposed in [25]. The ML methods for automatically recognizing tomato-pollinating bees based on their buzzing sounds have been presented in [19].

N.-T. Nguyen et al. (Eds.): ICIT 2022, LNDECT 148, pp. 194–203, 2022.
https://doi.org/10.1007/978-3-031-15063-0_18

In the use of the standard ML methods, the input should not be the raw materials. That is, it is demanded to apply techniques to extract important features from the raw data before feeding them to the ML algorithms. The right method for feature extraction plays a very important role in enhancing the performance of these algorithms. It allows reducing significantly the dimension of original data without preserving physical meaning while providing the most essential features that best represent the data. Therefore, finding a feature extraction technique that is appropriate and compatible with a ML algorithm is a crucial task to have an efficient method for analyzing bee sound vibrations. This problem has also been the main concern in many studies in the literature, see, for example, [15, 21]. In this study, we investigate the performance in recognizing beehive audio of different feature extraction methods combined with classical ML algorithms. A number of feature extraction methods have been applied and compared in terms of classification performance, including the Mel frequency cepstral coefficients (MFCC), the Short-time Fourier transform (STFT), the Constant-Q transform (CQT), the Spectral contrast (SC), and the Chroma. We also make a comparison of the classification accuracy of using these features and using a combination of many different types of features as suggested in [14]. The obtained results show that by choosing the right feature extraction method, the performance of ML algorithms in classifying bee sounds has been improved significantly.

The rest of the paper is organized as follows. Section 2 presents a brief description of the chosen feature extraction methods and the ML algorithms. The details of our experiment, from the data description to the use of the proposed approach, and the obtained results and a comparison with a previous study have been provided in Sect. 3. Section 4 is for some concluding remarks and perspectives of the study.

2   Methodology

2.1   Feature Extraction

In order to apply ML techniques, the first step is to extract features from the raw signals into feature vectors. Feature vectors are numerical vectors that represent a sample of data and usually used in many computations or classifications.

Mel Frequency Cepstral Coefficient (MFCC) is one of the most popular feature extraction methods in sound recognition. It was initially firstly proposed for human speech perception but now is widely used in many different types of sound recognition such as animal sound [18]), music [10], or acoustic event detection [22]. The MFCC converts the raw audio data into a vector of coefficients representing the original signal. Several steps for extracting MFCC features have been shown in Fig. 1. More details of the method explanation for each step can be seen in [24].

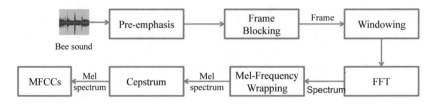

Fig. 1. The block diagram of extracting MFCC features

Short-Time Fourier Transform (STFT) is an extension method of Fourier transform (FT) of a windowed signal [7]. This is a effective tool for audio signal processing to analyze non-stationary signals. STFT points out that the signal contains which frequencies and they appear in which time intervals while the FT only gives the information of frequency over the whole signal and hides the information of time. The STFT is presented by the equation as follows:

$$X_{STFT}[m, n] = \Sigma_{k=0}^{N-1} x[k]g[k - m]e^{-j\pi nk/N} \tag{1}$$

where $x[k]$ denotes a input signal and $g[k]$ is a N-point window function.

Constant-Q Transform (CQT) is firstly proposed for music processing [4]. This method converts the original data in the time domain into the time-frequency domain using a logarithmic spaced frequency axis and ensuring that the Q- factors (rates of the center frequencies to bandwidths) of all frequency bins are equal. The CQT could be considered as a wavelet transform because it increases time resolution towards higher frequencies and the frequency resolution is better for low frequencies. This mechanism is similar to the human auditory system in that we can hear audio having an approximately "constant Q" frequency range. The original of the method can be seen in [4].

Spectral Contrast (SC) is initially devoted to audio genre classification in [12]. This is a measure indicates that the energy of frequency at each timestamp. Thus, it allows measuring the energy variation of an audio signal.

This method computes spectral peaks, valleys, and their differences in each sub-band separately. These features process relative spectral characteristics where strong spectral peaks often correspond with harmonic components and non-harmonic components (noise sounds) usually present in spectral valleys.

Chroma (a Chroma Feature) is a set of 12 elements indicating how much energy of each pitch class {C, C#, D, D#, E, F, F#, G, G#, A, A#, B} presenting in the signal [13]. This feature is a strong tool, widely used for analyzing music whose pitches can be meaningfully categorized and should be robust to noise [20]. The objective of this feature type is to represent the harmonic content (for example chords) of a short-time window of audio. The feature vector is computed the magnitude spectrum by using a STFT, CQT, etc. Figure 2 presents the general procedure of extracting Chroma features.

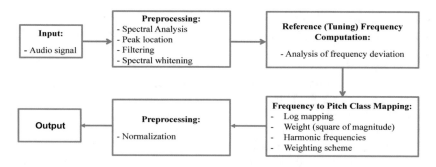

Fig. 2. The schema of Chroma features extraction process [20]

2.2 Machine Learning Methods

Logistic Regression (LR) is well-known in statistical analysis for binary classification. In LR, a logistic function is used to predict a binary dependent variable by analyzing the relationship between one or more existing independent variables.

Also, LR can be applied for binomial, ordinal, or multinomial cases. In the binomial case, the observed outcome for a dependent variable can have only two possible types, coded by "0" or "1". When the outcome contains more than three types, it refers to multinomial LR. Finally, ordinal LR is for the case when dependent variables are ordered. LR has been widely used as an important tool in the discipline of machine learning [1].

K-Nearest Neighbors (KNN) is a non-parametric statistical learning method because there is no requirement of assumption about the distribution of data. Due to using all the instances for training, it is also known as a lazy learning algorithm. To classify a new instance, KNN applies the distance from the input to the points in the training set. The use of KNN in classification has been discussed broadly in [11].

Support Vector Machine (SVM) is a classification algorithm that aims to find out optimal hyper-plane to maximize the margin or the support vectors [23]. As a result, it can maximize the distance between data points of both classes.

The separated hyper-plane, which is in the form of a linear regression function, can be demonstrated by the formula:

$$f(x) = w^T x + b \tag{2}$$

where w is the weight vector, b is the bias and x is the input vector. To overcome the non-linear problem, SVM uses Kernel functions to transform original data into a new space.

The new space, called feature space, is usually high dimensional so that the classes become linearly separable.

Extreme Gradient Boosting (XGBoost) is an ensemble learning technique and an implementation of gradient boosted decision trees designed for speed and performance [6]. XGBoost provides parallel boosting trees, smoothing training loss, and regularization (normalization of errors, coefficients, and variables). It combines the predictive power of all these base trees, uses a variation of the loss function, and gives a single model with the best performance. The idea of this algorithm is to rectify mistakes performed by the model in the previous step, learn from it, and then in the next step XGBoost improves the performance.

To reduce over-fitting and faster training process, randomization techniques are also deployed in XGBoost. These techniques are applied to random select sub-samples to create individual trees and random choose features at tree and tree node levels. Besides, XGBoost uses percentiles of the data where only a subset of candidates of the split is tested and the gain is calculated applying aggregated statistics to avoid scanning all possible candidate splits. This makes the XGBoost algorithm the ability of solving many data science problems in a fast and accurate way.

Random Forest (RF) introduced in [2], is also an ensemble learning algorithm. It includes multiple single trees, where each tree in the forest is created on a random sample of the training data. RF generates fully grown trees without pruning in order to keep the low bias. Each tree is trained on a bootstrap sample set, and for splitting at each node, a random subset of the attributes is considered. In this way, the randomness creates diversity among the trees and the low correlation between trees is controlled in the forest. As a result, they are more accurate and stable as more trees are added. A discussion for the choice of the number of trees in the RF can be seen in [17].

Extremely Randomized Trees (ET) is another ensemble ML algorithm that we applied in this study. Like RF, it is also many un-pruned decision trees and samples of the features at each split point of a decision tree. However, the degree of randomization in ET is stronger than that in RF as it selects a split point at random rather than using an algorithm to achieve an optimal split point as in RF.

There are some parameters to be considered in ET such as the number of decision trees, the number of input features to randomly select for each split point, and the minimum number of samples required in a node to create a new split point. In theory, the larger the number of decision trees ET creates, the better performance it can result in. However, it also requires more time for the calculation while the obtained result may be negatively affected once it is beyond a critical number of trees. The rationale behind the ET method for the other parameters from both the bias-variance and computational points of view have been discussed in detail in [9].

3   Experiments

3.1   Data Description

To investigate the performance of the methods of extracting features, we use two datasets introduced in [14]. Both two datasets include the audio data captured from beehives at different times and in different places in the U.S. The data have been classified into one of the three categories, i.e. bee buzzing, cricket chirping, and ambient noise.

The first dataset, BUZZ1, comprises 10,260 audio samples, separated into a train/test dataset of 9110 samples and a validation dataset of 1150 audio samples. As described by the authors, the samples in the validation dataset were separated from the ones in the training/testing dataset by beehive and location. All the samples have been labeled and the number of samples in each category has been shown in Table 1.

Due to the use of audio data from the same beehives, there is an overlap between the training and testing datasets in BUZZ1. To avoid this problem, the BUZZ2 has been created that separated the training data from the testing data in terms of beehive and location. This dataset contains 12,914 audio samples in total, classified into the same three categories as in BUZZ1. Table 1 shows the number of samples in each category. More details on the process of collecting samples for BUZZ1 and BUZZ1 and related information can be seen in [14].

Table 1. The number of samples in each category in two datasets

Type of samples	BUZZ1		BUZZ2		
	Training & Testing	Validation	Training	Testing	Validation
Bee	3000	300	2402	898	1000
Cricket	3000	500	3000	500	1000
Noise	3110	350	2180	934	1000

3.2   The Performance of the Feature Extractions Methods

In this study, five different methods for extracting features from BUZZ1 and BUZZ2 datasets have been investigated, including the MFCC features, the STFT features, the CQT features, the Chroma features, and the Spectral contrast features. After being extracted from raw audios, these important features have been fed to several ML algorithms to classify if the sounds are bee buzzing, cricket chirping, or ambient noise. We also apply the technique of tuning hyperparameters for the ML algorithms. The performance of the applied feature extraction methods based on the different types of ML algorithms on the test set and validation set has been presented in Tables 2 and 3 (BUZZ1 dataset) and Tables 4 and 5 (BUZZ2 dataset), respectively.

As can be seen from Table 2, on the test set of BUZZ1, the ensemble ML-based methods (ET, RF, and XGB) lead to a better result compared to the other classical ML algorithms like KNN, LR, and SVM, especially when using CQT features. When it comes to each type of feature, MFCC ones provide the best representation of the data as the classification performance using these features as input is higher than the others with an accuracy of more than 99%, regardless of the utilized ML algorithms. However, on the validation set of BUZZ1, the performance of MFCC is not as good as on the test set. This is also true for all the ML algorithms on this test set as the accuracy shown on Table 3 is lower than the ones on Table 2. The highest accuracy belongs to the XGB method when using the input of the feature extracted by CQT. Although the MFCC features give very good results on the test set, the classification results are no longer impressive with the validation set. In general, the results on Table 3 show that the performance of MLs when using features extracted by Chroma method is superior compared to other methods. In comparison with the method applied in [14] where the authors used a combination of many features extracted from each raw audio sample, i.e. 40 MFCC's, 12 Chroma coefficients, 128 Mel spectrogram coefficients, 7 spectral contrast coefficients and 6 tonnetz coefficients, the Chroma method only with 24 features in this study give a more or less result. In particular, using LR and RF algorithms the accuracy of Chroma features in this study is lower than the combination of 193 features in [14], which are 89% and 84.5% compared to 94.6% and 93.21%, while the accuracy of KNN and SVM with Chroma features is higher than the one with a combination of 193 features in [14], i.e. 88.7% and 88.4% compared to 85.47% and 83.91%.

Table 2. The performance of machine learning algorithms using various types of features on the test set of BUZZ1

Method	MFCC	STFT	CQT	Chroma	Spectral contrast
ET	99.70%	98.20%	98.90%	96.50%	98.30%
RF	99.80%	99.70%	99.70%	96.40%	98.00%
XGB	99.80%	99.80%	99.70%	97.00%	98.40%
KNN	99.50%	98.50%	85.20%	96.20%	98.60%
LR	99.60%	96.00%	85.60%	94.50%	95.00%
SVM	99.00%	98.50%	78.30%	94.50%	98.60%

The advantage of using different kinds of feature extraction methods can see more clearly on the BUZZ2 dataset. Similar to the case of BUZZ1, the classification performance of the ML methods with MFCC features in the test set shown in Table 4 are still the best with high accuracy. The accuracy of using features extracted from STFT method, however, is poor. Using this method, even in the test set, only ET and RF result can achieve an accuracy of higher than 84%, while the others give quite low results, which are less than 40%. On

Table 3. The performance of machine learning algorithms using various types of features on the validation set of BUZZ1

Method	MFCC	STFT	CQT	Chroma	Spectral contrast
ET	75.60%	83.00%	81.30%	87.00%	54.80%
RF	76.90%	69.50%	86.00%	84.50%	54.00%
XGB	54.90%	60.50%	93.70%	88.40%	42.70%
KNN	66.30%	56.00%	59.80%	88.70%	30.40%
LR	69.60%	35.20%	69.00%	89.00%	26.00%
SVM	76.30%	56.00%	67.90%	88.40%	30.40%

the validation set, not only the STFT features but also the CQT and the Spectral contrast features are also not a good choice due to the poor results presented in Table 5. By contrast, the Chroma features show an impressive accuracy as in the case of BUZZ1. Using these features, the performance of all the ML algorithms have been enhanced significantly, leading to a much better result compared to the obtained results in [14]. For example, the accuracy in classifying bee sounds from cricket sounds and the ambient noise of KNN, LR, SVM, and RF based on Chroma features in this study are 89.8%, 89.3%, 87.8% and 87.8%, which are notably higher than the values of 37.42%, 68.53%, 56.6%, and 65.8% from the same ML algorithms based on a combination of many different types of features as mentioned above (see Table 23 in [14]). The ML algorithms using the input of the features extracted from Chroma even outperform several deep learning methods in [14] like ConvNet 1, ConvNet 2, and ConvNet 3. According to a discussion in [14], one should consider that BUZZ2 is a more challenging dataset compared to BUZZ1, and the deep learning (DL) methods are preferable for the situation. However, the finding from this study shows that by choosing the right feature extraction method, the ML algorithms can still bring a better result. Finally, we can see from Table 3 and Table 5 that the values of accuracy based on the Chroma features are very close, around 85% to 90%, regardless of the ML algorithms or the datasets (i.e. validation sets of BUZZ1 or BUZZ2). That is to say, the method can extract essential features that represent well the data, allowing the ML algorithms to best classify the datasets.

Table 4. The performance of machine learning algorithms using various types of features on the test set of BUZZ2

Method	MFCC	STFT	CQT	Chroma	Spectral contrast
ET	99.50%	94.30%	99.20%	88.20%	94.80%
RF	98.80%	83.60%	98.50%	87.80%	94.20%
XGB	98.80%	40.30%	98.20%	89.70%	93.10%
KNN	99.20%	39.10%	77.10%	89.70%	92.80%
LR	99.44%	39.90%	80.10%	88.20%	88.50%
SVM	95.50%	40.00%	75.30%	87.40%	88.10%

Table 5. The performance of machine learning algorithms using various types of features on the validation set of BUZZ2

Method	MFCC	STFT	CQT	Chroma	Spectral contrast
ET	80.40%	25.00%	35.50%	89.50%	76.00%
RF	82.40%	23.10%	23.20%	87.80%	83.20%
XGB	76.90%	34.50%	50.60%	90.50%	77.30%
KNN	54.50%	32.60%	47.80%	89.80%	71.60%
LR	79.00%	34.80%	56.60%	89.30%	63.30%
SVM	72.60%	33.90%	53.10%	87.80%	77.20%

4   Concluding Remarks and Perspectives

In this study, we have investigated the performance of the ML algorithms based on many different feature extraction methods for identifying audio samples. The obtained results show that the Chroma features are the best to represent the two datasets BUZZ1 and BUZZ2. Using these features allows enhancing the classification performance of the ML algorithms, making them outperform several DL methods in a previous study in the literature. Feeding these features to the DL algorithms or combining them with other features from different methods to find a better feature extraction method would be our concern for further research.

References

1. Bennouna, G., Tkiouat, M.: Scoring in microfinance: credit risk management tool-case of Morocco. Procedia Comput. Sci. **148**, 522–531 (2019)
2. Breiman, L.: Random forests. Mach. Learn. **45**(1), 5–32 (2001). https://doi.org/10.1023/a:1010933404324
3. Bromenshenk, J.J., Henderson, C.B., Seccomb, R.A., Rice, S.D., Etter, R.T.: Honey bee acoustic recording and analysis system for monitoring hive health. US Patent 7,549,907, 23 June 2009
4. Brown, J.C.: Calculation of a constant Q spectral transform. J. Acoust. Soc. Am. **89**(1), 425–434 (1991)
5. Cejrowski, T., Szymański, J., Mora, H., Gil, D.: Detection of the bee queen presence using sound analysis. In: Nguyen, N.T., Hoang, D.H., Hong, T.-P., Pham, H., Trawiński, B. (eds.) ACIIDS 2018. LNCS (LNAI), vol. 10752, pp. 297–306. Springer, Cham (2018). https://doi.org/10.1007/978-3-319-75420-8_28
6. Chen, T., Guestrin, C.: XGBoost: a scalable tree boosting system. In: Proceedings of the 22nd ACM SIGKDD International Conference on Knowledge Discovery and Data Mining, pp. 785–794 (2016)
7. Dennis, G.: Theory of communications. J. Inst. Electr. Eng. **93**, 429–457 (1946)
8. Ferrari, S., Silva, M., Guarino, M., Berckmans, D.: Monitoring of swarming sounds in bee hives for early detection of the swarming period. Comput. Electron. Agric. **64**(1), 72–77 (2008)
9. Geurts, P., Ernst, D., Wehenkel, L.: Extremely randomized trees. Mach. Learn. **63**(1), 3–42 (2006). https://doi.org/10.1007/s10994-006-6226-1

10. Ghosal, A., Chakraborty, R., Dhara, B.C., Saha, S.K.: Music classification based on MFCC variants and amplitude variation pattern: a hierarchical approach. Image Process. Pattern Recogn. **5**(1), 21 (2012)
11. Guo, G., Wang, H., Bell, D., Bi, Y., Greer, K.: KNN model-based approach in classification. In: Meersman, R., Tari, Z., Schmidt, D.C. (eds.) OTM 2003. LNCS, vol. 2888, pp. 986–996. Springer, Heidelberg (2003). https://doi.org/10.1007/978-3-540-39964-3_62
12. Jiang, D.N., Lu, L., Zhang, H.J., Tao, J.H., Cai, L.H.: Music type classification by spectral contrast feature. In: Proceedings of the IEEE International Conference on Multimedia and Expo, vol. 1, pp. 113–116 (2002)
13. Kattel, M., Nepal, A., Shah, A.K., Shrestha, D.: Chroma Feature Extraction, pp. 1–9. Springer, Berlin (2019)
14. Kulyukin, V., Mukherjee, S., Amlathe, P.: Toward audio beehive monitoring: deep learning vs. standard machine learning in classifying beehive audio samples. Appl. Sci. **8**(9), 1573 (2018)
15. Mutlag, W.K., Ali, S.K., Aydam, Z.M., Taher, B.H.: Feature extraction methods: a review. In: Journal of Physics: Conference Series, vol. 1591, p. 012028. IOP Publishing (2020)
16. Nolasco, I., Terenzi, A., Cecchi, S., Orcioni, S., Bear, H.L., Benetos, E.: Audio-based identification of beehive states. In: 2019 IEEE International Conference on Acoustics, Speech and Signal Processing (ICASSP), ICASSP 2019, pp. 8256–8260. IEEE (2019)
17. Oshiro, T.M., Perez, P.S., Baranauskas, J.A.: How many trees in a random forest? In: Perner, P. (ed.) MLDM 2012. LNCS (LNAI), vol. 7376, pp. 154–168. Springer, Heidelberg (2012). https://doi.org/10.1007/978-3-642-31537-4_13
18. Ramirez, A.D.P., de la Rosa Vargas, J.I., Valdez, R.R., Becerra, A.: A comparative between Mel Frequency Cepstral Coefficients (MFCC) and Inverse Mel Frequency Cepstral Coefficients (IMFCC) features for an automatic bird species recognition system. In: 2018 IEEE Latin American Conference on Computational Intelligence (LA-CCI), Gudalajara, Mexico, pp. 1–4. IEEE (2018)
19. Ribeiro, A.P., da Silva, N.F.F., Mesquita, F.N., de Cássia Souza Araújo, P., Rosa, T.C., Mesquita-Neto, J.N.: Machine learning approach for automatic recognition of tomato-pollinating bees based on their buzzing-sounds. PLoS Comput. Biol. **17**(9), e1009426 (2021)
20. Serrá, J., Gómez, E., Herrera, P., Serra, X.: Chroma binary similarity and local alignment applied to cover song identification. IEEE Trans. Audio Speech Lang. Process. **16**, 1138–1151, p. 336 (2008)
21. Sharma, G., Umapathy, K., Krishnan, S.: Trends in audio signal feature extraction methods. Appl. Acoust. **158**, 107020 (2020)
22. Socoró, J.C., Alías, F., Alsina-Pagès, R.M.: An anomalous noise events detector for dynamic road traffic noise mapping in real-life urban and suburban environments. Sensors (Basel, Switzerland) **17**(10), E2323 (2017)
23. Vapnik, V.N.: The Nature of Statistical Learning Theory. Springer, Heidelberg (1995)
24. Han, W., Chan, C.F., Choy, C.S., Pun, K.P.: An efficient MFCC extraction method in speech recognition. In: 2006 IEEE International Symposium on Circuits and Systems, Island of Kos, Greece, p. 4. IEEE (2006)
25. Zgank, A.: IoT-based bee swarm activity acoustic classification using deep neural networks. Sensors **21**(3), 676 (2021)

Indoor Localization Simulation Based on LoRa Ranging

Thanh Danh Pham[✉][iD], Nhut Quang Tran[iD], and Trong Nhan Le[iD]

Ho Chi Minh City University of Technology, Ho Chi Minh City, Vietnam
danh.phamthanh@hcmut.edu.vn

Abstract. Recently, indoor localization has played a vital role in many applications such as route guidance, device tracking, and disaster relief. Many different wireless technologies (WiFi, ultra-wideband (UWB), Bluetooth low energy (BLE), radio-frequency identification (RFID), and LoRa) have been used for indoor localization [1]. Standing out from other wireless technologies, LoRa with its low-power and long-range advantages in transmission shows that it can be a potential candidate for several indoor positioning applications. Additionally, modeling and simulation systems have been developed a very long time ago to reduce time and cost for experiments in numerous fields, including quantum physics, industry, and astronomy. These days, thanks to the powerful computation ability of computers, the simulation is more accurate and precise. Therefore, in this study, we propose a model for indoor localization with LoRa wireless technology on the OMNeT++ simulation.

Keywords: Indoor localization · LoRa · Simulation · OMNeT++

1 Introduction

Positioning has lately become a popular study topic, with much of the attention focused on exploiting existing technology to solve the difficulty of determining locations, positions of people, equipment, and other items. Positioning can be difficult due to the dependencies on the environment in which the positioning is performed. There are several Position Estimation Techniques (PET) [2] and they are principally divided into two types based on their location: outdoor and indoor [3]. Outdoor positioning is done outside a building. On the other hand, indoor positioning is carried out inside buildings such as warehouses, hospitals, and malls. Each type of position has its own different strengths and weaknesses. However, different applications may require different positioning methods to meet their requirements. For instance, GPS, which is the most popular technology nowadays, gives the location of an object or person on the earth's surface with satisfactory accuracy. However, GPS would behave with limitations in indoor environments as the lack of line of sight and being blocked by solid things, obstacles. In buildings with multiple floors, the electricity will go off due to unexpected fires or disasters. Systems that operate depending on the power

© The Author(s), under exclusive license to Springer Nature Switzerland AG 2022
N.-T. Nguyen et al. (Eds.): ICIT 2022, LNDECT 148, pp. 204–214, 2022.
https://doi.org/10.1007/978-3-031-15063-0_19

supply cannot last long. Therefore, the demand for investigating and developing a new indoor positioning system that can meet the requirements of high accuracy, long-range operation, and low-energy consumption is essential. The advantages of LoRa technology are promising when it is applied to this indoor positioning system.

The Indoor Positioning Mechanism (IPM) continuously determines the position of an object in a physical space. IPM can be performed with various approaches [4] and has requirements that differentiate it from outdoor positioning. There are five main quality metrics of indoor positioning systems: system accuracy and precision; coverage and its resolution; latency in making location updates; the building's infrastructure impact; and the effect of random errors on the system, such as errors caused by signal interference and reflection. Therefore, to keep these metrics at a satisfactory level, we promote an approach that first evaluates them at a simulation level before eventually deploying them in real-world settings. Likewise, the simulation must be configured with the same algorithm and hardware specifications, also including the errors from the environmental impacts.

Problem Formulation: In a certain indoor tracking environment, 2D-view from the top, we determine the position of an unknown position wireless sensor node called Unknown Node (UN). This node is randomly deployed in the environment with the assumption that we already know the position of the stationary anchor node (AN) and that at least 3 of those ANs are needed to estimate that UN. Also, the UN must be inside the triangle area generated from those 3 ANs to prevent the false estimation generated when the actual position of the UN is outside the area [5]. After that, the estimation process is described in Fig. 1. The process is divided into three main phases. The first one is known as the estimating phase, where each UN estimates its distance from ANs depending on the time of arrival (ToA) of the signal. The second phase applies the IPM to the 3 ToA values yielded from the first process and returns the estimated uncleaned position UN. The last step is to filter the uncleaned position and return the highest optimized position value.

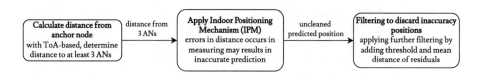

Fig. 1. Three-phase process of localization

The paper is organized as follows. In Sect. 2, we briefly introduce related works related to this study. Section 3 deals with the implementation of indoor localization by applying our IPM model in the OMNeT++ simulation environment. Finally, the results and discussion of our work are presented in Sect. 4.

2   Related Works

2.1   PET in the Indoor Environment

The fundamental limits of localization in the indoor environment are the minimum requirement of the number of ANs, which is three, and the 2D-fix map generated. Generally, positioning strategies for indoor environments consist of a ranging phase where nodes use distance-dependent or range-based signal relations. Examples of this technique are RSS, ToA, AoA, etc. A quick comparison among ranged-based techniques is displayed in Table 2. Ranging quality is affected by the impact of the noisy ToA, AoA, or RSS measurements. There are bounds to reduce the error [6,7]. A precision of 2m could be achieved with these bounds, assuming that map-based prior knowledge and map-aware localization are used [8,9]. However, in this paper, we will not use these bounds for better accuracy but instead apply a state-of-the-art hybrid technique between ToA - a range-based, and a range-free technique called the Centroid scheme [6], which has the lowest communication overhead and is easily implemented [10].

2.2   OMNeT++ Simulator

OMNeT++ is an open-source, discrete event simulator that has an extensible, modular, component-based, C++ simulation library and framework, primarily used to build network simulators [11]. OMNeT++ has the ability to describe the operation of Wireless Sensor Network precisely with the help of MiXim framework. MiXiM framework was designed and created for mobile Wireless Sensor Network as it provides various models such as TCP/IP layers, radio wave propagation, interference estimation, radio transceiver power consumption, mobility, and numerous IEEE standard wireless protocols.

2.3   Localization with LoRa Approach

Considering indoor localization applications, various types of factors can affect the precision of measured results. Multipath effect signals are reflected and attenuated by walls, furniture, human and noise interference [12]. In this study, we choose LoRa wireless technology because of its advantages which can overcome listed difficulties. Firstly, with relatively low cost and power consumption, the LoRa's signal is resilient to multipath effect or noise [1]. Besides, its good through-wall penetration ability shows that LoRa has great potential for indoor localization applications. Table 1 summarizes the characteristics of LoRa compared to other wireless technologies.

Table 1. Indoor wireless technologies characteristic

	Wifi	GSM	A-GPS	Bluetooth	Zigbee	UWB	LoRa
Accuracy	1–5 m	10–50 m	5–10 m	1–5 m	1–10 m	6–10 cm	1–5 m
Operation range	50 m	40 km	20 km	10 m	300 m	45–300 m	12 km
Penetration ability	Moderate	Poor	Poor	Poor	Moderate	Good	Moderate
Power consumption	High	High	High	Low	Low	High	Low
Cost	Low	Low	Low	Low	Low	High	Low

Table 2. Localization techniques overview.

Technique	Characteristics	Strength & Weakness	Applicability
TOA	Distance timestamp between UN and AN	Need synchronization as ranging one-way	Common in cellular networks & wireless sensor networks (WSN)
TDOA	Utilized ToAs in several ANs	Needs highly precise synchronization between UNs instead of between ANs and UNs	Common in wireless sensor networks
AOA	Angle between & intersection the UN and ANs	Requires new hardware, additional costs, and larger node sizes	More appropriate for ANs rather than UNs
RSSI	Attenuation signal by propagation from AN to UN	Need accurate propagation model, low cost as mostly RX used to estimate RSS, errors with mobility	Common in Wireless local area networks

3   Indoor Localization in OMNeT++ with LoRa Ranging

3.1   OMNeT++ Configuration

Besides the parameters attenuation, path loss, interference, and wave propagation in the simulation environment, we also set up a wireless network with UN and ANs that matches the physical locations of these nodes in reality because we use the data measured in real life to simulate the errors which can happen in the process of determining the position of the UN.

3.2   Errors Filtering

Given that, in some specific applications, we have a fixed number of times to measure the distance before determining the position of the UN. The filter that we propose is the mean distance of residuals.

$$\overline{D}_i = \sum_{j=1,\ i\neq j}^{N} \frac{d_{ij}}{N-1} \tag{1}$$

where,

$\overline{D}_i$ is the average distance of position i to others

d_{ij} is the distance between position i and j can be calculated by Eq. 3

N is the number of predicted positions that we have

After that, the mean of these average distances is calculated.

$$\overline{M} = \frac{\sum_{i=1}^{N} \overline{D}_i}{N} \qquad (2)$$

where, $\overline{M}$ is the mean of average distances i to other positions.

In the final step, we define a threshold (T) to filter errors. When the difference between $\overline{D}_i$ and $\overline{M}$ exceeds this threshold, e.g. $|\overline{D}_i - \overline{M}| > T$, the position i^{th} is rejected. If there are many rejected positions, the actual position of the UN cannot be determined in this prediction.

3.3   Indoor Positioning Mechanism in Simulator

In two-dimensional coordinates and a set of UNs. In order to calculate the distance UN and AN, Euclidean equation could be formulated as shown in Eq. 3

$$d_{ij} = \sqrt{(x_i - x_j)^2 + (y_i - y_j)^2} \qquad (3)$$

where,

$\quad i \in ANs$

$\quad j \in UNs$

$\quad (x_i, y_i)$ are coordinates of ANs

$\quad (x_j, y_j)$ are coordinates of UNs

The estimation error, E_j, of the jth non-anchor node is given by the Eq. 4

$$E_j = \sum_{i=1}^{n} \left(d_{ij} - \hat{d}_{ij}\right)^2 \qquad (4)$$

where,

$\quad d_{ij}$ is the real distance

$\quad \hat{d}_{ij}$ is the estimated distance

$\quad n$ is the number of UNs

LoRa antennas in the real-world communicate with each other via radio signal, and following the range-based technique, we can deduce the location geometrically in the Oxy-plane. As referred to phase 1 in Fig. 1, the output will be 3 distance values from the ANs, or equivalently, the 3 radii with the center of 3 ANs. This mechanism in simulation is described as follows.

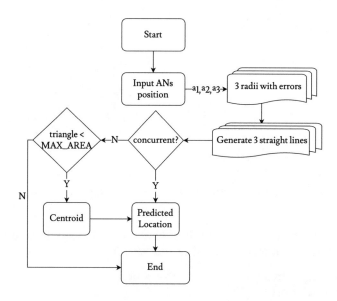

Fig. 2. Flow chart of Indoor Positioning Mechanism (IPM)

1. The position of ANs are known in preparation step before.
2. The UN send packet to these ANs, then ANs will reply the point of time that they receive the packet to the UN.
3. Based on time, the UN can calculate the radii to 3 ANs. The 3 perpendicular lines (with blue color in Fig. 4) will be generated depending on the relative position of each pair of circles.
4. If these three lines are concurrent, then location is predicted. If not, these three lines will form a triangle.
5. If the area of this triangle is less than a pre-defined threshold, the predicted location is the centroid of the triangle. Otherwise, the predicted location is discarded.

Ideally, the UN would be located at the intersection of at least three circles with centers at the anchor nodes and radii equal to the distance to each of these anchor nodes, as shown in Fig. 3a. However, it is highly unlikely that a single point of intersection is obtained. **Centroid Scheme**: Centroid is a range-free algorithm used to estimate the location of the UK, which depends on the coordinates of anchor nodes. Referring to step Centroid in Fig. 2, when 3 straight lines are generated and not concurrent, they will form a triangle with 3 vertices h, j, and k. Then, the centroid of this triangle will be the predicted location, $(predict_x, predict_y)$. An example is the map displayed in Fig. 5.

$$(predict_x, predict_y) = \left(\frac{h_x + j_x + k_x}{3}, \frac{h_y + j_y + k_y}{3}\right) \tag{5}$$

Possibility Errors with Relative Position: The error in the first phase results in falsely measuring the radius of each AN. In the second phase, we

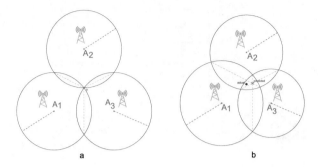

Fig. 3. a. Ideal position without errors. b. Predicted position with errors

will apply IPM to all relative positions of the two circles, which are produced when errors occur, all possible positions of two circles are:

(a) Intersecting with no circle cover other center
(b) Without intersecting
(c) Intersecting a circle cover other center
(d) Intersecting a circle lying over other circle

For the rest of formulas for all cases, without losing generality, we assume the will calculate the straight line which is perpendicular to a_1a_2 from circle (a_1).

Case a. Intersecting with no circle cover other center

In this case, the position is shown in Fig. 4a, the sum of distances d_{a_1} and d_{a_2} in Eq. 6 is greater than the sum of the Euclidean distances of AN a_1 and AN a_2

$$\begin{cases} d_{a_1a_2} \geqslant d_{a_1} \\ d_{a_1a_2} \geqslant d_{a_2} \\ d_{a_1a_2} \leqslant d_{a_1} + d_{a_2} \end{cases} \tag{6}$$

$$d_{a1p} = \frac{1}{2} * \frac{d_{a_1}^2 - d_{a_2}^2 + d_{a_1a_2}^2}{d_{a_1a_2}} \tag{7}$$

Case b. Without intersecting

As shown in the Fig. 4b, the sum of distances d_{a_1} and d_{a_2} in Eq. 6 is smaller than the sum of the Euclidean distances of AN a_1 and AN a_2

$$\begin{cases} d_{a_1} \geqslant d_{a_2} \\ d_{a_1} + d_{a_2} \leqslant d_{a_1a_2} \end{cases} \quad \text{or} \quad \begin{cases} d_{a_2} \geqslant d_{a_1} \\ d_{a_1} + d_{a_2} \leqslant d_{a_1a_2} \end{cases} \tag{8}$$

$$d_{a1p} = d_{a_1} + (d_{a_1a_2} - d_{a_1} - d_{a_2}) * \frac{d_{a_2}}{d_{a_1}} \tag{9}$$

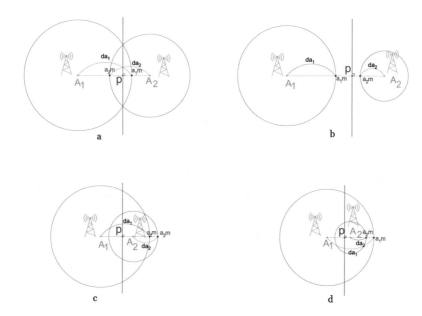

Fig. 4. All relative positions when error occurs (Color figure online)

Case c. Intersecting a circle cover other center

This case is described in Fig. 4c, occurred when the radius of one circle is large enough to pass the other center but not completely lie over that circle.

$$\begin{cases} d_{a_1} \geqslant d_{a_1 a_2} \\ d_{a_1 a_2} + d_{a_2} \geqslant d_{a_1} \end{cases} \quad \text{or} \quad \begin{cases} d_{a_2} \geqslant d_{a_1 a_2} \\ d_{a_1 a_2} + d_{a_1} \geqslant d_{a_2} \end{cases} \tag{10}$$

$$d_{a_1 p} = d_{a_1} * \frac{d_{a_1 a_2}}{d_{a_1} + d_{a_2}} \tag{11}$$

Case d. Intersecting a circle lying over other circle

Figure 4d shows the case when a is large enough to pass other center but not completely lying over that circle.

$$\begin{cases} d_{a_1} > d_{a_1 a_2} \\ d_{a_1 a_2} + d_{a_2} < d_{a_1} \end{cases} \quad \text{or} \quad \begin{cases} d_{a_2} > d_{a_1 a_2} \\ d_{a_1 a_2} + d_{a_1} < d_{a_2} \end{cases} \tag{12}$$

$$d_{a_1 p} = d_{a_1 a_2} * \frac{d_{a_1}}{d_{a_2}} \tag{13}$$

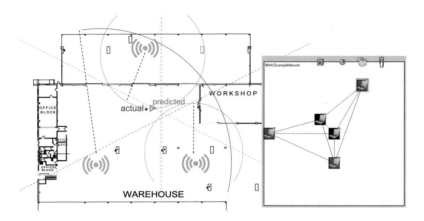

Fig. 5. An example when applying IPM in real-world and OMNeT++

4   Result and Discussion

4.1   Result

After running the simulation multiple times, the statistic is collected with two main factors number of measurements in each prediction and threshold (T). Tables 3 and 4 show the relationship between these two factors. Assume that each measurement costs one second.

Table 3. Error(%) based on number of pings and thresholds(T)

Number of measurements	Thresholds(T)				
	10 m	8 m	6 m	4 m	2 m
3	2.26	2.17	2.09	2.01	1.91
4	2.01	1.99	1.97	1.95	1.95
5	1.93	1.92	1.9	1.88	1.86
7	1.85	1.84	1.82	1.79	1.78
9	1.8	1.79	1.78	1.76	1.76
10	1.78	1.77	1.76	1.75	1.74

Table 4. Response time(s) based on number of pings and thresholds(T)

Number of measurements	Thresholds(T)				
	10 m	8 m	6 m	4 m	2 m
3	3.03	3.12	3.29	3.68	5.19
4	4	4	4	4.02	4.15
5	5	5	5.04	5.13	5.53
7	7.03	7.12	7.34	7.84	9.6
9	9	9	9.03	9.15	9.73
10	10	10.02	10.09	10.32	11.39

4.2 Discussion and Future Work

From Table 3, 4 above, the accuracy can be improved by increasing the number of measurements in each prediction or decreasing the threshold. However, choosing a small threshold can lead to the increment of response time. Therefore, depending on the requirements of specific indoor localization applications, these two factors should be suitably balanced and selected.

In our work, we investigate indoor localization in 2D environment. However, for 3D environment (offices, malls, apartments, buildings with multiple floors), this approach can be applied in the vertical axis first, and then in the horizontal axis to obtain the location.

We cannot deny that the chosen mechanism, TOA-based, does not accurately sum up the whole picture related to indoor positioning. We hope that more mechanisms will be tested out to thoroughly evaluate the simulation. One of them is device-based mechanisms, such as smartphone positioning with built-in WiFi and Bluetooth. We also believe that this paper introduces the idea not only restricted to the OMNeT++ tool but also another simulation tool [13].

4.3 Conclusion

This study researches the application of simulation in indoor localization using LoRa ranging technology. We proposed a model for simulating indoor localization and two mechanisms to filter errors that can happen in measurement. In short, the results of this study give a general relationship between accuracy and responsiveness. Furthermore, we hope that our results can provide necessary information to others when they want to use indoor localization in a specific application.

References

1. Kim, K., et al.: Feasibility of LoRa for smart home indoor localization. Appl. Sci. **11**, 415 (2021)

2. Hasan, H., Hussein, M., Saad, S.M., Dzahir, M.A.M.: An overview of local positioning system: technologies, techniques and applications. Int. J. Eng. Technol. (UAE) **7**, 1–5 (2018)
3. Ghorpade, S., Zennaro, M., Chaudhari, B.: Survey of localization for Internet of Things nodes: approaches, challenges and open issues. Future Internet **13**(8) (2021). https://www.mdpi.com/1999-5903/13/8/210
4. Yassin, A., et al.: Recent advances in indoor localization: a survey on theoretical approaches and applications. IEEE Commun. Surv. Tutor. **19**(2), 1327–1346 (2017)
5. Kwak, J., Sung, Y.: Beacon-based indoor location measurement method to enhanced common chord-based trilateration. J. Inf. Process. Syst. **13**, 1640–1651 (2017)
6. Saad, E., El-Hosseini, M., Haikal, A.: Recent achievements in sensor localization algorithms. Alex. Eng. J. **57** (2018)
7. Liu, H., Darabi, H., Banerjee, P., Liu, J.: Survey of wireless indoor positioning techniques and systems. IEEE Trans. Syst. Man Cybern. Part C (Appl. Rev.) **37**(6), 1067–1080 (2007)
8. Montorsi, F., Mazuelas, S., Vitetta, G., Win, M.: On the performance limits of map-aware localization. IEEE Trans. Inf. Theory **59**, 5023–5038 (2013)
9. Chandrasekaran, G., et al.: Empirical evaluation of the limits on localization using signal strength. In: 2009 6th Annual IEEE Communications Society Conference on Sensor, Mesh and Ad Hoc Communications and Networks, pp. 1–9 (2009)
10. He, T., Huang, C., Blum, B.M., Stankovic, J.A., Abdelzaher, T.: Range-free localization schemes for large scale sensor networks. In: Proceedings of the 9th Annual International Conference on Mobile Computing and Networking, MobiCom 2003, pp. 81–95. Association for Computing Machinery, New York (2003). https://doi.org/10.1145/938985.938995
11. Varga, A.: The OMNeT++ discrete event simulation system. In: Proceedings of the ESM 2001, vol. 9 (2001)
12. Curran, K., Furey, E., Lunney, T., Santos, J., Woods, D., McCaughey, A.: An evaluation of indoor location determination technologies. J. Locat. Based Serv. **5**, 61–78 (2011)
13. Xiao, J.: A survey on wireless indoor localization from the device perspective. ACM Comput. Surv. **49** (2016)

Mobile Applications Testing Based on Bigraphs and Dynamic Feature Petri Nets

Thanh Binh Nguyen[1]([✉]), Thi Thanh Binh Le[2], Oum-El-Kheir Aktouf[3], and Ioannis Parissis[3]

[1] The University of Danang - Vietnam Korea University of Information and Communication Technology (VKU), Da Nang, Vietnam
ntbinh@vku.udn.vn
[2] The University of Danang - University of Science and Education, Da Nang, Vietnam
lttbinh@ued.udn.vn
[3] University Grenoble Alpes, Grenoble INP, LCIS, Valence, France
oum-el-kheir.aktouf@lcis.grenoble-inp.fr,
ioannis.parissis@grenoble-inp.fr

Abstract. Nowadays, mobile smartphones are being widely used. They allow users to access a variety of services provided by mobile applications (mobile apps). These services are location-based services, meaning that a user's location is taken into consideration for service provision. Testing these mobile apps is challenging due to the complexity of context variability (i.e., a user's location). Current testing approaches cannot efficiently handle dynamic variability of mobile apps. To solve this problem, this paper introduces a model-based testing approach of mobile apps that uses a combination of a Bigraphical Reaction System (BRS) model and a Dynamic feature Petri net (DFPN) for automatic generation of test cases. Our model addresses the mobile app testing challenges related to the context of mobile apps, and especially to changes in the context location.

Keywords: Mobile applications testing · Bigraphs · Dynamic feature Petri nets

1 Introduction

In recent years, more and more mobile apps have been developed to support different needs in news, social, tourism, business, health, and other domains. Hundreds of new apps are released and downloaded daily. By the end of year 2023, it is expected that the global revenue from the mobile app market will be $935 billion whereas almost 299 billion downloads of mobile apps will be performed [1].

Due to the increasing impact of mobile apps, there is an urgent need for a sound testing process. Indeed, testing mobile apps is a challenging research topic as testing a mobile app induces to validate it on many different contexts including different hardware platforms, operating systems, web browsers, many geographical locations for location-dependent apps, etc. All these varying features are referred to as the context of mobile apps.

N.-T. Nguyen et al. (Eds.): ICIT 2022, LNDECT 148, pp. 215–225, 2022.
https://doi.org/10.1007/978-3-031-15063-0_20

This paper develops new testing techniques for automatic generation of test cases using models for context-aware mobile apps. We will mainly focus on addressing the context location variability.

The remaining parts of this paper are organized as follows: In Sect. 2, we present main features of context-aware mobile apps. In Sect. 3, we discuss related work. In Sect. 4, we propose a test model for context-aware mobile apps. This model is illustrated on a case study in Sec. 5. Finally, Sect. 6 concludes our approach and outlines future work.

2   Context-Aware Mobile Apps

Modern mobile devices, with their mass of sensors, are capable of identifying many of the contexts, which surround the user. These sensors offer the opportunity to gain more information about the user's environment than before. Mobile apps that have these features are called context-aware.

Context awareness has become a trendy topic with mobile apps. When dealing with context, entities can be classified in three groups: people (groups, individuals), places (buildings, rooms, etc.) and things (computer components, physical objects, etc.). Each entity can be described by different attributes, which can be distinguished into four categories: location (an entity's position, co-location, etc.), identity (each entity has a unique identifier), status (the properties of an entity, e.g., lighting or temperature for a room) and time (used to accurately define situation or ordering events, etc.) [2].

Mobile apps are increasingly incorporating functionalities which behaviors not only depend on explicit user input but also on the state of the surrounding environment. This type of mobile apps is known as context-aware mobile apps.

Context-aware mobile apps often consist of a middleware and a collection of services, and adapt their behavior according to the changing context [3]. They provide adaptive services responding to the dynamically changing contexts in the environment [4]. This means that, context-aware mobile apps use current context to provide proper services to the user.

3   Related Work

Testing is extremely important and costly in the software development life cycle. To test context-aware apps, the tester must not only verify the system functions by providing user interactions, but he/she also has to provide these interactions according to different combinations of the system context.

For most test automation tools, test engineers develop test scenarios manually, only the execution of test cases is automated. Furthermore, model-based testing (MBT) is used to automatically generate test cases from the system according to a test model. Model based system engineering aims to focus efforts on modeling the solution instead of writing the code [3]. There are two key elements related to MBT approaches [5]: (i) Selecting proper models to describe the software behaviors (UML diagrams and finite state machines are often used to build models); and (ii) selecting criteria to generate test cases.

Currently, there are research works targeting context-aware applications and context-aware mobile apps testing. In particular, Siqueira et *al.* [6] made an exhaustive survey about this test field, where they describe the main papers on related techniques for automatic generation of test cases using model-based testing approaches. Achilleos et *al.* [7] defined three modelling components: Presentation model, Petri Net model, Context model. These components have been generated and integrated into the model-driven environment as Eclipse plug-ins. Seiger and Schlegel [8] created a test model based on DFPNs. Then, they defined context rules and integrated them into their model. Yu et *al.* [4] used two modelling notations, bigraphical labelled transition system and finite state machine. Test cases were generated by tracing interactions between the environment model and the middleware model. Griebe and Gruhn [9] defined a development model based on a meta model. Then, they transformed the development model into Petri net. Mirza and Khan [10] defined a development model as a UML activity diagram. Then, the development model is transformed into a function net.

On the basis of the above overview, we can see that test models based on adding context data directly to the behavioral model, as in the work of Griebe and Gruhn [9], Mirza and Khan [10] are complex and not suitable to model large mobile apps. Another test model where test cases are generated for changing environment and that uses data-flow-based coverage criteria, is adopted by Yu et *al.* [4] and seems to be an interesting idea. Indeed, separate models for the context and the behavior of the context-aware mobile apps, leads to models that are more manageable and simpler to modify and to extend. However, context aware mobile apps use the user's context to provide adaptive services. A service in service-oriented architecture may be an atomic service or a composite service [11]. Yu et *al.* [4] proposed a model-based testing approach based on bigraphical modeling for the context-aware mobile apps with an atomic service. The authors experiment on an airport application and an atomic illumination service. Nevertheless, this model is not implemented for context-aware mobile apps with a composite service. In fact, Yu et *al.* [4] used an extended finite state machine to model an atomic service. But, when the number of services in context-aware applications grows, then the number of states also increases and using such extended finite state machine model could be impractical. To overcome this limitation, we investigate in our work the use of DFPN instead of finite state machine, in combination with bigaph model.

4   A Test Model for Context-Aware Mobile Apps

Based on the above analysis, we propose a novel test model for context-aware mobile apps with both atomic and composite services. Our model comprises two separate diagrams: a bigraph for modeling the context and a DFPN for modeling the behavior of the context-aware mobile apps.

4.1   Bigraph

Bigraphs are graphs with nodes and edges. The nodes may be nested, representing locality and the edges link the nodes. Details about the notion of bigraph can be found in [12–14].

Example: The TripAdvisor application provides a set of services (Tourist Spots, Hotels, Restaurants) which fit to the user's location. Let's consider an example of a simulated environment, modeled as a bigraph in Fig. 1.

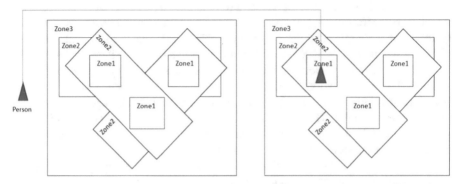

Fig. 1. A bigraph for modeling in a simulated environment of the TripAdvisor application

There are two structures on the nodes of a bigraph, they may be nested, and they may also be connected by links. The linkage is independent of the nesting, so links often cross node boundaries [12]. We can observe the particular bigraph in Fig. 1, as follows:

Three regions (rectangles): Zone 3 is a city of a country. Zone 2 is a district belonging to the city. Zone 1 is a ward belonging to the district.
A person: a human. A person may enter or leave another Zone.

Each model using bigraphs is represented by two parameters: a signature that makes the bigraph represent formal entities of the model and a set of reaction rules to represent their behaviors [14]. Reaction rules in a bigraph model are used to present the dynamics in the form r -> r', meaning that r can change its state to r' in suitable contexts. These changes may involve both placing and linking. The reaction rules of the TripAdvisor application are presented in Table 1.

4.2 Dynamic Feature Petri Net Model

Dynamic feature Petri net is an extension to Petri nets. A basic Petri net consists of places which are represented by circles. Each place may be marked by tokens that are represented by dots. Transitions (rectangles) and arcs (arrows) connect transitions and places. A marking may be created by putting tokens into places. A transition is activated when all places are marked with tokens. The state of a Petri net can be changed by letting a transition consume tokens from input places and produce new ones in output places. Multiple places can be marked in the same state [15].

In DFPN, transitions are additionally annotated in the following way [15]:

$$\text{Transitions} = (\text{application condition})/(\text{update expression}) \tag{1}$$

An application condition is a Boolean logical formula on a set of features. An update expression describes how the feature selection evolves after the transition.

The execution of a DFPN based on the feature selection is called application condition and the mechanism to update the feature selection is called update expression.

DFPN extends Petri net by adding a feature selection to the state of the Petri net marking transitions with application conditions and update expressions.

4.3 Interaction Among Models

Middleware is in between the environment and a set of services. Get and Put are two interfaces which are defined between middleware and context environment. The middleware uses Get interface to get context information from the environment, and Put interface to take action on the environment. The middleware sends a request to a service and acquires return results if any through a call interface between the middleware and services. The interaction between the middleware and the environment is modeled as the Cartesian product of models.

5 Case Study

To provide a proof of concept of our proposed approach for context-aware mobile apps testing, we applied it to test the TripAdvisor application which is introduced in Sect. 4.1.

TripAdvisor provides location-based travel services, which helps travelers to search tourist Spots, Hotels and Restaurants based on their locations. We consider the following scenario: There is 1 service (e.g., Tourist Spots service or Hotels service or Restaurants service) in Zone 1, there are 2 services e.g., (Tourist Spots service and Hotels service) or (Hotels services and Restaurants service) or (Tourist Spots services and Restaurants service) in Zone 2 and there are 3 services (e.g., Tourist Spots service and Hotels service and Restaurants service) in Zone 3.

5.1 Modeling the Environment with a Bigraph

There are 15 reaction rules shown in Table 1, describing the behaviors of the user and interactions with the environment.

Reaction rule r0 (see Fig. 2): The user is outside of the Zone (C0) and moves into Zone 1 having one service {Tourist Spots service (C1) or Hotels service (C2) or Restaurants service (C3)}.

Reaction rule r1: User moves from Zone 1 having one service {Tourist Spots service (C1)} moves into Zone 2 having two services (including pre-existing service) {Tourist Spots service and Hotels service (C4)} or {Tourist Spots service and Restaurants service(C5)}.

Reaction rule r2: User moves from Zone 1 having one service {Tourist Spots service (C1)} moves into Zone 2 having two services (pre-existing services are not included) {Hotels service and Restaurants service (C6)}.

Reaction rules r3: User moves from Zone 2 having two services {Tourist Spots service and Hotels service (C4)} into Zone 3 having 3 services {Tourist Spots service, Hotels service and Restaurants service (C7)}.

Table 1. Descriptions of 15 reaction rules of the TripAdvisor application.

No.	Description
1	User enters Zone 1 having one service (Tourist Spots service)
2	User enters Zone 1 having one service (Hotels service)
3	User enters Zone 1 having one service (Restaurants service)
4	User moves from Zone 1 having one service (Restaurants service) to Zone 2 having two services (Tourist Spots service and Hotels service)
5	User moves from Zone 1 having one service (Restaurants service) to Zone 2 having two services (Restaurants service and Hotels service)
6	User moves from Zone 1 having one service (Restaurants service) to Zone 2 having two services (Restaurants service and Tourist Spots service)
7	User moves from Zone 1 having one service (Tourist Spots service) to Zone 2 having two services (Restaurants service and Hotels service)
8	User moves from Zone 1 having one service (Tourist Spots service) to Zone 2 having two services (Tourist Spots service and Hotels service)
9	User moves from Zone 1 having one service (Tourist Spots service) to Zone 2 having two services (Tourist Spots service and Restaurants service)
10	User moves from Zone 1 having one service (Hotels service) moves to Zone 2 having two services (Tourist Spots service and Restaurants service)
11	User moves from Zone 1 having one service (Hotels service) moves to Zone 2 having two services (Hotesl service and Restaurants service)
12	User moves from Zone 1 having one service (Hotels service) moves to Zone 2 having two services (Hotels service and Tourist Spots service)
13	User moves from Zone 2 having two services (Hotels service and Tourist Spots service) to Zone 3 having 3 services (Tourist Spots service, Hotels service and Restaurants service)
14	User moves from Zone 2 having two services (Hotels service and Restaurants service) to Zone 3 having 3 services (Tourist Spots service, Hotels service and Tourist Spots service)
15	User moves from Zone 2 having two services (Tourist Spots service and Restaurants service) to Zone 3 having 3 services (Tourist Spots service, Hotels service and Restaurants service)

5.2 Modeling the Middleware with a DFPN

Context-aware mobile apps include the environment and the system under test (SUT). There are two components in the SUT: middleware and services. The environment obtains context information. The SUT obtains a set of services for the middleware to invoke. The interactions between the environment and the SUT are implemented through the get and put interfaces. We add these two interfaces to the transitions of DFPN as in [8]. We model the TripAdvisor application with a DFPN as follows:

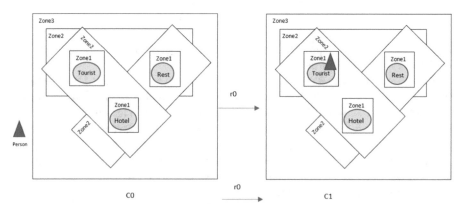

Fig. 2. Reaction rule r0

We describe this application with a set of services: Tourist Spots service (T), Hotels service (H), Restaurants service (R) and the location changing of the user among Zones. Each service has 2 states and 2 transitions respectively: Tourist Spots service (T) has 2 states of no service (T0) and service (T1) and transitions t0 and t1. For instance, Fig. 3 shows the DFPN for Tourist Spots service.

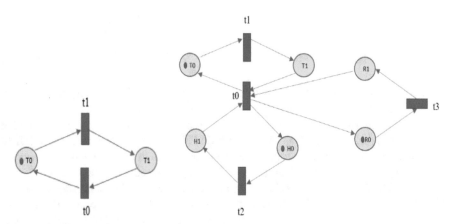

Fig. 3. DFPN for tourist spots service.

Fig. 4. DFPN for the TripAdvisor application

Basically, a transition is activated if its input states are marked. In the beginning, the token may be consumed by two transitions t0 and t1, if transition t1 fires, the token moves from state T0 to state T1, and Tourist Spots service is enabled. Symmetrically, when token moves from state T1 to state T0 (t0 fires), Tourist Spots service is disabled.

Similarly, Hotels service (H) has 2 states of no service (H0) and service (H1) and transitions t0 and t2. Restaurants service (R) has 2 states of no service (R0) and service (R1) and transitions t0 and t3. Three services in TripAdvisor application can be modeled with DFPN (see Fig. 4).

5.3 Assessments

In order to evaluate our approach on the TripAdvisor app, we use three different criteria (all paths, all-uses, and all-defs) as in [4].

We implement our test model with test situation 1: The user is outside of the Zone and moves into Zone 1 having one service (Tourist Spots service); then he/she moves into Zone 2 having two services (Tourist Spots service and Hotels service); then he/she moves into Zone 3 having 3 services (Tourist Spots service, Hotels service and Restaurants service).

An interesting bigraph in Fig. 5 and the reaction rules in Table 1 constitute an example of a bigraphical reactive system. Applying the matching of these reaction rules, a bigraph label transition system is generated.

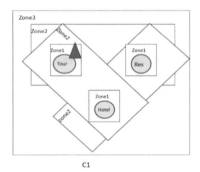 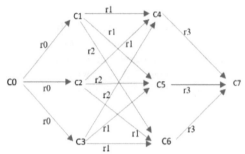

Fig. 5. A bigraph of interest.   **Fig. 6.** Bigraph label transition system

We model this test situation 1 with bigraph as shown in Fig. 6 where reaction rules as labels indicated as r0 to r3 and bigraph as states indicated as C0 to C7.

We continue to model this test situation 1 with DFPN as shown in Fig. 7, where transition as labels indicated as t0 to t3 and DFPN as states indicated as P0 to P7. In particular, the user is outside of the Zone (P0). The user is in Zone 1 having one service {Tourist Spots service (P1) or Hotels service (P2) or Restaurants service (P3)}. The user is in Zone 2 having two services {Tourist Spots service and Hotels service (P4)} or {Tourist Spots service and Restaurants service (P5)} or {Hotels service and Restaurants service (P6)}. The user is in Zone 3 having 3 services {Tourist Spots service, Hotels service and Restaurants service (P7)}.

Then, we provide the combination between BLTS and DFPN with test situation 1. This synchronization model is modeled as a Cartesian product between them. We have the following result:

$$C0P0 \xrightarrow{t1} C0P1 \xrightarrow{r0} C1P1 \xrightarrow{t2} C1P4 \xrightarrow{r1} C4P4 \xrightarrow{t3} C4P7 \xrightarrow{r3} C7P7$$

The BLTS in Fig. 6 may contain a large number of reaction paths. There are too many possible test-case sequences in the synchronized model. So, we use testing strategies

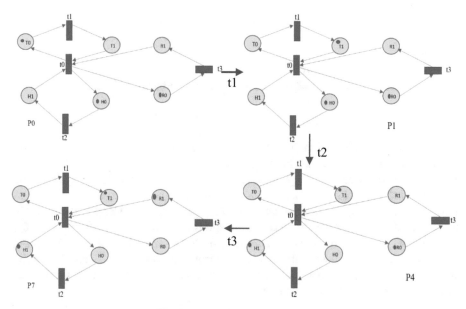

Fig. 7. DFPN for test situation 1

based on pattern-flow based testing in [4] to select test cases by selecting a set of reaction paths such that the set of paths cover all-defs or all-uses.

Regarding pattern-flow based testing, r0 is an example of pattern def of the bigraph of interest as shown in Fig. 5 where the user enters Zone 1 with Tourist Spots service; r1 is a pattern def-use, r2 is a pattern use, r3 is a pattern def-use. Test cases are generated from test situation 1 as follows:

1. $C0 \xrightarrow{r0} C1 \xrightarrow{r1} C4$
2. $C0 \xrightarrow{r0} C1 \xrightarrow{r1} C4 \xrightarrow{r3} C7$
3. $C0P0 \xrightarrow{t1} C0P1 \xrightarrow{r0} C1P1 \xrightarrow{t2} C1P4 \xrightarrow{r1} C4P4 \xrightarrow{t3} C4P7 \xrightarrow{r3} C7P7$

Test situation 2: The user is outside of the Zones and moves into Zone 1 having one service (Tourist Spots service), then he/she moves into Zone 2 having two services (Restaurants service and Hotels service), then he/she moves into Zone 3 having 3 services (Tourist Spots service, Hotels service and Restaurants service).

Modeling this test situation 2 with bigraph is also shown in BLTS in Fig. 6.

We model this test situation 2 with DFPN as follows:

$$C0P0 \xrightarrow{t1} C0P1 \xrightarrow{r0} C1P1 \xrightarrow{t0} C1P0 \xrightarrow{r2} C6P0 \xrightarrow{t2} C6P2 \xrightarrow{t3} C6P6 \xrightarrow{t1} C6P7 \xrightarrow{r3} C7P7$$
$$\searrow_{t3} C6P3 \nearrow_{t2}$$

Test cases are generated from test situation 2 as follows:

1. $C0 \xrightarrow{r0} C1 \xrightarrow{r2} C6$
2. $C0P0 \xrightarrow{t1} C0P1 \xrightarrow{r0} C1P1 \xrightarrow{t0} C1P0 \xrightarrow{r2} C6P0 \xrightarrow{t2} C6P2 \xrightarrow{t3} C6P6 \xrightarrow{t1} C6P7 \xrightarrow{r3} C7P7$

$$\xrightarrow{t3} C6P3 \xrightarrow{t2}$$

Context-aware mobile applications provide adaptive services in response to the dynamically changing contexts in the environment. In the TripAdvisor application, its environment comprises of a range of moving user and physical facilities. Physical facilities include Zone 1, Zone 2, Zone 3 and moving user may refer to the user who may move through Zones. Providing suitable services to end user according to change of the environment is hard. So, testing these applications is challenging. This paper handles this problem by using the proposed approach on the test situations of the TripAdvisor application. That means, when the user moves through Zones (Zone 1, Zone 2, Zone 3), the adaptive services (Tourist Spots service, Hotels service and Restaurants service) are provided according to the user's movement.

6 Conclusion and Future Work

Context-aware mobile applications interact with the environment in real time, which keeps on changing. Therefore, one challenge of context-aware mobile applications testing is to generate test cases for the changing environment. This paper proposes to use BRS as an environmental model and DFPN as a middleware model. By synchronizing the BRS and DFPN, test cases are generated to verify the interactions between the environment and the middleware. Our novel test model for context-aware mobile apps that uses a combination of BRS model and DFPN solves the challenge of changing context, especially location context. In the future, we will implement our test model with testing tools to assess our approach.

References

1. https://www.statista.com/statistics/269025/worldwide-mobile-app-revenue-forecast/. Last Accessed 20 Nov 2021
2. Dey, A.K., Abowd, G.D.: A conceptual framework and a toolkit for supporting rapid prototyping of context-aware applications. J. Hum. Comput. Interact. 16(2–4), 7–166 (2001)
3. Baldauf, M., Rosenberg, F., Dustdar, S.: A survey on context-aware systems. In: International Journal of Ad Hoc and Ubiquitous Computing, pp. 236–277. Geneva, Switzerland (2007)
4. Lian, Y., Tsai, W.-T., Perrone, G.: Testing context-aware applications based on bigraphical modeling. IEEE Trans. Reliab. 65(3), 1584–1611 (2016). https://doi.org/10.1109/TR.2016.2575444
5. Mehmood, M.A., Khan, M.N.A., Afzal, W.: Transforming context-aware application development model into a testing model. In: 8th International Conference on Software Engineering and Service Science, pp. 177–182. IEEE Press, Beijing, China (2017)

6. Siqueira, B.R., Ferrari, F.C., Souza, K.E., Camargo, V.V., Lemos, R.J.S.T.: Testing of adaptive and context-aware systems: approaches and challenges. Verification and Reliability **1772**, 1–46 (2021)

7. Achilleos, A.P., Kapitsaki, G.M., Papadopoulos, G.A.: A framework for dynamic validation of context-aware applications. In: 15th International Conference on Computational Science and Engineering, pp. 532–539. IEEE, Cyprus (2012)

8. Puschel, G., Seiger, R., Schlegel, T.: Test modeling for context-aware ubiquitous applications with feature petri nets. In: 2nd Workshop on Model-Based Interactive Ubiquitous Systems, pp. 37–40. ACM, Copenhagen, Denmark (2012)

9. Griebe, T., Gruhn, V.: A model-based approach to test automation for context-aware mobile applications. In: 29th Annual ACM Symposium on Applied Computing, pp. 420–427. ACM, New York, USA (2014)

10. Mirza, A.M., Khan, M.N.A.: An automated functional testing framework for context-aware applications. IEEE Comput. **6**(1), 46568–46583 (2018)

11. http://docs.oasisopen.org/wsbpel/2.0/CS01/wsbpel-v2.0-CS01.html. Last Accessed 21 Oct 2021

12. Milner, F.: The Space and Motion of Communicating Agents, 2nd edn. Cambridge Univesity Press, UK (2009)

13. Yu, L., Gao, J.: Generating test cases for context-aware applications using bigraphs. In: 8th International Conference on Software Security and Reliability, pp. 137–146. IEEE Press, San Francisco, USA (2014)

14. Milner, R.: Pure bigraphs: structure and dynamics. Inf. Comput. **204**(1), 60–122 (2006)

15. Muschevici, R., Clarke, D., Proenca, J.: Feature petri nets. In: 14th International Conference, pp. 13–17. Jeju, Korea (2010)

Recommendation System with Artificial Intelligence for Welding Quality Improvement

Trung Tin Tran[1] ⓘ, Jang Hee-Dong[2], and Anh Quang Nguyen Vu[3(✉)]

[1] FPT University, Da Nang 550000, Vietnam
tintt19@fe.edu.vn
[2] Dong-A University, Busan 49315, South Korea
[3] Vietnam-Korea University of Information and Communication Technology, Da Nang 550000, Vietnam
nvaquang@vku.udn.vn

Abstract. This paper proposes the recommendation system (RS) to support the welding quality improvement system (WQIS) with artificial intelligence (AI). The proposed RS is applied to make suggestions for beginners and experienced workers about the welding quality predictions based on weld button size and the welding quality improvement by increasing welding current. The goal of the paper aims to develop an RS that has the ability to learn, analyze, predict and make these suggestions to humans through AI. Support Vector Machines (SVM) have been employed to predict the welding quality with impact parameters such as instantaneous (IHR), electrode tip diameter (D_e), and the status of welding current (I_w) during the operation of the resistance spot welding (RSW) machine. The practical experiments are set up with an RSW machine using an AC inverter on Galvanized (GI) steel to collect the dataset for the SVM model. Through the experimental result, the effectiveness of the utility application is validated. In addition, these experimental results should be helpful for developing the high-performance RSW machine with AI applications in practice.

Keywords: Resistance spot welding · Recommendation system · SVM · The welding quality prediction and improvement

1 Introduction and Related Work

The recommendation system (RS) is known as a part of information filtering system which helps the users seek the prediction of rating or preference that users would give to an item or the service recommendations [1]. Currently, the RS has been upgraded with several machine learning algorithms to provide users the suggestion for their purposes in [2] or build the framework for RS as shown in [3]. The synthesis of a state-of-the-art review and perspectives is implemented by Baicun Wang et al. [4] to collect these approaches of techniques for making welding systems intelligent through signal processing, feature extraction, modeling, decision-making, and learning. Resistance spot weld (RSW) is an important equipment to melt the components and body of the car with

N.-T. Nguyen et al. (Eds.): ICIT 2022, LNDECT 148, pp. 226–235, 2022.
https://doi.org/10.1007/978-3-031-15063-0_21

more than a thousand weld points [5]. A the welding quality improvement system (WQIS) is a meaningful method to smartly track, classify the welding quality and suggest some ways to improve the welding quality.

As the given of the welding quality requirements, among the weld attributes, weld size, in terms of nugget width or button diameter, are the most frequently measured and most meaningful in determining a weld's strength as represented in [6]. The welding quality depends on many impact factors such as welding current, welding time, compression force, and the contact area of electrode tip diameter (or called electrode tip diameter) as shown in [7]. A study by R. J Bowers et al. [8] investigated the effect of electrode geometry on welding current distribution. The distribution of welding current at the contact area depends largely on the electrode tip diameter size. That means the quality of the welding nugget depends on the distribution of welding current focusing on the contact area between the electrode tip and the metal sheet. In this case the contact surface diameter of the electrode tip increases, the welding current distribution is scattered causing poor quality weld points. A study that was implemented by M. Zhou et al. [9] discovered the relationship between the quality and attributes of spot welding. An online qualitative nugget classification had been proposed by Mahmoud El-Banna et al. [10] based on linear vector quantization (LVQ) neural network for estimation of the weld button size class-based a small number of dynamic resistance patterns.

Recent studies have mentioned new methods for classifying the welding quality using several of machine learning and deep learning techniques. A proposed system, called SemML was introduced by Yulia Svetashova et al. in [11] to address these challenges by empowering ML-based quality monitoring methods with semantic technologies. They evaluated SemML on the Bosch use-case of electric resistance welding with very promising results. Baifan Zhou et al. [12] developed 12 machine learning pipelines with a novel feature-engineering approach of 4 feature settings and 3 machine learning methods (linear regression, multi-layer perception, and support vector regression) to predict quality of upcoming the welding operations before they happen. A proposed study was implemented by Lei Zhou et al. in [13] to collect the state information in the welding process and combine it with the unsupervised learning method to predict the final welding quality.

As mentioned above, most studies have focused on predicting the welding quality based on machine learning methods. Furthermore, previous approaches have not used particular machine learning methods to recommend or suggest some methods for improving the welding quality by using some impacted parameters from the welding machine. In contrast to previous studies, the authors focus on developing a recommendation system to diagnose and predict the welding quality. At the same time, the proposed recommendation system will give suggestions to improve the welding quality of the RSW machines. Accordingly, SVM model is utilized for training, diagnosing, and predicting the welding quality and suggesting these methods to improve the welding quality in this study.

In fact, the estimation of weld point quality depends very much on the experiences and welding techniques of the workers. The workers who can make a preliminary assessment of the welding quality based on their long-term working experiences. They can decide to adjust the welding current value and determine the suitable time to replace the electrode tip for improving the welding quality. However, these actions depend entirely on the skills and experiences of senior workers.

The purpose of this study is to design the RS that will evaluate, predict the welding quality, and suggest the method to improve the welding quality to the workers. In the proposed RS, the authors use SVM model on several parameters that affect weld button formation for building a predictive model in the welding quality. In addition, the proposed RS reflects the prediction of the welding quality for suggesting the method to help the workers select the suitable set welding current of RSW machine with the particular welding conditions.

The rest of the paper is organized as follows. In Sect. 2, the authors introduce the overview of WQIS and RS. Section 3 defines the requirements for data collection based on the experiments in practice and implements the methodology in detail of RS to support the users in making their decisions to improve the welding quality. Some preliminary experimental results are discussed in Sect. 4, and the authors conclude with our plan for future work in Sect. 5.

2   Background of RS with AI

2.1   The Welding Quality Improvement System (WQIS)

The WQIS is the assistance system designed to support the workers to determine the welding quality and recommend the methods for improving the welding quality. The proposed RS uses for WQIS is a system that combines with the machine learning (ML) model. It plays a role as the professional checker to give the evaluation instructions of welding works for workers based on predictability and data analysis to provide the appropriate suggestions with welding conditions.

ML algorithm by using Support Vector Machine (SVM) in regression helps RS improve the ability of learning and predicting the welding quality from the real work.

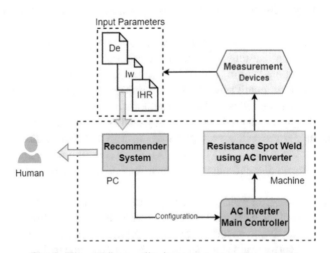

Fig. 1. The welding quality improvement system concept.

Particularly, it supports the workers to perform the welding works for each welding condition. The proposed WQIS is illustrated in Fig. 1. As shown in Fig. 1, the measured parameters including welding current (I_w), instantaneous heating rate (IHR), electrode diameter (D_e), and weld button size (D_w) are gathered by data acquisition device and monitoring software to compute IHR values.

2.2   RS Architecture of WQIS

The structure of RS applied for WQIS is illustrated in Fig. 2. In order to build the RS with ML, SVM model for regression has been applied to predict and give the recommendation. In Fig. 2, the structure of RS is composed of two modules: (i) predicting the welding quality; (ii) welding current prediction as to the improvement method. Data classification and regression are the core component of module (i). However, in the current implementation, the main task of the module (i) aims to predict the welding quality based on the initial parameters of the RSW machine (i.e., I_w, IHR, D_e, D_w).

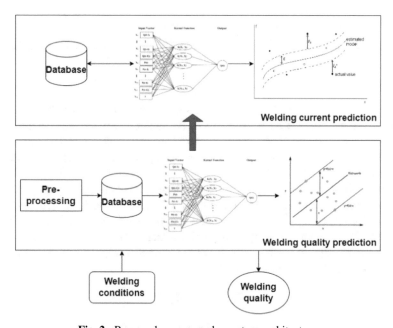

Fig. 2. Proposed recommender system architecture.

It is completely impossible to measure the weld button size in practice during the welding process. Therefore, it will be difficult for workers to accurately perceive the quality of the next weld point. As can be seen from Fig. 2, to perform the welding quality assessment, the proposed RS with the SVM model will predict the quality of weld points based on the change of IHR. In the next stage, the predicted weld button size will be the input of the model to accurately predict the value of the corresponding welding current, thereby giving suggestions to the worker for increasing the welding current.

3 Methodology

3.1 Data Acquisition

In order to build the proposed RS, the historical data is collected and the input parameter values are measured for the training and test dataset of the proposed RS. The training dataset must be collected through specific experiments. To produce the training dataset, the setup of the initial test conditions is presented in Table 1.

The practical experiments are set up with Galvanized steel to collect the dataset aims to validate the performance of the proposed RS. An RSW machine with an AC inverter will be used for making the practical experiments. The power source uses 3-phase 380–440 VAC with frequency 60 Hz, the type of electrode tip is truncated cone (type T) made of copper.

Table 1. Initial configuration of RSW machine for the experiment.

Mateiral	Thickness (mm)	Electrode force (kN)	Welding current (kA)	Welding time (cycle)	Electrod diameter
GI	1.6	4.53	10	13	6

To determine the impacts of these factors on the welding quality, the experiments are performed with the different diameters of the electrode tip, including 6 mm, 7 mm, 8 mm, 9 mm, and 10 mm. Then, for each more than 150 weld points, the survey of the weld button diameter and electrode tip diameter is repeated. Approximately 6,000 weld points are procedure.

In order to verify the weld strength, the destructive evaluation (e.g. peel, chisel tests, and a metallographic test) and non-destructive evaluation with methods such as ultrasonic A-scan, B-scan are utilized as shown in [6, 10]. In this study, the test method used to check the welding quality is the peel test technique to check the weld strength of the specimen. Figure 3 shows the formula (1) applied for the tested specimens to calculate the weld button diameter in [14].

$$d_w = (d_1 + d_2)/2 \tag{1}$$

where, d_w denotes the average diameter of the weld button with symmetrical and asymmetrical buttons, d_1 and d_2 are the measured diameter of the weld button of maximum and minimum diameters, respectively. A weld button is considered in the case of the weld button size is calculated as a formula $d = 3\sqrt{t}$ and $d = 6\sqrt{t}$ (t is the thickness of the metal sheets). The feature extraction data of the welding quality are collected based on the weld button diameter.

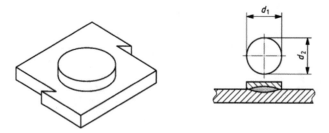

Fig. 3. Symmetrical weld button.

3.2 SVM with Regression for RS

As mentioned above, the collections of weld button size cannot be done in parallel with welding work in progress. Therefore, SVM model for regression takes the first step of predicting the welding quality based on several input parameters as mentioned above. The result of these predictions will then be used as an input of the next step in predicting the change of welding current at the next welding points.

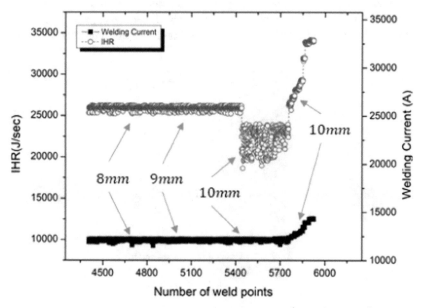

Fig. 4. Relationship between IHR, I_w, and D_e from 4,300th point to 5,924th point.

Figure 4 illustrates the relationship between IHR, I_w, and D_e from 4,300th point to 5,924th point during the whole weld process. Therefore, IHR will decrease when the electrode tip diameter increases, meanwhile the welding current is still stable at 10 kA. Figure 5 shows the electrode tip diameter and the number of weld points.

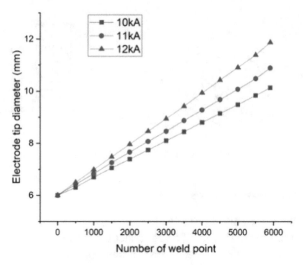

Fig. 5. Relationship between D_e and I_w.

The SVM model for regression (Fig. 6) is responsible for predicting the welding quality of each weld point and welding current for beginner and experencied workers in the case. As shown in [15], the prediction function is expressed as following the Eq. (2).

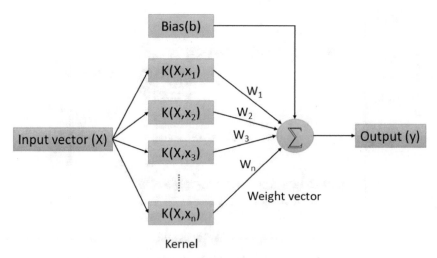

Fig. 6. SVM architecture

$$y = f(X) = \sum_{K=1}^{n} W_k.K(X, x_k) + b \qquad (2)$$

The metric methods are used to evaluate the performance of the model are mean square error (MSE), mean absolute error (MAE) và R^2. The MSE score is described below.

$$MSE(X, h) = \frac{1}{m} \sum_{i=1}^{m} (h(x_i) - y_i)^2 \tag{3}$$

The MAE score is written below.

$$MAE(X, h) = \frac{1}{m} \sum_{i=1}^{m} |h(x_i) - y_i| \tag{4}$$

where, m denotes the number of instances in the dataset, X is the matrix containing all the feature values of all instances in the dataset, y plays a role of desired output value, h denotes the prediction functions (called a hypothesis), $\hat{y}_i = h(x_i)$ is the predicted value.

The kernel function represents the scalar product in a high-dimensional space in which the data become linearly separable as known as the "kernel trick" as shown in [16, 17]. In this model, the Radial Basis Function (RBF) is used for kernel function with hyperparameter C is introduced to balance margin size and classification errors (C = 100). Also, tolerated deviations from the observed and predicted values of training data are at most ε, and larger errors are penalized ($\varepsilon = 0.1$).

4   Result

Accordingly, the experiment is set up to test and validate the performance of the proposed RS for predicting the welding quality and adjusted welding current. The dataset is extracted 5,924 samples from practical experiments as mentioned in Sect. 3, with 4,739 samples being used as a training set and 1,184 samples as a test set.

As can be seen from Figs. 4 and 5, Table 2 shows the impact of IHR and D_e on D_w rapidly. The welding current is gradually increased from 100 A to 500 A to verify the change of IHR corresponding to I_w value.

Table 2. Extracted feature between IHR, I_w, D_e, D_w.

IHR (J/Sec)					I_w(kA)
6 mm	7 mm	8 mm	9 mm	10 mm	
24656	25046	25437	21406	18480	10.0
25570	25835	25785	22082	19339	10.5
26382	26554	26945	21304	19667	11.0
26929	27128	27390	29316	30019	11.5
27390	27832	28187	30308	30652	12.0
28300	28523	28726	31304	31921	12.5

As mentioned above, the SVR model is used to predict the welding quality, thereby serving as a reference for the evaluation and classification of the welding quality. At the same time, the SVR model is also used to predict the adjusted welding current for the welding quality improvement. In Table 3, the evaluation metrics are listed to evaluate the performance of the regression model for predicting the welding quality and predicting the adjusted welding current for improving the welding quality. As shown in Table 2, the SVR model predicts the welding quality and welding current with several metrics evaluations as mentioned above.

Table 3. Evaluation metrics for SVR.

	MSE	MAE	R^2
The welding quality	0.0089	0.0736	0.9325
Welding current	0.0085	0.0735	0.9356

5 Conclusion and Future Work

In this study, the authors present the proposed RS and novel method for the welding quality improvement with the machine learning model. Form the describing details on the proposed RS's structure and its performance above, the authors developed the RS with SVM model for regression. This model supports the evaluation of the welding quality based on predicting the weld button size. Then, it predicts the welding current with predicted the welding quality to give the suggestion of improvement method (such as increasing welding current). The SVM model implements the prediction of the welding quality with the best accuracy of 0.0089 of MSE, 0.0736 of MAE, and 0.9325 for R^2, respectively. Meanwhile, the result of the prediction of the adjusted welding current as the quality improvement method is shown with 0.0085 of MSE, 0.0735 of MAE, and 0.9356 of R^2. Once the proposed RS is able to give the better choice for workers to improve the welding quality.

In the future work of this study, the author plans to focus on improving the machine learning model in the proposed RS with the agent to recommend several parameters such as the electrode diameter for ready condition selection. So, the proposed RS will be played the role of the professional examiner for a worker in the future.

References

1. Adomavicius, G., Tuzhilin, A.: Toward the next generation of recommender systems: a survey of the state-of-art and possible extensions. IEEE Trans. Knowl. Data Eng. **17**, 734–749 (2005)
2. Portugal, I., Alencar, P., Cowan, D.: The use of machine algorithms in recommender systems: a systematic review. Expert Syst. Appl. **97**, 205–227 (2018)
3. Aguliar, J., Valdiviezo-Díaz, P., Riofrio, G.: A general framework for intelligent recommender systems. Appl. Comput. Inform. **13**, 147–160 (2017)

4. Wang, B., Hu, S.J., Sun, L., Freiheit, T.: Intelligent welding system technologies: State-of-art review and perspectives. J. Manuf. Syst. **56**, 373–391 (2020)
5. Aluminum automotive manual website: automotive aluminum joining manual 2015, chapter 11. URL: https://www.european-aluminum.eu/resource-hub/aluminum-automotive-manual/. Last Accessed 01 Mar 2021
6. AWC1.1: Recommended Practices for Resistance Welding, 6th edn. American Welding Society (2019)
7. Pouranvari, M., Marashi, S.P.H.: Factors affecting mechanical properties of resistance spot welds. Mater. Sci. Technol. **26**, 1137–1144 (2010)
8. Bowers, R.J., Sorensen, C.D., Eagar, T.W.: Electrode gometry in resistance spot welding. Weld. J. **45-s** (1990)
9. Zhou, M., Zhang, H., Hu, S.J.: Relationship between quality and attributes of spot weld. Weld. J. **72-s** (2003)
10. El-Banna, M., Filev, D., Chinnam, R.B.: Online qualitative nugget classification by using linear vector quantization neural network for resistance sot welding. Int. J. Adv. Manf. Technol. (36), 237–248 (2008)
11. Svetashova, Y., et al.: Ontology-enhanced machine learning: a Bosch use case of welding quality monitoring. In: Pan, J.Z., et al. (eds.) ISWC 2020. LNCS, vol. 12507, pp. 531–550. Springer, Cham (2020). https://doi.org/10.1007/978-3-030-62466-8_33
12. Zhou, B., Pychynski, T., Reischl, M., Karlamov, E., Mikut, R.: Machine learning with domain knowledge for predictive quality monitoring in resistance spot welding. J. Intell. Manuf. **33**, 1139–1163 (2022)
13. Zhou, L., Zhang, T., Zhang, Z., Lei, Z., Zhu, S.: A new online quality monitoring method of chain resistance upset butt welding based on isolation forest and local outlier factor. J. Manuf. Process. **68**, 843–851 (2021)
14. French Standard: Resistance Welding – Testing of Welds – Peel and Chisel Testing of Resistance Spot and Projection Welds, pp. 1–12. AFNOR-French Standard Institute (2015)
15. Smola, A.J., Bernhard, S.: A tutorial on support vector machine. Stat. Comput. (14), 199–222 (2004)
16. Drucker, H., Burges, C.J., Kaufman, L., Smola, A., Vapnik, V.: Support vector regression machines. Adv. Neural. Inf. Process. Syst. **9**, 155–161 (1996)
17. Boser, B.E., Guyon, I.M., Vapnik, V.N.: A training algorithm for optimal margin classfiers. In: Proceedings of the Fifth Annual Workshop on Computational Learning Theory, pp. 144–152. Pittsburgh Pennsylvania USA (1992)

Sentence-Level Sentiment Analysis Using a CNN Model on Contextualized Word Representations

Huyen Trang Phan[1] , Ngoc Thanh Nguyen[2] , and Dosam Hwang[1]([⊠])

[1] Department of Computer Engineering, Yeungnam University,
Gyeongsan, South Korea
dshwang@yu.ac.kr

[2] Department of Applied Informatics, Wroclaw University of Science and Technology,
Wroclaw, Poland
Ngoc-Thanh.Nguyen@pwr.edu.pl

Abstract. The progress in Internet-based applications such as websites and social networks has resulted in the generation of significant amounts of reviews and opinions associated with policies, influencers, services, and products. The emotion in these opinions becomes a significant source for politicians, celebrities, servicers, and producers who intend to improve their activities and make better decisions. Sentiment analysis (SA) is an essential tool for extracting and analyzing these emotions. Various state-of-the-art SA methods with good performance have been developed. This paper presents a model, called context-based convolutional neural network (CNN), for sentence-level SA. Unlike previous methods, the proposed model focuses on the role of context information expressed in a sentence for SA. The proposed method comprises the following procedures: (i) Sentences are converted into vectors based on the BERT model; (ii) contextualized word representations are extracted using the BiLSTM model over word embeddings; (iii) sentence-level sentiment is analyzed using the CNN model over the contextualized word representations. Experiments on benchmark datasets prove that the proposed method can improve the performances of previous methods.

Keywords: Sentiment analysis · BERT-BiLSTM · BERT-CNN

1 Introduction

Numerous comments are posted daily on Internet-based applications such as social networks, websites, and blogs. These opinions are a significant source of data; in this regard, e-commerce and e-governments have effectively exploited this data source. Politicians, celebrities, and producers can utilize these opinions to provide more suitable policies and create better products. Sentiment analysis (SA) is an essential tool for extracting and analyzing the emotions expressed in these opinions. SA comprises three primary levels: documents, sentences, and

N.-T. Nguyen et al. (Eds.): ICIT 2022, LNDECT 148, pp. 236–245, 2022.
https://doi.org/10.1007/978-3-031-15063-0_22

aspects. Document-level SA determines the sentiment polarities of the entire text based on the overall sentiment of users regarding entities. Sentence-level SA is used to classify a sentence as objective or subjective or to determine whether it expresses negative, neutral, or positive emotions. Meanwhile, aspect-level SA is used to determine the sentiment regarding the specific aspects of each entity. Herein, we refer to sentence-level SA, which classifies the sentiment in an entire sentence, as negative, neutral, or positive.

Various high-performance sentence-level SA methods have been developed, in particular deep-learning-based approaches, such as CNNs [6], LSTMs, BiLSTMs, and most recently BERTs. Although deep-learning-based SA methods achieve better performance than lexicon-based and machine learning-based methods, they do not consider the context of sentences for analyzing sentiments. This motivated us to propose a sentence-level SA method, known as context-based CNN, by focusing on the role of contextualized word representations. Contextualized word representation [12] involves building a vector for each word conditioned on its context that directly captures both (1) word syntax and semantics and (2) word contexts in various linguistics. In this study, the combination of BERT and BiLSTM was used to create a contextualized word representation.

BERT is an embedding model that uses attention mechanisms as transformers to establish relationships between words via an encoder at the input and a decoder at the output. Unlike other embedding models that use the input as one word at a time, BERT can use the entire sentence at once based on transformers. Therefore, BERT can learn the real meanings between words [4]. Unlike basic grammar-based methods, only statistical characteristics that disregard the real meaning between words are considered based on the context information. BiLSTM can learn contextual information that is suitable for the logic of human language [1]. CNNs have demonstrated superior performance in various tasks and applications, including sentence SA and document classification [11]. The proposed method includes the following procedures: (i) Sentences are converted into vectors based on the BERT model; (ii) contextualized word representations are extracted using the BiLSTM model over word embeddings; (iii) sentence-level sentiment is analyzed using the CNN model over contextualized word vectors. The main contribution of this study is to propose a method that can capture more effectively the contextual knowledge by combining BiLSTM and BERT models for improving the performance of the sentence-level sentiment analysis.

The remainder of this paper is organized as follows: Sect. 2 provides a review of studies associated with deep-learning-based SA. Section 3 describes the research problem. Section 4 presents the proposed method. Section 5 provides the experimental results and evaluations. The final section presents the conclusions and future work.

2   Related Works

Deep-learning-based approaches are correctly oriented in SA. In this section, we list and discuss some approaches associated with SA, such as CNNs, LSTM, and BERT.

(i) The CNN model was first proposed by Collobert for semantic role-labeling tasks [3]. In another attempt, Collobert [2] used a CNN model that served as a syntactic parse. Kim [6] developed a CNN-based SA method is based on a simple CNN model with one convolution layer for creating feature maps, and it is often used as a baseline for various SA methods. Based on Kim's CNN-based model, various variants have been introduced, such as a densely connected CNNs [20] for text classification. In addition, Poria et al. [15] applied a CNN for document feature extraction and then input it into a multiple-kernel learning model for SA. In another study [14], an extended LSTM was used for context information extraction from surrounding sentences. All CNN-based methods proved that CNNs can achieve superior performance in terms of SA.

(ii) LSTMs are the most state-of-the-art methods for recurrent neural networks and are often used in an intermediate step for extracting and representing features in SA methods. Tai et al. [19] proposed an LSTM-based tree to improve semantic representations. Ruder et al. [16] introduced hierarchical BiLSTM to exploit language independence. Wang et al. [21] introduced LSTM with an attention mechanism for the critical section of a sentence. Liu et al. [8] focused on extracting left and right context information using different attention mechanisms. Yang et al. [22] presented an SA model using a BiLSTM with two attention mechanisms. Ma et al. [9] developed a network of interactive attention to integrate information from both the target and context. The LSTM-based SA methods proved that LSTM can extract context information and is suitable for human language logic. (iii) Since 2019, the performance of SA methods has improved significantly owing to the introduction of the BERT model [4], the performance of sentiment analysis methods has improved significantly. Various SA methods have applied BERT to create contextualized embeddings for input sentences before classifying sentiments [5,18]. BERT can use an entire sentence at once based on transformers. Therefore, BERT can learn the real meanings between words [4].

The analysis above suggests that favorable results can be achieved if all three models, i.e., BERT, BiLSTM, and CNN, are combined for sentence-level SA—this is a hypothesis to be proven in this study.

3   Research Problems

3.1   Problem Definition

For a finite set of n sentences $S = \{s_1, s_2, ..., s_n\}$ and for a specific sentence s_i, let c_i be the contextualized word embeddings of sentence s_i. The objective of this proposal is to construct a context-based CNN for sentence-level SA. This objective can be formalized by obtaining a mapping function $F : (c_i) \rightarrow \{0, 1, 2\}$ such that:

$$F(c_i) = \begin{cases} 1, & \textit{if sentiment expressed in } s_i \textit{ is positive,} \\ 0, & \textit{if sentiment expressed in } s_i \textit{ is neutral,} \\ 2, & \textit{if sentiment expressed in } s_i \textit{ is negative} \end{cases} \tag{1}$$

3.2   Research Questions

The main aim of this study is to propose a context-based CNN model for sentence-level SA. Therefore, we attempted to answer the following research questions:

- How to extract the contextualized word representations by combining the BERT and BiLSTM models?
- How to construct the context-based CNN model using CNN over the contextualized word representations?
- How to use the context-based CNN model for sentence-level SA?

4   The Proposed Method

In this section, we present a methodology to improve the performance of previous CNN-based methods. The workflow of the method is shown in Fig. 1, which includes the following four procedures. First, BERT is used to convert words in a sentence into word vectors (called sentence embeddings or sentence vectors). Second, BiLSTM is used over sentence embedding to create contextualized word representations. Third, a CNN is used over contextualized word representations to classify sentence-level sentiments. The details of these procedures are explained in the following subsections.

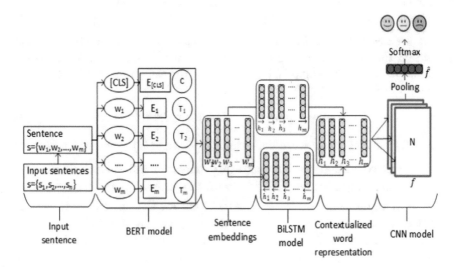

Fig. 1. A framework of the proposed method.

4.1   Creating Sentence Embeddings

Given a sentence $s \in S$ with $s = \{a_1, a_2, ..., a_m\}$, we use the BERT model to create the vectors of the words as follows:

First, a sequence of words is created as

$$\hat{s} = [CLS] + s \tag{2}$$

where [CLS] is a symbol that is added at the beginning of a sentence to indicate the beginning of each sentence.

Second, the sequence $\hat{s}$ is fed into the pre-trained BERT model[1] [4] to create the hidden representations as

$$\hat{C} = BERT(\hat{s}) \in R^{m \times d_w} \tag{3}$$

where $\hat{C}$ is the hidden representation matrix, in which each row $c \in \hat{C}$ indicating a word vector. d_w is the dimension of word vector.

Finally, the final sentence vectors $C \in R^{m \times d_c}$, where d_c is the dimension of vectors C, is created by using the average pooling on matrix $\hat{C}$ to integrate the inner information extracted from words in $\hat{C}$.

4.2   Creating Contextualized Word Representations

Because LSTM can effectively extract contextual information, it is suitable for human language logic [1]. In this study, we used BiLSTM over BERT embeddings to extract contextualized word representations in the following phases:

Input Layer: Therefore, from the input sentence s, based on the BERT model, we obtain the sentence vectors C. These vectors are used to create the input vectors A of the BiLSTM model [13] by adding the weight matrix $W^a \in R^{d_a \times m}$. This layer is formulated as follows:

$$A = Sigmoid(W^a.C + b^a) \in R^{d_a \times d_c} \tag{4}$$

where b^a is the bias vector with dimension d_a, $Sigmoid$ is the activation function.

BiLSTM Layer: This layer is to aggregate contextual information via two word directions including a forward LSTM ($\overleftarrow{lstm}$) to encode the sentence from left to right ($\overrightarrow{h}_i$), and a backward LSTM ($\overrightarrow{lstm}$) to encode the sentence in the reverse direction ($\overleftarrow{h}_i$). Thus, BiLSTM converts the input vectors $A = \{a_1, a_2, .., a_m\}$ to contextualized vectors $H = \{h_1, h_2, ..., h_m\}$ by concatenating two hidden vectors $\overrightarrow{h}_i$ and $\overleftarrow{h}_i$. This layer has a significant role in representing words following context. This layer is formulated as follows:

$$\overrightarrow{h_i} = \overrightarrow{lstm}(a_i) \in R^{d_h}, i = [1, m] \tag{5}$$

$$\overleftarrow{h_i} = \overleftarrow{lstm}(a_i) \in R^{d_h}, i = [m, 1] \tag{6}$$

---

[1] https://github.com/google-research/bert.

$$h_i = [\overrightarrow{h_i}, \overleftarrow{h_i}] \tag{7}$$

Output Layer: From the input vectors $A \in R^{d_a \times d_c}$, we obtain the contextualized word representation matrix $H = (h_1, h_2, ..., h_m) \in R^{m \times d_h}$, where d_h is the dimension of the contextualized word vector.

4.3   Building Sentence-Level Sentiment Classifier

In this study, we used the CNN model which is presented in our previous paper [13] over a contextualized word representation matrix H to convert elements in matrix H into the low-dimensional matrix in the following phases:

Input Layer: The contextualized word representations matrix $H \in R^{m \times d_h}$ created by the BERT-BiLSTM model is the input layer of the sentiment classifier.

Convolutional Layer: This layer is to capture important information from the contextualized word representations. This layer uses a filter $N \in R^{q \times d_h}$ of length q from i to $i + q - 1$ to slide over matrix H to create features map f as follows:

$$f = [f_1, f_2, .., f_m] \tag{8}$$

where f_i, $1 \leq i \leq m$, is the vector obtained after sliding filter N each time and is computed as

$$f_i = ReLU(N \ominus H_{i:i+m-1} + b) \tag{9}$$

where $\ominus$ is the convolution operator; $ReLU$ is a activation function. b is a bias.

Max-Pooling Layer: Because the size of feature vectors $f_i \in f$ will be dissimilar if we have the different lengths of contextualized representations and the different sizes of the filters. Therefore, this layer is to convert feature map vector f into new feature map vector $\hat{f}$ with the fixed size by computing the maximum number from vectors f_i. This layer is formulated as follows:

$$\hat{f} = [\hat{f}_1, \hat{f}_2, .., \hat{f}_m] \tag{10}$$

where $\hat{f}_i = Max(f_i)$. Then vector $\hat{f}$ is fed into next layer.

Fully Connected Layer: This layer is to analyze the sentence-level sentiment by fine-tuning the sentiment features of the previous layer as follows:

$$\hat{y} = Softmax(E \cdot \hat{f} + b) \tag{11}$$

where $E \in R^{l \times m}$ and $b \in R^l$ are a learnable matrix and a bias, respectively. l is the number of class labels.

4.4   Training Model

The context-based CNN model is trained by minimizing the cross-entropy loss of the predicted and real label distributions as follows:

$$L = -\sum_{i}^{l} y_i log(\hat{y}_i) + \lambda \|\theta\|^2 \tag{12}$$

where l is the number of class labels in the training set, y_i is the i-th real distribution of label, and $\hat{y}_i$ is the i-th predicted distribution label. λ represents the coefficient of L_2 regulation. θ represents the parameter set from the previous layers.

5   Experiment

5.1   Data Acquisition

In this study, to demonstrate the performance of our model and ensure an unbiased comparison of the proposed method with other methods, we used benchmark datasets such as IMDB[2] and FinancialPhraseBank[3] (FPB) are respectively from Kaggle[4]. Detailed information regarding the databases is shown in Table 1.

Table 1. Databases used in experiments

Class	IMDB-train	IMDB-validation	IMDB-test	FPB-train	FPB-validation	FPB-test
Positive	4140	1035	5176	2558	639	320
Neutral	–	–	–	4862	1215	608
Negative	4134	1033	517	773	193	97
Total	8274	2068	5693	8193	2047	1025

The following hyperparameters were suggested for the proposed model. For the BERT model, we used pre-trained BERT[5] with dimensions of 768. All the model weights were initialized with a uniform distribution. The dimensions of the hidden state vectors were set to 512. We used Adam [7] as the optimizer, with a learning rate of 2e−5. The coefficient of L_2-regularization was 10^{-5} and batch size was 8. Moreover, for the CNN model, the sequence length was set to 54, the filter size was (3, 4, 5), and the number of filters was 16.

The experimental results were obtained by averaging three trials with random initialization, where the *Precision, Recall, F_1, Loss,* and *Accuracy* [17] were adopted as the evaluation metrics. Furthermore, we performed paired tests on these metrics to verify whether the improvements achieved by our models over the baselines were significant.

---

[2] https://www.kaggle.com/lakshmi25npathi/imdb-dataset-of-50k-movie-reviews/version/1.
[3] https://www.kaggle.com/ankurzing/sentiment-analysis-for-financial-news/version/5?select=all-data.csv.
[4] https://www.kaggle.com/.
[5] https://github.com/google-research/bert.

5.2   Baseline Methods

To prove that the performance of our model is better than those of other models, we deployed three different methods, including our proposed method and two baselines, on two datasets.

– The ensemble of LSTM and CNN [10] is an SA model where LSTM captures temporal information, whereas the CNN extracts the local structure from the data.
– CNN + GloVe, LSTM + GloVe [10] are variants of the CNN and LSTM using GloVe embeddings as the input layer.

5.3   Evaluation Results

The performance of the SA methods over the specified datasets is presented in Tables 2, 3, and 4.

Table 2. Performance of our proposal on train and validation (val) datasets

Epoch	Train loss	Train F_1	Val loss	Val accuracy	Val precision	Val recall	Val F_1
IMDB							
1	0.363	0.834	0.218	0.925	0.939	0.925	0.924
2	0.206	0.926	0.219	0.923	0.941	0.923	0.941
3	0.176	0.936	0.216	0.926	0.926	0.926	0.924
4	0.151	0.946	0.223	0.923	0.941	0.923	0.923
FPB							
1	0.569	0.765	0.393	0.848	0.878	0.848	0.839
2	0.302	0.896	0.305	0.894	0.910	0.894	0.889
3	0.251	0.919	0.285	0.899	0.915	0.899	0.894
4	0.207	0.939	0.270	0.906	0.919	0.906	0.900

Table 3. Performance of our proposal on test dataset

Dataset	Test loss	Test accuracy	Test precision	Test recall	Test F_1
IMDB	0.179	0.917	0.931	0.917	0.916
FPB	0.295	0.891	0.902	0.891	0.884

As shown in Tables 2 and 3, although the number of samples in the two datasets was the same, the proposed method performed slightly better on the IMDB dataset than the FPB dataset. This is primarily because the samples in the IMDB dataset were classified into two classes, whereas samples in the FPB were categorized into three classes. Therefore, the samples in the IMDB were

Table 4. Performance comparison of models on IMDB dataset (%)

Method	Accuracy
CNN + GloVe	89.3
LSTM + GloVe	89.0
Ensemble of LSTM and CNN	90.0
Proposed method	91.7

sparser than those in the FPB. The proposed method can significantly improve this result by constructing a dataset that ensures a better balance between the sentiment classes.

Table 4 presents the performance comparison of the models. As shown, the proposed method yielded better results than the baseline methods for the IMDB dataset. The proposed method improved the accuracy by at least 1.7% and up to 2.4% compared with the ensemble of LSTM, CNN, and LSTM + GloVe. How did the proposed method enhance the accuracy of the baseline methods? In this study, the BERT model captured the semantics of the text effectively, and the BiLSTM model accurately extracted the context of the sentence. Moreover, the CNN model is currently one of the algorithms that achieves good accuracy for SA. Therefore, the results confirm that using a CNN for contextual information representation significantly affects the accuracy of SA methods.

6   Conclusion and Future Works

In this study, a method to improve the performance of sentence-level SA using CNN over BERT-BiLSTM was introduced. Experimental discussions showed that the proposed method significantly improved the performance of sentence-level SA over two benchmark datasets. However, the proposed method did not simultaneously consider all contextual factors, semantic relations, and emotional knowledge when building the classifier model. In the future, we will focus on developing a model that simultaneously represents contextual factors, semantic relations, and sentimental knowledge to improve the performance of SA methods.

References

1. Cai, R., et al.: Sentiment analysis about investors and consumers in energy market based on BERT-BiLSTM. IEEE Access **8**, 171408–171415 (2020)
2. Collobert, R.: Deep learning for efficient discriminative parsing. In: Proceedings of the Fourteenth International Conference on Artificial Intelligence and Statistics, pp. 224–232. JMLR Workshop and Conference Proceedings (2011)
3. Collobert, R., Weston, J., Bottou, L., Karlen, M., Kavukcuoglu, K., Kuksa, P.: Natural language processing (almost) from scratch. J. Machine Learn. Res. **12**(ARTICLE), 2493–2537 (2011)

4. Devlin, J., Chang, M.W., Lee, K., Toutanova, K.: BERT: pre-training of deep bidirectional transformers for language understanding. arXiv preprint arXiv:1810.04805 (2018)

5. Huang, B., Carley, K.M.: Syntax-aware aspect level sentiment classification with graph attention networks. arXiv preprint arXiv:1909.02606 (2019)

6. Kim, Y.: Convolutional neural networks for sentence classification. CoRR **abs/1408.5882** (2014). http://arxiv.org/abs/1408.5882

7. Kingma, D.P., Ba, J.: Adam: a method for stochastic optimization. arXiv preprint arXiv:1412.6980 (2014)

8. Liu, J., Zhang, Y.: Attention modeling for targeted sentiment. In: Proceedings of the 15th Conference of the European Chapter of the Association for Computational Linguistics: Volume 2, Short Papers, pp. 572–577 (2017)

9. Ma, D., Li, S., Zhang, X., Wang, H.: Interactive attention networks for aspect-level sentiment classification. arXiv preprint arXiv:1709.00893 (2017)

10. Minaee, S., Azimi, E., Abdolrashidi, A.: Deep-sentiment: sentiment analysis using ensemble of CNN and Bi-LSTM models. arXiv preprint arXiv:1904.04206 (2019)

11. Niepert, M., Ahmed, M., Kutzkov, K.: Learning convolutional neural networks for graphs. In: International Conference on Machine Learning, pp. 2014–2023. PMLR (2016)

12. Peters, M.E., et al.: Deep contextualized word representations. CoRR **abs/1802.05365** (2018). http://arxiv.org/abs/1802.05365

13. Phan, H.T., Nguyen, N.T., Hwang, D.: Convolutional attention neural network over graph structures for improving the performance of aspect-level sentiment analysis. Inf. Sci. **589**, 416–439 (2022)

14. Poria, S., Cambria, E., Hazarika, D., Majumder, N., Zadeh, A., Morency, L.P.: Context-dependent sentiment analysis in user-generated videos. In: Proceedings of the 55th Annual Meeting of the Association for Computational Linguistics (volume 1: Long papers), pp. 873–883 (2017)

15. Poria, S., Chaturvedi, I., Cambria, E., Hussain, A.: Convolutional MKL based multimodal emotion recognition and sentiment analysis. In: 2016 IEEE 16th International Conference on Data Mining (ICDM), pp. 439–448. IEEE (2016)

16. Ruder, S., Ghaffari, P., Breslin, J.G.: A hierarchical model of reviews for aspect-based sentiment analysis. arXiv preprint arXiv:1609.02745 (2016)

17. Schütze, H., Manning, C.D., Raghavan, P.: Introduction to Information Retrieval, vol. 39. Cambridge University Press, Cambridge (2008)

18. Sun, C., Huang, L., Qiu, X.: Utilizing BERT for aspect-based sentiment analysis via constructing auxiliary sentence. arXiv preprint arXiv:1903.09588 (2019)

19. Tai, K.S., Socher, R., Manning, C.D.: Improved semantic representations from tree-structured long short-term memory networks. arXiv preprint arXiv:1503.00075 (2015)

20. Wang, S., Huang, M., Deng, Z., et al.: Densely connected CNN with multi-scale feature attention for text classification. In: IJCAI, pp. 4468–4474 (2018)

21. Wang, Y., Huang, M., Zhu, X., Zhao, L.: Attention-based LSTM for aspect-level sentiment classification. In: Proceedings of the 2016 Conference on Empirical Methods in Natural Language Processing, pp. 606–615 (2016)

22. Yang, M., Tu, W., Wang, J., Xu, F., Chen, X.: Attention based LSTM for target dependent sentiment classification. In: Proceedings of the AAAI Conference on Artificial Intelligence, vol. 31 (2017)

Stock Price Prediction in Vietnam Using Stacked LSTM

Nguyen Trung Tuan[1] (ID), Thu Hang Nguyen[2(✉)] (ID), and Thanh Thi Hien Duong[2] (ID)

[1] National Economics University, Hanoi, Vietnam
tuannt@neu.edu.vn
[2] Hanoi University of Mining and Geology, Hanoi, Vietnam
{nguyenthuhang,duongthihienthanh}@humg.edu.vn

Abstract. The direction of the stock market is always complex, stochastic, and highly volatile. In addition to traditional forecasting models such as linear regression and Automatic Regression Integrated Moving Average (ARIMA) models, analysts are now trying to apply modern deep learning models to predict trends direction of the stock market to achieve more accurate forecasting. In this conducting research, we have investigated and applied the state-of-the-art deep learning sequential model, namely the Stacked Long Short-Term Memory Model (Stacked LSTM) to the prediction of stock prices the next day. The experimental result on three benchmark datasets: stocks of Apple Inc. (AAPL), stocks of An Phat Bioplastic JSC (AAA), and stocks of Bank of Foreign Trade of Vietnam (VCB) has shown the effectiveness of the predictive model. Furthermore, we discovered that the suitable quantity of hidden layers is two, and when we continue to increase the quantity of hidden layers to three or four, the Stacked LSTM model does not improve the predictive power, even though it has a more complex model structure.

Keywords: Stock price prediction · Stock forecasting · Time series forecasting · Stacked LSTM

1 Introduction

Stock prices have been a popular topic in the modern economy. In the stock market, because of high volatility, the stock price prediction has an important impact on the decisions in trading and investing. Hence, stock price forecasting has been an interesting field for researchers, traders, investors, and corporate.

Financial time series with flexible and complex variables are not easy for forecasting. Data of the stock market are numerous and almost nonlinear, so we must develop models which can dissect many hidden layers. Traditional statistical analysis models have been widely used for a long time in economics and finance data analysis, including stock forecasting problems, such as exponential smoothing and Autoregressive Integrated Moving Average (ARIMA) [1–3]. Recent studies have shown that deep learning models are capable of exploring hidden and dynamics patterns in the data by learning themselves, so these models can give better forecasting results than statistical [4].

© The Author(s), under exclusive license to Springer Nature Switzerland AG 2022
N.-T. Nguyen et al. (Eds.): ICIT 2022, LNDECT 148, pp. 246–255, 2022.
https://doi.org/10.1007/978-3-031-15063-0_23

Furthermore, Recurrent neural networks (RNN) are powerful types of neural networks designed to resolve sequence dependence. RNN uses not only the input data, but also the previous outputs for predicting the current output, so it is considered the network for sequential data. RNN association of the nodes makes a directed diagram and internal memory themselves are used to deal with flexible input sequences. The state of each node is time-varying by the real-valued activation function. The learning model in RNN is determined by transition between states so it always has the same input size. Otherwise, the same transaction function which has the same parameters at each step was used in the system [4].

A type of RNN is the Long Short-Term Memory network (LSTM) which has large structures and is used to train data with a successful outcome. An issue RNN faces are having vanishing and exploding gradients. To address these issues, LSTM networks operate "computational gates" which help handle data and keep more needed information. Hence, LSTM models have outperformed with time-series data, compared to other sequences [5]. Furthermore, the price forecasting in the stock market has been implemented with more highly accurate by using LSTM networks. Di Persio and Honchar performed three kinds of RNNs, namely: a basic RNN, an LSTM, and a Gated Recurrent Network (GRU), and included that the accurate results of the other RNNs were not good as the LSTM which was at 72% [6]. Pang et al. suggested two LSTM models to forecast the stock market: one had an embedding layer, and another had an autoencoder. The outcomes of the LSTM with embedding are more quality, with the result of accuracy being 57.2, compared to 56.9% of another [7].

There are some crucial factors to improve the performance of DNN architecture for prediction. Hiransha et al. showed that large of data is the first factor according to the bigger of the quantity of data, the higher the quality result of the model [4]. Otherwise, Adding or reducing the quantity of hidden layers is also affected to perform of models. The research of Karsoliya confirmed that the model can be had issues in training after the fourth layer. Moreover, the size of hidden notes in each layer can be followed by the thumb regulations that the hidden layer has a quantity of nodes is 2/3 the quantity of nodes in the input layer [8]. Hossain et al. performed two models LSTM and GRU for prediction. First, the features were passing the LSTM to work for forecasting. And then this forecast result was got to the GRU model to perform one more forecasting. The outcome of this model was better than the performance of the LSTM model or GRU model when they worked independently [9]. In recent years, combining models for forecasting have become popular, so the optimization of each model is considered. Assunta et al. predicted stock prices by implementing the multilayer perceptron (MLP) model. It found that in each layer, the optimization of the size of hidden layers and hidden nodes performed differently in each case and must be investigated through trial and error [10].

Many models have been used to predict stock price, but LSTM still has been one of the most common choices for experiments with successful results. Motivated by this trend, in our study, we investigate the Stacked LSTM models by considering the number of hidden layers, hidden nodes in each layer, and size of data aiming to find out the optimized model for forecasting stock price. The models have been verified with three datasets collected from the daily stock prices in the past of three companies, namely:

Apple Inc. (AAPL), An Phat Bioplastic JSC (AAA), and Bank of Foreign Trade of Vietnam (VCB).

The structure of the paper is arranged as follows. In Sect. 2, we introduce brief presentations of the LSTM and then the Stacked LSTM model - the model that we will optimize architecture for stock price prediction. We then provide some empirical analysis to point out the params settings for the Stacked LSTM model to achieve the best prediction outcomes in the next part - Sect. 3. Finally, the last section - Sect. 4 is a conclusion.

2 Methodology

2.1 Long Short-Term Memory

A particular model of RNN is the LSTM network which introduces an internal cell state or memory state and gating mechanisms. It can retain short-term memory while capturing long-range dependencies in data [11]. The LSTM models were applicated in the financial domain [9, 10], sequence learning domain [13] and have achieved a lot of results.

Each cell of LSTM works with gates: Input gate (i_t), output gate (o_t) and forget gate (f_t) (see in Fig. 1) [5, 14, 15].

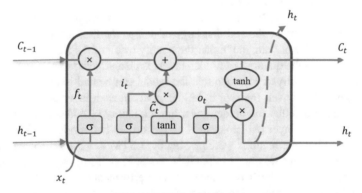

Fig. 1. LSTM cell [14, 12]

The forget gate will get which candidate data from the previous cell state and clear out. It takes the inputs and output of the previous hidden state (h_{t-1}), and input of current state (x_t), and input them through the sigmoid activation function (σ) which each value output is a vector between 0 and 1. The output is 0 means that the information is removed while 1 indicates that the information is kept.

The input gate with another sigmoid layer will choose the value to update in the cell state. A memory cell $\left(\tilde{C}_t\right)$ is also created which uses tanh activation function on the same inputs, but the output is between -1 and 1. The information from the cell state is dropped with the negative result and is added with the positive value. The result was defined by how much each cell state should be updated, and times the output from tanh and input gate sigmoid activation:

$$i_t = \sigma\left(W_i.\left[h_{t-1}, x_t\right] + b_i\right) \tag{1}$$

$$f_t = \sigma\left(W_f.\left[h_{t-1}, x_t\right] + b_f\right) \tag{2}$$

$$\widetilde{C}_t = tanh\left(W_C.\left[h_{t-1}, x_t\right] + b_C\right) \tag{3}$$

Finally, a new cell state was obtained by the forget vector (f_t) times the previous cell state (C_{t-1}) and the result is added with the multiplication between input gate and tanh vector:

$$C_t = f_t * C_{t-1} + i_t * \widetilde{C}_t \tag{4}$$

The output gate (o_t) (a sigmoid layer) will be used to filter output for the next step, and return to the hidden state (h_t):

$$o_t = \sigma\left(W_o.\left[h_{t-1}, x_t\right] + b_o\right) \tag{5}$$

$$h_t = o_t * tanh(C_t) \tag{6}$$

2.2 Stacked LSTM

The foundational LSTM model includes only one hidden layer. The network was expanded by adding hidden LSTM layers called Stacked LSTM. Each hidden layer of this network includes multiple cell states which are piled on top of each other. Furthermore, we receive one outcome after each input time step through each layer. Therefore, each hidden layer of LSTM can have a sequence output for all input time steps, instead of one output like other models [16].

A Multilayer Perceptron becomes deeper when more hidden layers are added. The learned performances through training process from previous layers were collected by the expanding hidden layer and create a new performance of networks which was developed to high levels of abstraction. Hence, the approach for prediction was improved with more accuracy based on the multi-layer model [17, 18].

Nowadays, Stacked LSTM is a strong method to handle the complex sequence data for forecasting as of result of outperforming [19, 20].

3 Experiments and Results

3.1 Datasets

The conducting research was done for three datasets of stock price, namely: AAPL stock was obtained from Yahoo Finance which is used widely in research for forecasting stock price; AAA stock and VCB stock which were collected from Vietstock [21]. The daily history prices of AAA stock and VCB stock are from 5 January 2015 to 31 December 2021, including 1744 and 1749 data points, respectively. AAPL stock is between 31 January 2015 to 29 December 2019, with 1236 data points. The attribute categories of

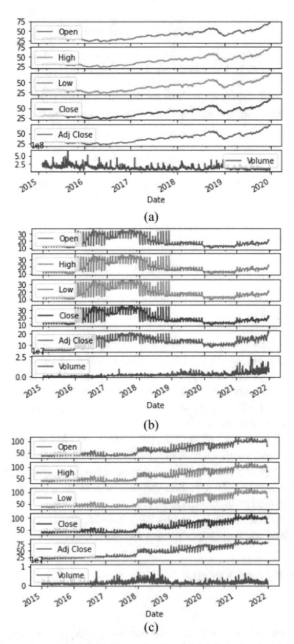

Fig. 2. Stock price performance: (a) AAPL stock price, (b) AAA stock price, (c) VCB stock price.

these data have six types, namely: Open, High, Low, Close, Adj Close, Volume (see in Fig. 2).

3.2 Parameter Settings and Evaluation Metrics

We use Keras library for Python to implement DNN model and Jupyter Notebook was used to conduct the analysis. To define the LSTM dataset, the data of the model was divided into two sets: the training data was indicated for 80% of the dataset and the test data was set for the rest with 20%. The quantity of layers in the Stacked LSTM model was designed from 2 to 4 hidden layers. A dropout layer was set before the Dense layer to avoid overfitting. The regulation of thumb method was applied so the quantity of nodes in the hidden layers is 2/3 of the size of the input layer [8]. The detail of the quantity of LSTM layers for each case is shown in Table 1.

Table 1. The quantity of layers in the stacked LSTM model

Quantity of LSTM layers	Nodes of LSTM layers in order
2	(45, 30)
3	(67, 45, 30)
4	(100, 67, 45, 30)

While training data, we used mean squared error (MSE) which is popular and applied the most as a loss function. Besides that, the Stacked LSTM network was optimized by applying the Adam algorithm [14, 20, 22]. Furthermore, with big data, the model was performed with 30 batches, 1000 epochs, and put sliding window size was 7 days.

Finally, the prediction models are evaluated by three used widely metrics in the research community, which are Root Mean Squared Error (RMSE), Mean Absolute Error (MAE) and Mean Absolute Percentage Error (MAPE) where the lower value the better. Let N be the number of samples, x_t and $\hat{x}_t$ are the ground-truth and prediction results, respectively, the evaluation metrics RMSE, MAE, and MAPE are defined by Eqs. 7, 8, and 9, respectively [19, 23].

$$RMSE(x, \hat{x}) = \sqrt{\frac{1}{N} \sum_{t=1}^{N} (x_t - \hat{x}_t)^2} \tag{7}$$

$$MAE(x, \hat{x}) = \frac{1}{N} \sum_{t=1}^{N} |x_t - \hat{x}_t| \tag{8}$$

$$MAPE(x, \hat{x}) = \frac{1}{N} \sum_{t=1}^{N} \left| \frac{x_t - \hat{x}_t}{x_t} \right| \tag{9}$$

3.3 Results and Discussion

The outcomes of evaluation metrics from Table 2 to Table 4 showed that the case with 2 LSTM layers in all models had the best performance with the lowest error measures. In the prediction models of AAPL stock prices, the RMSE is 0.725, the MAE is 0.503 and MAPE is 0.009. AAA stock prices with RMSE, MAE, and MAPE are 0.285, 0.219, and 0.014, respectively. The results of VCB stock are 1.793, 0.976, and 0.010 for RMSE, MAE, and MAPE, respectively. The prediction graphs for these works are displayed in Fig. 3. Otherwise, the value of error measures was increased when we added more the number of LSTM layers to 3 or 4. In the case of both AAPL and VCB stock prices, the error measures of 4 LSTM layers were about double compared with 3 LSTM layers.

Compared with other LSTM works, our performance has an important contribution that we can improve the model by optimizing the parameters which can make the model become the most efficient only with two hidden layers of architecture.

Table 2. The performance of evaluation metrics for AAPL stock predictions

Quantity of LSTM layers	Metrics		
	RMSE	MAE	MAPE
2	**0.725**	**0.503**	**0.009**
3	1.280	0.949	0.017
4	2.695	1.464	0.024

Table 3. The performance of evaluation metrics for AAA stock predictions

Quantity of LSTM layers	Metrics		
	RMSE	MAE	MAPE
2	**0.285**	**0.219**	**0.014**
3	0.380	0.278	0.018
4	0.352	0.261	0.016

Table 4. The performance of evaluation metrics for VCB stock predictions

Quantity of LSTM layers	Metrics		
	RMSE	MAE	MAPE
2	**1.793**	**0.976**	**0.010**
3	1.994	1.112	0.012
4	4.063	3.078	0.031

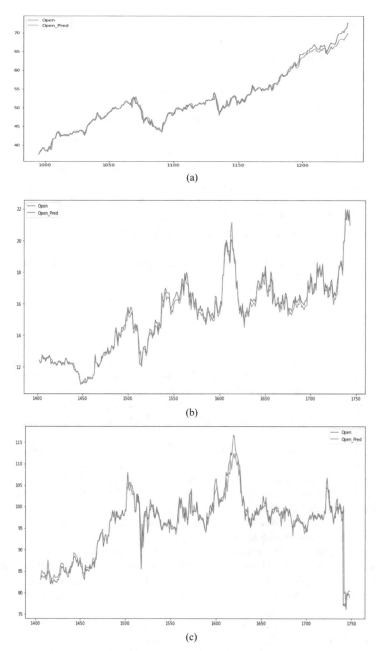

Fig. 3. Performance of stock price prediction for 2 LSTM layers: (a) AAPL stock price, (b) AAA stock price, (c) VCB stock price

4   Conclusion

In conducting research on stock price prediction, we recommend a neural network based on optimizing Stacked LSTM architecture. One of the approaches to improve performance is considering specific factors which are the quantity of data, the ratio of dataset divided into training and test set, the quantity of LSTM layers, and the nodes of each LSTM layer. Our implementation showed that we can achieve better outcomes after only 2 LSTM layers. The model has a lot of data that need to be trained so this performance helps to save the training time and complexity. Furthermore, our models were tested on three stock prices datasets of Apple Inc. (AAPL), An Phat Bioplastic JSC (AAA), and Bank of Foreign Trade of Vietnam (VCB), so it will help the traders and investors have more useful information to make good decisions and get more profit in Vietnam stock market. In future works, we will extend more crucial features that affect the stock price and could achieve real-time stock market forecasting in Viet Nam.

Acknowledgements. This work was funded by Hanoi University of Mining and Geology under grant number 65/QD-MDC.

References

1. Khashei, M., Hajirahimi, Z.: A comparative study of series arima/mlp hybrid models for stock price forecasting. Commun. Stat. – Simul. Comput. **48**(9), 2625–2640 (2019). https://doi.org/10.1080/03610918.2018.1458138
2. Kumar, M., Thenmozhi, M.: Forecasting stock index returns using ARIMA-SVM, ARIMA-ANN, and ARIMA-random forest hybrid models. IJBAAF **5**(3), 284 (2014). https://doi.org/10.1504/IJBAAF.2014.064307
3. Wadi, S.A., Almasarweh, M., Alsaraireh, A.A.: Predicting closed price time series data using ARIMA Model. MAS **12**(11), 181 (2018). https://doi.org/10.5539/mas.v12n11p181
4. Hiransha, M., Gopalakrishnan, E.A., Menon, V.K., Soman, K.P.: NSE stock market prediction using deep-learning models. Procedia Comput. Sci. **132**, 1351–1362 (2018). https://doi.org/10.1016/j.procs.2018.05.050
5. Olah, C.: Understanding lstm networks (2015). https://colah.github.io/posts/%202015%E2%80%9308-Understanding-LSTMs/
6. Di Persio, L., Honchar, O.: Analysis of recurrent neural networks for short-term energy load forecasting. Thessaloniki, Greece, p. 190006 (2017). https://doi.org/10.1063/1.5012469
7. Pang, X., Zhou, Y., Wang, P., Lin, W., Chang, V.: An innovative neural network approach for stock market prediction. J. Supercomput. **76**(3), 2098–2118 (2018). https://doi.org/10.1007/s11227-017-2228-y
8. Karsoliya, S.: Approximating number of hidden layer neurons in multiple hidden layer BPNN architecture. Int. J. Eng. Trends Technol. **3**(6), 4 (2012)
9. Hossain, M.A., Karim, R., Thulasiram, R., Bruce, N.D.B., Wang, Y.: Hybrid deep learning model for stock price prediction. In: 2018 IEEE Symposium Series on Computational Intelligence (SSCI), Bangalore, India, pp. 1837–1844 (2018). https://doi.org/10.1109/SSCI.2018.8628641
10. Salleh, H., Atrick Vincent, A.M.P.: An investigation into the performance of the multilayer perceptron architecture of deep learning in forecasting stock prices. UMT JUR, vol. 3, no. 2, pp. 61–68 (2021). https://doi.org/10.46754/umtjur.2021.04.006

11. Hochreiter, S.: The vanishing gradient problem during learning recurrent neural nets and problem solutions. Int. J. Unc. Fuzz. Knowl. Based Syst. **06**(02), 107–116 (1998). https://doi.org/10.1142/S0218488598000094

12. Heaton, J.B., Polson, N.G., Witte, J.H.: Deep learning for finance: deep portfolios. Appl. Stochast. Models Bus. Ind. **33**(1), 3–12 (2017). https://doi.org/10.1002/asmb.2209

13. Luong, M.-T., Sutskever, I., Le, Q.V., Vinyals, O., Zaremba, W.: Addressing the rare word problem in neural machine translation (2015). http://arxiv.org/abs/1410.8206. Accessed 1 Mar 2022

14. Aryal, S., Nadarajah, D., Rupasinghe, P.L., Jayawardena, C., Kasthurirathna, D.: Comparative analysis of deep learning models for multi-step prediction of financial time series. J. Comput. Sci. **16**(10), 1401–1416 (2020). https://doi.org/10.3844/jcssp.2020.1401.1416

15. Hochreiter, S., Schmidhuber, J.: Long short-term memory. Neural Comput. **9**(8), 1735–1780 (1997). https://doi.org/10.1162/neco.1997.9.8.1735

16. Brownlee, J.: Stacked long short-term memory networks (machinelearningmas-tery.com), 18 August 2017. https://machinelearningmastery.com/stacked-long-short-term-memory-net works/

17. Pascanu, R., Gulcehre, C., Cho, K., Bengio, Y.: How to construct deep recurrent neu-ral networks (2014). http://arxiv.org/abs/1312.6026. Accessed 24 Feb 2022

18. Graves, A., Jaitly, N., Mohamed, A.: Hybrid speech recognition with deep bidirectional LSTM. In: 2013 IEEE Workshop on Automatic Speech Recognition and Understanding, Olomouc, Czech Republic, pp. 273–278 (2013). https://doi.org/10.1109/ASRU.2013.670 7742

19. Koenecke, A.: Applying deep neural networks to financial time series forecasting. Institute for Computational & Mathematical Engineering, Stanford, California, USA (2020). https://web.stanford.edu/~koenecke/files/Deep_Learning_for_Time_Series_Tutorial.pdf

20. Al Ridhawi, M.: Stock market prediction through sentiment analysis of social-media and financial stock data using machine learning (2021). https://doi.org/10.20381/RUOR-27045

21. "Vietstock" Vietstock. https://vietstock.vn/

22. Kingma, D.P., Ba, J.: Adam: a method for stochastic optimization (2017). http://arxiv.org/abs/1412.6980. Accessed 02 Mar 2022

23. Terna, P.: A Deep Learning Model to Forecast Financial Time-Series (2015)

Intelligence Service Experiments

A High-Performance FPGA-Based Feature Engineering Architecture for Intrusion Detection System in SDN Networks

Tran Hoang Quoc Bao[1,2], Long Tan Le[3], Tran Ngoc Thinh[1,2(✉)] (iD), and Cong-Kha Pham[4]

[1] Ho Chi Minh City University of Technology (HCMUT), 268 Ly Thuong Kiet Street, District 10, Ho Chi Minh City, Vietnam
{thqbao.sdh20,tnthinh}@hcmut.edu.vn

[2] Vietnam National University Ho Chi Minh City (VNU-HCM), Linh Trung Ward, Thu Duc District, Ho Chi Minh City, Vietnam

[3] The University of Sydney, Sydney, Australia
lole4878@uni.sydney.edu.au

[4] University of Electro-Communications (UEC), Tokyo 182-8585, Japan
phamck@uec.ac.jp

Abstract. Software-defined networking (SDN) has been seen as a next-generation networking architecture that separates the control plane and data plane to make networks agile and flexible. However, the nature of software-based centralized control makes this emerging architecture vulnerable to cyber security issues. In this paper, we propose to build and integrate a high-performance machine learning-based Network Intrusion Detection System (NIDS) for SDN. Particularly, we leverage co-design HW/SW techniques to accelerate and improve the performance of NIDS. We design and implement an FPGA-based feature engineering processor for data dimensionality reduction based on AutoEncoder for NIDS, and deploy in a high-speed NetFPGA-SUME platform. The experiment results show that the proposed design approximately occupies 16% of LUTs, FFs, BRAMs, and 23% DSPs hardware resources. The maximum frequency of the design prototype is 233 MHz. The performance of Autoencoder on NSL-KDD feature data is presented through bandwidth and the packet processing time with the reconstruction loss from hardware encode data with original data is 0.00077.

Keywords: Hardware acceleration · Machine learning · FPGA · NIDS · SDN

1 Introduction

Software-Defined Networking (SDN) has emerged as a novel network architecture approach enabling the underlying network devices to be abstracted from an application point of view. The idea of SDN is to decouple the data plane from

© The Author(s), under exclusive license to Springer Nature Switzerland AG 2022
N.-T. Nguyen et al. (Eds.): ICIT 2022, LNDECT 148, pp. 259–268, 2022.
https://doi.org/10.1007/978-3-031-15063-0_24

the control plane to shift network control into a centralized controller, helping address a set of critical problems in conventional networks and can be widely applicable to a range of domains such as cloud computing, Internet of Things, edge computing, smart grid, cognition-based networks, etc. Although SDN offers immense benefits to the networking industry, the centralized nature of the control plane leads to many critical challenges. Above all, securing this type of network from cyber-security attacks is one of the most vital tasks attracted the attention of many researchers.

Meanwhile, machine learning (ML) has recently been seen as one of the most powerful weapons to deal with security issues. ML provides cognitive capabilities for identifying potential security threats and vulnerabilities thereby can be applied in solving the most common tasks of cyber-security including regression, prediction, and classification. With the advent of ML technologies, many security solutions have been developed to build a secure SDN environment. Notwithstanding the existing approaches have shown efficiency in protecting SDN from network anomalies, there are several weaknesses that need to be concerned. First, the majority of ML-based security solutions in SDN are software applications deployed in SDN controllers, while ML techniques in networking may require a very large number of training and testing data from many networking devices. Second, the feature learning process of the previous ML solutions is mostly based on feature selection methods, which lacked meaning in representing network traffic data.

In this paper, we propose to build a high-performance and efficient ML-based network intrusion detection system (NIDS) to enhance security for SDN. In particular, we work toward the goal which designs and implements an FPGA-based high-speed feature engineering mechanism for learning the meaning of data representation and speeding up the processing speed for machine learning classification tasks. Our main contributions are as follows

- We propose a high-performance architecture for ML-based NIDS in SDN networks using the co-design HW/SW techniques.
- We design and implement a high-speed feature engineering process based on FPGA and Autoencoder model for improving the feature extraction performance of the NIDS.
- We evaluate the proposed system on a practical testbed built upon NetFPGA-10G and NetFPGA-SUME platforms. Experimentally, we demonstrate that our approach has shown efficiency and low latency in detecting cyber-attacks under high-speed network traffic.

The remainder of the paper is organized as follows. Section 2 presents the background and related works. The proposed methodology with the detailed design and implementation is described in Sect. 3. Experimental evaluations are shown in Sect. 4, and we conclude our research works in Sect. 5.

2   Background and Related Work

2.1   Software-Defined Networking

Software-Defined Networking (SDN) [1] is an emerging network technology promising to overcome the limitations faced by traditional TCP/IP networks.

By decoupling the data plane and control plane, SDN enables the network to be programmable, intelligently, and centrally controlled, thereby providing an abstract layer managing the entire network consistently and holistically, regardless of the underlying network technology. Basically, an SDN model consists of a forwarding layer, a control layer, and an application layer. Despite bringing many benefits for network configuration and management, the centralized control layer shows to be vulnerable to anomaly activities in a network such as cyber-attacks and network penetration, therefore, it is necessary to integrate security solutions into this kind of system.

2.2 Network Intrusion Detection Systems

Network Intrusion Detection Systems (NIDSs) [2] are well-known for detecting network anomalous intrusions in traditional network systems. NIDS is capable of analyzing network traffic behavior and internal access to help the system identify unauthorized activities. Based on detection mechanisms, NIDSs are often divided into two main approaches: signature-based and anomaly-based. While signature-based detection is effective at finding sequences and patterns that may match a particular pre-determined rule, anomaly-based detection uses ML as well as other statistical methods to analyze data on a network to detect malicious behavior patterns linked to cyber attacks. With the emergence of ML, anomaly-based NIDS promises to become more accurate and efficient solutions in reducing the security risks of the current network systems from the effects of cyber-security, especially in tackling the problem of network security on SDN architecture.

2.3 AutoEncoder

AutoEncoder [3] has recently been seen as one of the dominant neural network models which efficiently learns to decode and encode data representation in an unsupervised manner. Autoencoder has shown efficiency in many applications such as data encoding, dimensionality reduction, and variability reduction. Typically, an Autoencoder consists of two parts: Encoder and Decoder. Accordingly, the Encoder tries to reduce data features through a neural network model by inputting original data $x \in \mathbb{R}^n$ to a hidden space layer $h \in \mathbb{R}^m$ such that $m < n$ using the following mapping function f:

$$h = f(W * x + b) \tag{1}$$

while W and b are the parameters of the model neural, meaning the weights and bias correspondingly.

From the encoded data, Autoencoder then tries to reconstruct the original spatial dimension in Decoder by using another mapping function f':

$$\tilde{x} = f'(W' * h + b') \tag{2}$$

while W' and b' are the weights and biases of the neural network in Decoder.

The ultimate goal of training an AutoEncoder is to minimize the following reconstruction loss:

$$\tilde{MSE} = \frac{1}{n} \sum_{i=1}^{n} (Y_i - \hat{Y}_i)^2 \tag{3}$$

while MSE is mean square error, n is the number of data points, Y_i and $\hat{Y}_i$ are observed values and predicted values respectively.

2.4   Related Work

The authors in [4] do an extensive review on the existing approaches on applying machine learning, deep learning, and artificial intelligence for NIDS. NetFPGA hardware implementation with P4 compiler in data plane using Snort Rule [5]. In [6], the controller system of SDN is expanded to include NIDS processing blocks, which are ensemble multiple layers with implementation autoencoder method - Griffin. Aiming to improve the anomaly detection system with 98% accuracy, they tested on Yisroel Mirsky's dataset. The result in [7] achieves 90% precision experiment in UNSW-NB15 dataset with a deep learning method. This contribution is using CNN and DNN for anomaly detection and combining PCA techniques for lower data dimensions. The research [8] uses GAN for generating malicious data and retraining them by machine learning technique, the data which is tested on the proposed system is CIC-IDS2017 dataset.

In addition to developing anomaly detection on SDN, a number of studies have applied a spatial reduction to decrease the system's processing time. Basant et al. [9] use PCA for reducing the feature, saving performance resource and time processing, and securing classification accuracy. The decrease of dimensionality data using AutoEncoder and combining with some technique machine learning for IDS in contribution [10]. The paper [11] proposed Feature Selection which is an implementation of the Pearson correlation technique with high accuracy. The article [12] shows that the AutoEncoder algorithm brings significantly higher efficiency as compared to PCA in terms of data spatial reduction ability. The research in Long et al. [13] proposes balance datasets that use the WGAN algorithm and Sparse Autoencoder for testing NSL-KDD and UNSW-NB15. Those datasets have many rare classes and the contribution is acceleration hardware for NIDS. Execution time in software is limited with handling multiple algorithms.

3   Methodology

3.1   Overall Architecture

In this paper, we design and implement an ML-based NIDS on the data plane of SDN architecture to detect anomalous behaviors in network traffic. Figure 1 illustrates the overall architecture of the proposed system.

The proposed system aims to solve two critical problems of the existing NIDS. First, almost previous solutions employed feature selection techniques to do the

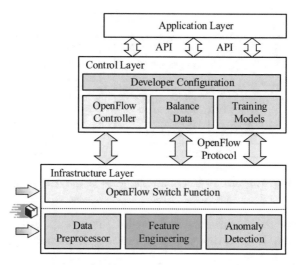

Fig. 1. The architecture of SDN with proposed design NIDS system

dimensionality reduction tasks which might not capture the meaning of network data. This results in low performance in detecting rare cyber attacks and causes overfitting. Secondly, the existing NIDS in SDN, software-based applications, are often deployed in the control plane. Due to the centralized manner of SDN controller, this deployment can lead to high latency and low-speed detection as well as exhaust SDN controller computing resources.

To deal with the aforementioned problems, we integrate into the SDN architecture some NIDS functions as shown in Fig. 1 including Balance Data, Training Models at the control plane; Data Preprocessor, Feature Engineering, and Anomaly Detection at the data plane. The Balance Data block is used to generate data to rebalance the class fields in the training set. Additional built-in blocks at the data plane accelerate the anomaly detection of network behavior performing the classification tasks of the machine learning model.

3.2 Data Plane

Figure 2 shows the structure of the function blocks at the SDN data plane. In particular, raw network packets are first preprocessed by tasks such as feature extraction and normalization, then will be reduced dimension by Feature Engineering through two phases. In the training phase, the data of reduced features will be processed in Balance Data. In the inference phase, data will be processed directly in Anomaly Detection. The operation progress in that block is to compute the neural network for data classification.

3.3 Feature Engineering

In order to accelerate the NIDS processing on the data plane, we employ Feature Engineering with the goal of reducing the data spatial dimension using AutoEn-

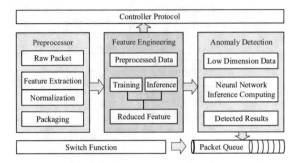

Fig. 2. Function blocks are designed at the data plane

coder, an unsupervised neural network model. Figure 3 shows the detailed architecture of Feature Engineering processor in the data encoding phase.

– **Input Parser** responsible for bitwise extraction of the packets into the input of the neural network, they have the same function as the input layer.
– **Encode Node** in Layers represents a set of nodes in the hidden layer, these nodes are used in the encoding phase. The module is implemented by calling the calculation submodules, including the following components:
  - The *Multiply* module is designed in the form of a set of Xilinx's multiplication core. They perform the multiplication between the input and the weight in a parallel manner.
  - The data after being calculated in the multiplication is immediately sent to the *Adder* module. They activate the signal with the calculation performed by the adder stratification method.
  - *Activation Relu* function returns the adder's output to a non-negative value. In synthetic hardware, the design set is realized consisting of a zero comparator and a data mux circuit.
  - *Memory* stores the parameter values of the data encoder model including the weight and bias. Since the model size is limited on hardware, this storage is implemented on memory-on-chip.

4   Experiment Results

To implement the proposed design, we use NetFPGA-SUME, a high-speed high-performance FPGA platform using Virtex 7 xc7vx690t chip produced by Xilinx. The dataset that we use to train and test the ML model is NSL-KDD [14] consisting of 41 features and reduced into 20 features. The input is represented based on the fix-point numbers suitable for FPGA computing. Input neurons are represented in 18 bit-width data frame which uses 2 bit-width for integer and 16 bit-width for fraction. Moreover, we use 32 bit-width to represent weight and bias with 4-bit integer and 28-bit fraction, respectively. The result of data in multiply block is 43 bit-width. The bit-width of input Adder is 32 bit, so they reduce bits of the output result in Multiply block.

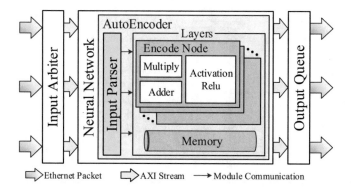

Fig. 3. Feature engineering: implementation AutoEncoder in hardware platform

4.1 Software Validation

To build the ML model for NIDS, we leverage TensorFlow, a popular ML library that supports configuring parameters for neural network models. We train multiple models with some configurations to ensure that the best model will be adapted to the hardware part for acceleration. The data reconstruction evaluation is based on the training set and the test set of NSL-KDD. The recommended configuration for applying ML to the hardware is described in Fig. 4. When considering data in the test set, the loss value fluctuates at 0.001. The results of Figs. 4b, 5a show that the configuration of the loss cross-entropy function and the softmax do not yield as good results as the relu-sigmoid and MSE configuration. Moreover, we conduct experiments with some different configurations in the number of nodes and layers. The results in Fig. 5b is close to the result in Fig. 4a, but cost more computational expense, which is not suitable for hardware implementation.

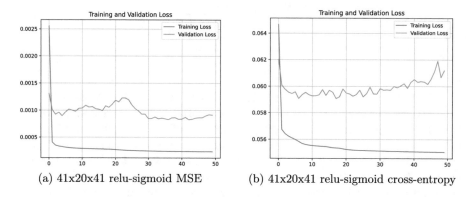

(a) 41x20x41 relu-sigmoid MSE   (b) 41x20x41 relu-sigmoid cross-entropy

Fig. 4. Model Autoencoder with different loss function

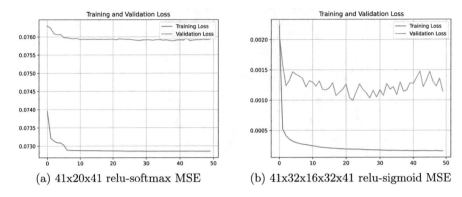

(a) 41x20x41 relu-softmax MSE   (b) 41x32x16x32x41 relu-sigmoid MSE

Fig. 5. Model Autoencoder with different configuration

The models of the algorithm have been tested and evaluated in several different configurations on software. In conclusion, we give the model $41 \times 20 \times 41$ with relu-sigmoid and use the MSE function as a measure of accuracy on hardware.

4.2 Resource Utilization

Table 1. Resource utilization for the implementation encoder core in hardware

Resources	Implementation
LUT	68055 (15.71%)
LUTRAM	1904 (1.09%)
FF	145870 (16.84%)
BRAM	214.5 (14.59%)
DSP	860 (23.89%)
BUFG	53.13%

Table 1 shows the results of the system's resource aggregation implemented on the network platform of NetFPGA hardware. Taking the largest share of the hardware implementation is the DSP, the product of the multiplications between input data and weight is Xilinx's IP-based multiply blocks, which is configured by using DSPs which take 23.89% available. The LUTs and FFs parts account for 15.71% and 16.84% number of resources because it is used to calculate the adder according to the formula 1. Parameters including weights and biases are stored on the on-chip memory so the BRAMs of 14.59%.

4.3   Performance Experiment

This section demonstrates the performance achieved by Feature Engineering blocks when deploying them on NetFPGA-SUME configurable hardware. With the implementation clock frequency requirement based on the NetFPGA-SUME project, the execution clock is 200 MHz. Through measured from the output data a total of 20 computational nodes with each node's output is 32 bits. Design according to the pipeline processing cycle of the processing block, where each clock corresponds to the output of 512 bits. As a result, the highest operating frequency of the design is 233 MHz.

4.4   Hardware Reconstruction Data

Figure 6 shows the loss value of the mean square error function when comparing reconstruction data from hardware and original data. Collected values are made from the data sample in the NSL-KDD test set. All exeperimental samples data give an approximate loss value of 0.00075–0.0079. So, it means that data encoding under hardware does not significantly change the properties of data features.

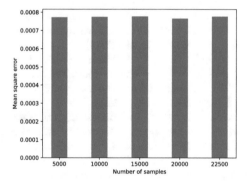

Fig. 6. MSE reconstruction feature NSL-KDD with encoded data in hardware

5   Conclusion

This research is the foundation for taking the NIDS system development to the next level. With hardware-acceleration, the system can respond promptly when an unauthorized intrusion occurs. The contribution of this work is integrating the dimension data reducer right in the dataplane. The design is implemented on the NetFPGA-SUME platform with Xilinx's Virtex7 processor. Realization resources per design take up an average of 16% LUTs, FFs, BRAMs, and 24% DSPs. The result of the performance achieved when measuring is 233 MHz. The encoded data from Encoder, the reconstruction of the data is not changed too much with the average loss value only equal to 0.00077 the original value.

Acknowledgment. This research is funded by Ho Chi Minh City University of Technology - VNU-HCM under grant number C2020-20-33. We acknowledge the support of time and facilities from Ho Chi Minh City University of Technology (HCMUT), VNU-HCM for this study. We acknowledge the donated NetFPGA-SUME board and licenses from XUP program of Xilinx company for this research.

References

1. The Open Networking Foundation: Software-Defined Networking (SDN) Definition. https://www.opennetworking.org/sdn-definition/
2. Axelsson, S.: Intrusion detection systems: a survey and taxonomy (2002)
3. Dor Bank, R.G., Koenigstein, N.: Autoencoders. arXiv (2021)
4. Hande, Y., Muddana, A.: A survey on intrusion detection system for software defined networks (SDN). In: Research Anthology on Artificial Intelligence Applications in Security, pp. 467–489. IGI Global (2021)
5. Tavares, K., Ferreto, T.C.: P4-ONIDS: a P4-based NIDS optimized for constrained programmable data planes in SDN. In: 2021 Anais do XXXIX Simpósio Brasileiro de Redes de Computadores e Sistemas Distribuídos, Brasil (2021)
6. Yang, L., Song, Y., Gao, S., Xiao, B., Hu, A.: Griffin: an ensemble of autoencoders for anomaly traffic detection in SDN. In: GLOBECOM 2020-2020 IEEE Global Communications Conference, pp. 1–6. IEEE (2020)
7. Banton, M., Shone, N., Hurst, W., Shi, Q.: Intrusion detection using extremely limited data based on SDN. In: 2020 IEEE 10th International Conference on Intelligent Systems (IS), pp. 304–309. IEEE (2020)
8. Qui, C.P.X., Quang, D.H., Duy, P.T., Pham, V.-H., et al.: Strengthening IDS against evasion attacks with GAN-based adversarial samples in SDN-enabled network. In: 2021 RIVF International Conference on Computing and Communication Technologies (RIVF), pp. 1–6. IEEE (2021)
9. Subba, B., Biswas, S., Karmakar, S.: Enhancing performance of anomaly based intrusion detection systems through dimensionality reduction using principal component analysis. In: 2016 IEEE International Conference on Advanced Networks and Telecommunications Systems (ANTS), pp. 1–6 (2016)
10. Srikanth Yadav, M., Kalpana, R.: Effective dimensionality reduction techniques for network intrusion detection system based on deep learning. In: Jacob, I.J., Kolandapalayam Shanmugam, S., Bestak, R. (eds.) Data Intelligence and Cognitive Informatics. Algorithms for Intelligent Systems, pp. 507–516. Springer, Singapore (2022). https://doi.org/10.1007/978-981-16-6460-1_39
11. Goyal, J.: Model based intrusion detection using data mining techniques with feature reduction (2022)
12. Singh, K., Kaur, L., Maini, R.: Comparison of principle component analysis and stacked autoencoder on NSL-KDD dataset. In: Singh, V., Asari, V.K., Kumar, S., Patel, R.B. (eds.) Computational Methods and Data Engineering. AISC, vol. 1227, pp. 223–241. Springer, Singapore (2021). https://doi.org/10.1007/978-981-15-6876-3_17
13. Le, L.T., Thinh, T.N.: On the improvement of machine learning based intrusion detection system for SDN networks. In: 2021 8th NAFOSTED Conference on Information and Computer Science (NICS), pp. 464–469 (2021)
14. Tavallaee, M., Bagheri, E., Lu, W., Ghorbani, A.A.: A detailed analysis of the KDD CUP 99 data set. In: 2009 IEEE Symposium on Computational Intelligence for Security and Defense Applications, pp. 1–6 (2009)

An Approach to Monitoring Solar Farms in Vietnam Using GEE and Satellite Imagery

Dung Nguyen$^{(\boxtimes)}$, Bao Ngoc Dinh, and Hong Anh Le

Faculty of Information Technology, Hanoi University of Mining and Geology,
18 Pho Vien, Bac Tu Liem, Hanoi, Vietnam
{nguyenthimaidung,dinhbaongoc,lehonganh}@humg.edu.vn

Abstract. Solar energy has been developed rapidly in Vietnam for three recent years. It becomes a promising energy sources when Vietnamese government shows the important factors of clean energy and commit to reduce thermal power stations as they cause side effects for environment. As the result, there is a need to manage solar stations that spread out the whole country. Satellite images are the data sources to observe the earth surface that can be used in many monitoring applications. For this purpose, the paper proposes a machine learning approach to automatically detect and calculate the solar farm area using Google Earth Engine which is a cloud-based platform for processing large scale spatial data. The method has been employed to solar stations in Dak Lak province and showed the potential solution for solar electricity management in Vietnam.

Keywords: Google Earth Engine · Solar farms · Satellite imagery · Machine learning

1 Introduction

Our planet is warmer year by year and climate change affects every country on the earth. Renewable energy shows that it is one of key factors to reduce the damage and harm of climate change side effects. Solar power is one of the fastest technology adopted not only in Vietnam but also worldwide to produce clean en. Solar power has been developed in Vietnam recently. Solar energy contributed nearly no part in energy sector in 2017 but became the largest PV installed capacity in South East Asian in 2020. Its strategic plan to produce up to 4000 Mega Watt in 2025 and 12000 MW in 2030. PV farms installed spread out from center to southeast of Vietnam, hence management tasks such as monitoring the development of the solar farm sites faces to many challenges. Electric of Vietnam is the cooperation that is responsible for managing and transferring to national electricity distribution network.

Satellite imagery contain essential information that can be used in many fields such as land use and land cover, natural hazard detection and monitoring, forest management, etc. These data are provided with various kind of satellite systems.

PlanetScope satellites lunched the first time in 2018. With approximately 130 satellites, Planet Scope is able to offer image of entire the earth surface every day. Maxar constellation also provides various kind of imagery such as GeoEye-1, QuickBird, Ikonos, WorldiView-1, WorldView-2, WorldView-3 and WorldView-4. Besides, we can also access to many other free satellite imagery sources such as Sentinel, Landsat.

From management perspective, satellite images are one of essential data sources can be used for monitoring [11–14, 16]. They provide data for electric providers regularly. For this reason, analyzing PV farms and the related information using remote sensing data received many attentions of researchers. One of approaches to detect solar farms is land user and land cover classification. Land cover change plays an important role in environment change. Land use indicates the physical land type such as vegetation, water, etc.that will be involved in natural management. GEE is a powerful cloud-based platform that provide to users computation libraries and various data-set catalog to develop and analyze the satellite images. This paper introduces an approach to monitoring solar stations as the land use classification by leveraging GEE and satellite imagery. As the best of our knowledge, even though the solar panel detection has been introduce by many work, this is the first report of solar monitoring study in Vietnam. The paper is organized as follows. Section 2 presents the study area and datasets that will be used in the research. In the Sect. 3, Google Earth Engine platform and algorithms used for the study is reported. Section 4 summarizes the related work. The implementation and results when applying the proposed method with solar stations in Dak Lak are presented in Sect. 5. Section 6 concludes the paper.

2   Study Area and Datasets

Dak Lak province is located in the Central Highlands of Vietnam with coordinates from $107°28'57''$ to $108°59'37''$ east longitude and from $12°9'45''$ to $13°25'06''$ north latitude. It has one city, one town and 13 districts occupying an area of around 13,000 square kilometers (Fig. 1). According the national power plan, Dak Lak has well natural conditions for solar energy development. It aims at 5,250 MW solar installation capacity in 2030 mainly located at M'Gar, Ea Sup, Buon Don and Ea HLeo districts.

In this study, we will do the experiment with Sentinel-2 level 2A and Maxar WorldView 3 with a resolution of 10 m and 0.5 m, respectively. The Sentinel-2 observation covers all continent surface with 13 bands spectrum [1] every 3–5 days. Sentinel-2 data, lunched in 2015 with level 2A and 2017 with level 2B, are free to access. Sentinel-2 level 2A tiles are $100 \times 100\,\mathrm{km}^2$ ortho-images in UTM/WGS84 projection. Maxar WorldView-3 [15], lunched in 2014, is a high resolution satellite providing imagery at 31 cm resolution. Table 1 demonstrates the specification of Sentinel-2 level-2A imagery.

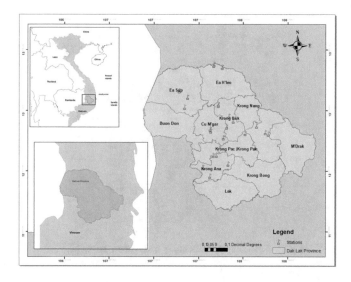

Fig. 1. The study area of Dak Lak province and solar stations

Table 1. Training classes

Product name	Sentinel-2 Level-2A
Number of bands	15
Cell size - X	10 m, 20, 60 m
Cell size - Y	10 m, 20, 60 m
Temporal resolution	Every 3–5 days

3   Google Earth Engine and Algorithms Used

3.1   Google Earth Engine

GEE is a cloud-based platform that provides a large public datasets, compute infrastructure and spatial APIs. The data catalog contains imagery and a lot of scientific datasets relating to climate, weather as well as the geophysical data such as Earth's terrain, land cover, and cropland. Figure 2 depicts the simplified architecture of GEE. Earth engine's imagery data catalog include Landsat and Sentinel satellite images. This study uses high-resolution optical images Sentinel-2A for monitoring. The computation layer not only provides support tools processing remote data but also leverages more than 20 machine learning classification algorithms. In this study, we will use two of them namely Random Forest and Classification and regression trees (CART).

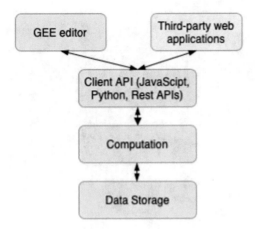

Fig. 2. GEE architecture (adapted from [2])

3.2 CART and Random Forest

Classification and regression trees (CART) were introduced in 1984 by Breiman et al. that can be applied in both classification and regression problems. These are techniques that construct the model by recursively partitioning the data and fitting a simple model with a partition. GEE provides the API to create an instance of CART classifier with a constructor as follows.

$$ee.Classifier.smileCart(maxNodes)$$

where, *maxNodes* is the maximum number of leaf nodes in each tree. The default value is null indicating that is no limit.

Random Forest is a simple and popular supervised learning technique that is based on the ensemble learning. It contains a number of tree decisions tree to predict the final output. GEE allows to initiate a Random Forest classifier as follow

$$ee.Classifier.smileRandomForest(numberOfTrees)$$

where, *numOfTrees* is the number of decision trees to create.

4 Related Work

Ioannou and Myronidis [3] proposed an approach to detect photo voltaic farms using CNN and Apple Maps image data service. Their approach consists of two steps: (i) creating high-resolution and low-resolution dataset my annotating Apple Map images and matching data provided Greek Regulatory Authority for Engergy; (ii) automatic detection of farm locations.

Yu *et al.* [10] presented a deep learning framework DeepSolar to extract locations and size of solar panels. The framework used images of Google Static Maps as the input develop CNN for automatic classification and segmentation.

They also offer a public database of US solar installation for researchers, solar developers to discover and build their own models. Based on DeepSolar, Mayer *et al.* [4] optimized the training model to efficiently handle with lower resolution imagery. They applied their proposed model to a large scale system at North-Rhine Westphalia, Germany.

Hou *et al.* [5] utilized UNet to develop SolarNet to perform segmentation on high resolution satellite imagery to map 430 solar farms in China that provides useful insights for local governance.

Matthias Zech and Joseph Ranalli [6] proposed a FCNN which is based on U-net architecture to spot solar farms locations with aerial images. They did the experiment with Tenser flow library and data collected from Google Maps in Oldenburg, Germany with 13,145 tiles.

Malof *et al.* [7] presented an algorithm to automatically detect of solar photovoltaic arrays using Randon Forest Classifier and aerial images. They experimented the algorithm with dataset of high-resolution aerial imagery over Fresno, California, US. They also proposed to use CNN to detect solar panels with 0.3m resolution aerial images at large scale [8]. They applied for 108 square km^2 in Boston and California with the correctness at 0.812 and 0.855 respectively.

5   Implementation and Result

5.1   General Flow

As mentioned above, there is a need to monitor the changes of solar stations specially the high volume generated one. Figure 3 shows the proposed solution. Instead of using high resolution imagery such as UAV, we proposed to use satellite imagery free sources such as Sentinel-2 that cover the whole country with high frequency in Vietnam. To monitor the specific areas that requires higher accuracy of the local policy makers, we apply high resolution imagery such as Maxar WolrdView. After detecting and calculating the solar farms at the specific location, the changes will be notified to the users.

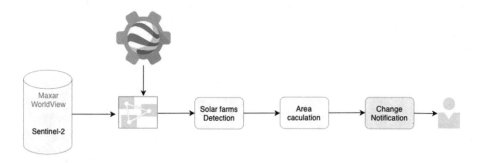

Fig. 3. General flow of monitoring solar farms

5.2   Solar Farms Extraction

In order to implement the analytic tasks, we follow the general approach for machine learning presented in [9]. The detail steps are described in Fig. 4:

1. Ingesting satellite images from many sources and providers. In this implementation, we use Maxar and Sentinel-2 imagery. Using GEE, we collect data directly on these sources. For the forthcoming machine learning model, the satellite image pre-processing must be executed. All the computational operator are implemented on cloud-based in GEE. In order to detect solar area, we prepare 4 classes namely solar, vegetation, bare, and urban illustrated in Table 2.

Table 2. Training data

Classes	No of samples
Solar	154
Urban	186
Bare	135
Vegetation	150

2. Initiating the appropriated classifiers provided by GEE machine learning library. In this implementation, we use RandomForest and CART by invoking constructors *ee.Classifier.smileRandomForest* and *ee.Classifier.smileCart* respectively. After creating an instance of classifier, we can customize the parameters for the classifier and train it. Training data were drawn by using geometry tools of GEE on top of Sentinel-2 true color composite image.
3. Detecting solar farms with input data using trained classifier.
4.1 Calculating the detected areas.
4.2 Assessing the model accuracy with input data.

5.3   Results

Figure 5 depicts the result on 04 datasets with Sentinel-2 in 04 solar stations. The solar sites are labeled as the black color, green color indicates the vegetation layer, yellow color is for bare layer, and pink color represents urban area. With higher resolution at 0.5 m, the classification result for Maxar Worldview is more accurate than Sentinel-2. Figure 6 shows the classification result predicting on the Maxar WoldView dataset. In case that the local managers want to know the more precise area of solar panels installations, satellite images with high resolution of 0.5 m (or above) is more suitable.

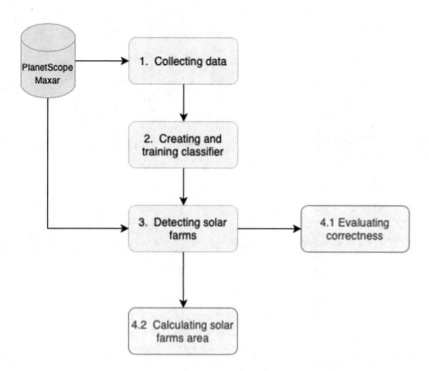

Fig. 4. Flow of solar farms analysis using GEE

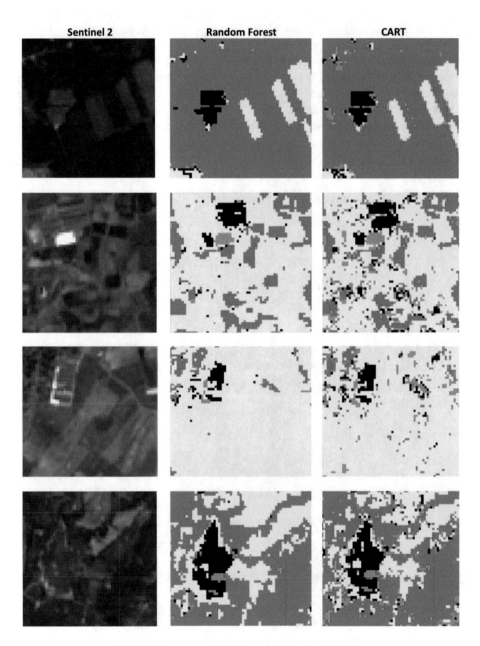

Fig. 5. Solar area detection with Sentinel-2 data

Fig. 6. Solar area detection with Maxar WorldView data

6 Conclusions

This paper presented a machine learning approach to monitor the solar farms in Vietnam using GEE and satellite imagery. We have constructed two classifiers with Random Forest and CART algorithms. Data is collected Sentinel-2 and Maxar WorldView images in 2021. The experiment in Dak Lak results shows that Random Forest algorithm achieves better detection with both of satellite images. With the processing at large-scale and big data-sets of GEE, the proposed approach shows it is the promising solution for monitoring solar stations development and operations in Vietnam. Currently, one limitation is that GEE does not provide built-in deep learning library. Instead, users have to use Google AI cloud services with payment per request. In the future, we will work out to integrate open source deep learning platform such as Tensor flow to work in GEE platform for predicting solar panels with CNN models.

Acknowledgment. This work is supported by the project no. B2022-MDA-01 granted by Ministry of Education and Training (MOET).

References

1. Sentinel-2: Sentinel Online (2022). https://sentinel.esa.int/web/sentinel/missions/sentinel-2. Accessed 19 Feb 2022
2. Gorelick, N., Hancher, M., Dixon, M., Ilyushchenko, S., Thau, D., Moore, R.: Google earth engine: planetary-scale geospatial analysis for everyone. Remote Sens. Environ. **202**, 18–27 (2017). Big Remotely Sensed Data: tools, applications and experiences. https://www.sciencedirect.com/science/article/pii/S0034425717302900
3. Ioannou, K., Myronidis, D.: Automatic detection of photovoltaic farms using satellite imagery and convolutional neural networks. Sustainability **13**(9) (2021). https://www.mdpi.com/2071-1050/13/9/5323
4. Mayer, K., Wang, Z., Arlt, M.-L., Neumann, D., Rajagopal, R.: Deepsolar for Germany: a deep learning framework for PV system mapping from aerial imagery. In: 2020 International Conference on Smart Energy Systems and Technologies (SEST), pp. 1–6 (2020)

5. Hou, X., Wang, B., Hu, W., Yin, L., Wu, H.: SolarNet: a deep learning framework to map solar power plants in china from satellite imagery (2019)
6. Zech, M., Ranalli, J.: Predicting PV areas in aerial images with deep learning. In: 2020 47th IEEE Photovoltaic Specialists Conference (PVSC), pp. 0767–0774 (2020)
7. Malof, J.M., Bradbury, K., Collins, L.M., Newell, R.G.: Automatic detection of solar photovoltaic arrays in high resolution aerial imagery. Appl. Energy **183**, 229–240 (2016). https://www.sciencedirect.com/science/article/pii/S0306261916313009
8. Malof, J.M., Hou, R., Collins, L.M., Bradbury, K., Newell, R.: Automatic solar photovoltaic panel detection in satellite imagery. In: 2015 International Conference on Renewable Energy Research and Applications (ICRERA), pp. 1428–1431 (2015)
9. Google: Google earth engine (2022). https://developers.google.com/earth-engine. Accessed 19 Feb 2022
10. Yu, J., Wang, Z., Majumdar, A., Rajagopal, R.: DeepSolar: a machine learning framework to efficiently construct a solar deployment database in the United States. Joule **2**(12), 2605–2617 (2018). https://www.sciencedirect.com/science/article/pii/S2542435118305701
11. Krapf, S., Kemmerzell, N., Khawaja Haseeb Uddin, S., Hack Vázquez, M., Netzler, F., Lienkamp, M.: Towards scalable economic photovoltaic potential analysis using aerial images and deep learning. Energies **14**(13) (2021). https://www.mdpi.com/1996-1073/14/13/3800
12. Phan, T.N., Kuch, V., Lehnert, L.W.: Land cover classification using google earth engine and random forest classifier-the role of image composition. Remote Sens. **12**(15) (2020). https://www.mdpi.com/2072-4292/12/15/2411
13. Basu, S., Ganguly, S., Mukhopadhyay, S., DiBiano, R., Karki, M., Nemani, R.: DeepSat: a learning framework for satellite imagery. In: SIGSPATIAL 2015. Association for Computing Machinery, New York (2015). https://doi.org/10.1145/2820783.2820816
14. Tamiminia, H., Salehi, B., Mahdianpari, M., Quackenbush, L., Adeli, S., Brisco, B.: Google earth engine for geo-big data applications: a meta-analysis and systematic review. ISPRS J. Photogram. Remote Sens. **164**, 152–170 (2020). https://www.sciencedirect.com/science/article/pii/S0924271620300927
15. Maxar Technologies: WorldView-3 Data sheet (2022). https://resources.maxar.com/data-sheets/worldview-3. Accessed 19 Feb 2022
16. Li, P., et al.: Understanding rooftop PV panel semantic segmentation of satellite and aerial images for better using machine learning. Adv. Appl. Energy **4**, 100057 (2021). https://www.sciencedirect.com/science/article/pii/S2666792421000494

An FPGA-Based Solution
for Convolution Operation Acceleration

Trung Pham-Dinh[1,2], Bao Bach-Gia[1,2], Lam Luu-Trinh[1,2],
Minh Nguyen-Dinh[1,2], Hai Pham-Duc[1,2], Khoa Bui-Anh[1,2],
Xuan-Quang Nguyen[1,2], and Cuong Pham-Quoc[1,2(✉)] (iD)

[1] Ho Chi Minh City University of Technology (HCMUT), Ho Chi Minh City,
Vietnam
{trung.pham.ktmt,bao.bachbbace12,lam.luu1602,minh.nguyen207bk,
hai.phamcse,khoa.bui140,nxquang,cuongpham}@hcmut.edu.vn
[2] Vietnam National University - Ho Chi Minh City (VNU-HCM),
Ho Chi Minh City, Vietnam

Abstract. Hardware-based acceleration is an extensive attempt to
facilitate many computationally-intensive mathematics operations. This
paper proposes an FPGA-based architecture to accelerate the convolution operation - a complex and expensive computing step that appears
in many Convolutional Neural Network models. We target the design to
the standard convolution operation, intending to launch the product as
an edge-AI solution. The project's purpose is to produce an FPGA IP
core that can process a convolutional layer at a time. System developers
can deploy the IP core with various FPGA families by using Verilog HDL
as the primary design language for the architecture. The experimental
results show that our single computing core synthesized on a simple edge
computing FPGA board can offer 0.224 GOPS. When the board is fully
utilized, 4.48 GOPS can be achieved.

Keywords: Convolution operation · FPGA · Hardware acceleration ·
IP core · Edge computing

1 Introduction

The prominence of Convolutional Neural Networks (CNN) has sparked computer
vision and AI technology. As the demand for these applications grows, scientists
have proposed many CNN models and fitted them to many aspects of the digital
world. Soon enough, embedded devices find themselves the urge to incorporate
the AI capability into their functionality to ease the exhausted computation on
the server. However, micro-controllers have limitations regarding the resources
needed to perform AI computation. Therefore, FPGAs (Field Programmable
Gate Arrays) dominate the battlefield, along with ASICs (Application Specific
Integrated Circuits) [2]. The FPGA technology has been rapidly growing, with
each new generation integrating more Intellectual Properties (IP) to expand the
chip's capability. These IP cores are the engineers' resources to integrate any
CNN functionality and inferences to the embedded devices.

This paper proposes our initial attempt to bring AI functionality to edge devices. Our work focuses on convolution operation acceleration using FPGA technology. Convolution is long known for its intensive and time-consuming computing process, and it occupies around 90% in nearly every CNN model. Therefore, the solution to this acceleration problem will undoubtedly lay the ground for further AI-at-the-edge developments.

The rest of the paper is organized as follows. The convolution operation and related work are presented in Sect. 2. We describe the overview of the proposed architecture in Sect. 3. Section 4 and 5 will continue to discuss the implementation of the IP core and its behavior under simulation environment, along with its synthesis report on different Xilinx's FPGA families. Finally, Sect. 6 concludes the paper.

2   Background and Related Work

In this section, we introduce an overview of the convolution operation that we are targeting as our hardware-based IP core. We then present related work for the FPGA-based CNN computing cores.

2.1   Convolution Operation

Convolution is an operation performing on two arguments referred as *input* (images) and *filters*. The output of the operation is known as the *feature map*. Consider our input image as a two-dimension matrix. The convolution operation is performed as shown in Eq. 1. We apply the element-wise multiplication of the Kernel matrix and a partial matrix of the input Image and take the sum of all the products. We continue to apply the same computation by "sliding" and "weighted-summing" the Kernel on the Image until we yield all the outputs of the Feature Map.

$$F(i,j) = (I \circledast K)(i,j) = \sum_m \sum_n I(i+m, j+n) \times K(m,n) \qquad (1)$$

In practical usage, most images come in RGB format, encoded to three matrices corresponding to each dimension of the picture - Red, Green, and Blue. In this case, the Kernel expands to a 3D tensor. Equation 2 describes the resulting Feature Map retrieved by performing the Convolution Operation subjecting to this change.

$$F(i,j) = (I \circledast K)(i,j,d) = \sum_d \sum_m \sum_n I(i+m, j+n, d) \times K(m,n,d) \qquad (2)$$

2.2   Related Work

In recent years, many studies in the literature have been proposed to accelerate CNN in FPGA. Some of them try to improve parallelism by exploiting the huge amount of hardware resources in FPGA such as research in [4,7–9]. Another approach is proposals that reduce the complexity of CNN by using lightweight models like BNN such as [3,5,10–14]. However, all these systems focus on high-performance FPGA computing platforms. Meanwhile, in this work, we target edge computing platforms with less amount of hardware resources and low energy consumption.

3   IP Core Architecture

Figure 1 conveys the primary functionality of our IP core. Because the core takes on one layer at a time, it always expects a set of feature maps containing C channels and the associated sets of K kernels (each kernel includes C channels) as inputs. After the accelerated computation, the IP core produces another set of feature maps, including K channels. These operations sum up the overall functionality of the architecture.

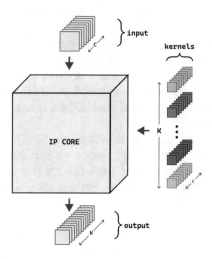

Fig. 1. IP core functionality

The proposed solution in this paper exploits the profound advantages of SoC architectural design. The integrated on-chip Processing System (PS) will efficiently transfer the input data down the stream. The IP initially loads the two sets of input data into its internal groups of BRAMs (Sect. 4.1 will further elaborate on the organizing of data among the RAM pool). Since the amount of data is typically large, we use a direct memory access controller, or DMA, to handle the transfer; hence cutting down the workload on the PS.

The IP has two groups of BRAMs to store the input image and the input kernel separately. The IP core's internal logic enables the Computing Core module when the DMA has finished transferring the input data. In the most abstract

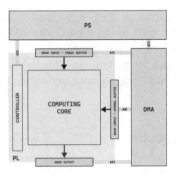

Fig. 2. Dataflow

sense, the Computing Core is a convolutional computation unit of the architecture. When fed with an image channel and its associated kernels, it spits out the resulting feature map after applying the convolutional operation. The detailed implementation to achieve this functionality is far more complicated and is discussed in Sect. 4.2.

After computing the feature map, the Computing Core will store it in another set of BRAMs. The DMA is continually responsible for transferring the calculated results back to the PS from this set of BRAM. All the communications between the DMA and the BRAMs mentioned so far are through AXI4 interfaces - a communication interface protocol that Xilinx has adopted as the standard for their IPs communication.

Figure 2 illustrates the dataflow discussed so far. The final sub-module to mention here is the Controller unit. To perform a correct convolution operation, it will receive the information needed from the PS (for example, the dimension of the input image and the input kernel). Section 4.2 will discuss the Computing core in depth.

4   Implementation

This section presents in detail the architecture of the IP core. We first start describing our memory management. Then, the introduction of the IP's central computing core, including sub-components, will be followed.

4.1   Memory Management

Block RAMs (BRAMs) in FPGA are common resources for storing data as they provide random access to memory elements. Xilinx provides IP Block Memory Generator (BMG) to help designers access BRAM efficiently under various configurations. However, BMG has only two ports for concurrently reading and writing; therefore, the architecture will distribute and organize data into multiple BMGs to exploit this concurrency behavior.

Input BRAMs. The architecture has a set of four BMGs for each input image (Fig. 3). Each BMG in the set will store one fourth of the image channels. Because the depth (the number of channels) per image varies, the size of each BMG must be large enough to hold the largest possible image. This also means that for small images, there will be redundant, or meaningless, slots in the BMGs. Here we use 4 BMGs but not some other arbitrary number because the majority of CNN inferences like the AlexNet [6] or the MobileNet [1] architecture has one common characteristic. For these inferences, all the produced feature maps are divisible by 4, except for the first input image. By separating the channels into four sets, we can process them in parallel, hence speeding up the computation of the operation.

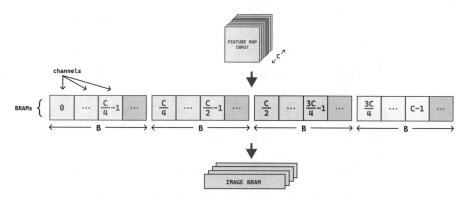

Fig. 3. The set of Image BRAMs. Here, B is the largest possible feature map size divided by 4, and C is the total number of channels of this layer's input image (or feature map).

The same divisible-by-4 idea applies to the kernel inputs. However, one standard convolutional step typically requires more than one multiple-channel kernel. Therefore, we organize the kernel data according to the number of kernels. Furthermore, because the number of channels is a multiple of 4, the number of kernels associated with each convolution layer must also be divisible by 4. Therefore, for each BMG that contains one-fourth of the channels of every kernel, we further divide it into four smaller BMGs, each holding one-fourth of the whole kernels. This distribution scheme guarantees that each BMG portion of the image has an associated set of four BMGs, breaking the computing process down to four separate tasks independent of each other.

Output BRAMs. Since an output feature map of one convolutional layer is the required input for the next one in a typical CNN model, the distribution of the computed feature map among the output BRAMs is identical to that of the input image BRAMs. It means that each one-fourth of the output channels will be stored in a separate BGM. These output BGMs will then be used to either read the result out (through the DMA) or compute the next convolution layer in the network.

4.2   Computing Core

Computing Core is the essential module of the proposed IP core architecture. As stated before, this module is responsible for producing the weighted sum (or partial sum) when feeding the input image and weight data from the BRAMs. Subsequent sections discuss the design and behavior of this module from the multi-channel view to the multi-kernel execution.

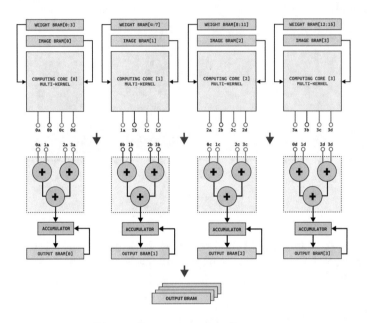

Fig. 4. Computing cores diagram

Multi-channel Architecture. The proposed architecture comprises four computing cores; each performs the convolution operation on one-fourth of the depth of the input feature maps. With the organization of the input data among the BRAMs discussed previously, the four computing cores can execute the operation totally in parallel, called multi-channel architecture. Each computing core has one associated image BRAM and four corresponding weight BRAMs - complying exactly with the number of BRAMs in the architecture. Figure 4 illustrates the proposed design. In this architecture, computed PSUM values of each core get accumulated continually into the output BRAMs until the processing depth of images is finished. The computing cores will continue to repeat the process but with another set of kernels. It keeps processing until all the kernels have been filtered.

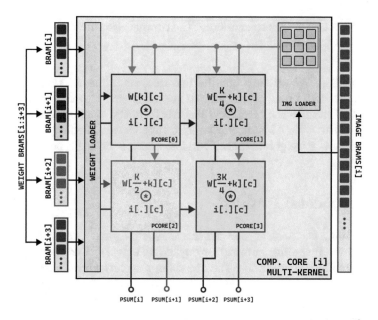

Fig. 5. A single computing core. In the diagram, $W[k][c]$ is the channel c^{th} of kernel k^{th}, $i[.][c]$ is the partial input image at the channel c^{th}, $\circledast$ is the convolution operator, $comp.core[i]$ is the ith computing core. In general, k iterates from 0 to $K/4 - 1$, where K is the total number of input kernels; and c iterates from $i*C/4$ to $(i+1)*C/4 - 1$, for each iteration of k, where C is total number of input kernels.

Multi-kernel Computing Core. Each computing core in the architecture produces four PSUM (Partial Sums) values due to four kernels received, as shown in Fig. 5. The produced PSUM values belong to the one-fourth output feature map channel, speeding up the computing process four times more. A single computing core contains four more sub-module called PCOREs. The transfer of data from the BRAMs to the PCOREs is through the intermediate loaders. Each PCORE computes a PSUM value according to the weight input it receives from the Weight Loader, while the Image Loader holds a set of nine pieces of input values for all the four PCOREs. This computing model is weight stationery since the weight data persists in the weight loader.

In contrast, the image loader continually fetches different input images after each computed set of PSUMs. The internal logic of a PCORE is simple. Each contains a set of MAC (Multiply and Accumulate) units and adder modules to perform a weighted-sum operation.

Pipeline. Essentially, the computing process of the computing cores can be divided into two stages. The first stage transfers sufficient data from the BRAMs to the loaders while the second stage performs the convolution operation and accumulates the computed PSUMs into the output BRAMs. The architecture pipelines these two stages to accelerate the computation further, effectively cutting down the wasted cycles.

Bias Handling. We utilize a handful feature from Xilinx's BMG IP to handle the bias input. The input bias is first to get initialized into the output BRAMs through the PS. Since the computed PSUMs from the computing core get accumulated into these BRAMs, it effectively causes the same effect as if the biases were added to the weighted sum. Therefore there is no logic needed to handle the bias.

5   Experimental Results

In this section, we present our synthesis results with three different Xilinx FPGA devices, including xc7z020clg400, xc7z020clg484, and xzcu3eg-sbva484-1. We then analyze simulation results to claim the throughput of our computing core.

5.1   Synthesis Results

Table 1 show the resources used on different FPGAs including the number of Look-up Table (LUTs), Flip-flop (FF), and percentage of these resources. Maximum working frequencies for operation are calculated based on "Data Path Delay" values of the synthesis. According to the table, our computing core consumes less than 5% hardware resources of the Pynq Z2 board, one of the most suitable FPGA boards for edge computing. In other words, we can deploy up to 20 cores concurrently to further improve computing performance.

Table 1. Synthesis result on different FPGAs

FPGA	#LUTs	#FF	Max frequency
xc7z020clg400-1	5027 (9.45%)	4959 (4.66%)	112 MHz
xc7z020clg484-1	5243 (9.86%)	5054 (4.75%)	93 MHz
xzcu3eg-sbva484-1-i	11917 (16.89%)	14522 (10.29%)	161 MHz

5.2   Simulation Results

Figure 6 describes a part of the simulation waveform of one Computing core on Vivado. Each feature signal (feature 0, feature1, feature2) concatenates three 8-bit feature entries to create a 3×3 tile of the feature. Each weight signal (weight0, weight1, weight2, weight3) links nine 8-bit weight entries, so each weight signal represents a channel of input weight. Because each Computing core can compute convolution on four different kernels simultaneously, four weight signals correspond to four different kernels. As a result, there are four partial sum signals (psum0, psum1, psum2, psum3), and each signal results from a single weight channel overlapping over a single 3×3 tile of the feature. The computing core needs eight clock cycles to compute four psum values and accumulate them to BRAM outputs. The system has four Computing cores to compute sixteen psum values for each eight clock cycles.

The system consists of four Computing cores; a single core can compute four psum values for each eight clock cycles. With the input feature of $[224 \times 224 \times 8]$ and input weight of $[8 \times 3 \times 3 \times 8]$, the system needs to compute 3,154,176 psum values. Based on Synthesis Result, the IP core can operate on 112 MHz clock frequency with the Pynq Z2 FPGA board. Therefore, we can deduce the theory time needed for computing this sample, which is 0.01408 s. In other words, the throughput of a single core is 0.224 GOPS. With the Pynq Z2 FPGA board for edge computing devices mentioned above, when 20 cores are deployed, our computing system can offer up to 4.48 GOPS.

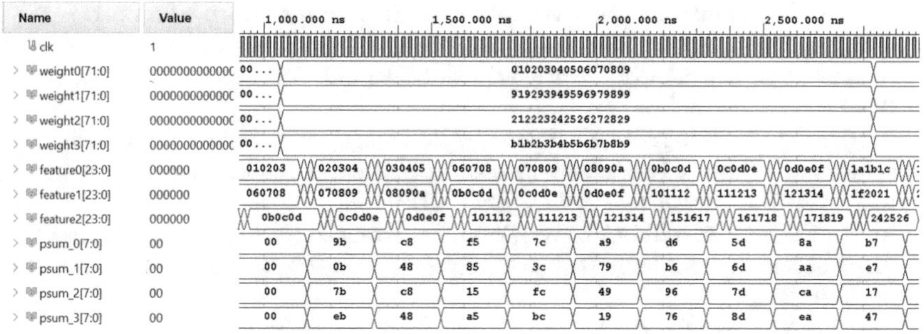

Fig. 6. One part of the waveform from the simulation of a single computing core

6   Conclusion

With the rapid growth in AI, the demand of real time processing is significantly high. Convolution operation is the core of convolution neural network and take up most of the computation time. Therefore, this paper proposes a FPGA-based architecture for accelerating convolution operation. The architecture is implemented by Verilog and be able to operate on different FPGA devices. This architecture can compute on multiple channels of feature and multiple kernels of filter simultaneously, which increases the parallelism on the computation. In addition, load stage and computation stage are pipelined, which significantly reduces the computation time and reachs 0.224 GOPS on a single core. This core is synthesized on different FPGAs and reachs maximum frequency of 161 MHz with reasonable used resources. Future optimization on this architecture is try to make it more flexible and able to operate on a higher frequency to prove the great prospect of FPGA in AI processing.

Acknowledgement. This research is funded by Vietnam National University - Ho Chi Minh City (VNU-HCM) under grant number B2021-20-02. We acknowledge the support of time and facilities from Ho Chi Minh City University of Technology (HCMUT), VNU-HCM for this study.

References

1. Howard, A.G., et al.: MobileNets: efficient convolutional neural networks for mobile vision applications (2017). https://doi.org/10.48550/ARXIV.1704.04861, https://arxiv.org/abs/1704.04861
2. DeMicheli, G., Sami, M.: Hardware/Software Co-design. Nato Science Series E. Springer, Netherlands (1996). https://www.springer.com/gp/book/9780792338833
3. Ghasemzadeh, M., Samragh, M., Koushanfar, F.: ReBNet: residual binarized neural network. In: 2018 IEEE 26th FCCM, pp. 57–64, May 2018
4. Guo, K., Han, S., Yao, S., Wang, Y., Xie, Y., Yang, H.: Software-hardware codesign for efficient neural network acceleration. IEEE Micro **37**(2), 18–25 (2017)
5. Jiao, L., Luo, C., Cao, W., Zhou, X., Wang, L.: Accelerating low bit-width convolutional neural networks with embedded FPGA. In: 2017 27th FPL, pp. 1–4 (2017)
6. Krizhevsky, A., Sutskever, I., Hinton, G.E.: ImageNet classification with deep convolutional neural networks. In: Advances in Neural Information Processing Systems, vol. 25. Curran Associates, Inc. (2012)
7. Li, H., Fan, X., Jiao, L., Cao, W., Zhou, X., Wang, L.: A high performance FPGA-based accelerator for large-scale convolutional neural networks. In: 2016 26th FPL, pp. 1–9 (2016)
8. Lin, X., Yin, S., Tu, F., Liu, L., Li, X., Wei, S.: LCP: a layer clusters paralleling mapping method for accelerating inception and residual networks on FPGA. In: 2018 55th ACM/ESDA/IEEE Design Automation Conference (DAC), pp. 1–6 (2018)
9. Liu, Z., Dou, Y., Jiang, J., Xu, J.: Automatic code generation of convolutional neural networks in FPGA implementation. In: 2016 FPT, pp. 61–68 (2016)
10. Moss, D.J.M., et al.: High performance binary neural networks on the Xeon+FPGA^{TM} platform. In: 2017 27th FPL, pp. 1–4 (2017)
11. Nakahara, H., Fujii, T., Sato, S.: A fully connected layer elimination for a binarizec convolutional neural network on an FPGA. In: 2017 27th FPL, pp. 1–4 (2017)
12. Nurvitadhi, E., Sheffield, D., Sim, J., Mishra, A., Venkatesh, G., Marr, D.: Accelerating binarized neural networks: comparison of FPGA, CPU, GPU, and ASIC. In: 2016 FPT, pp. 77–84 (2016)
13. Prost-Boucle, A., Bourge, A., Pétrot, F., Alemdar, H., Caldwell, N., Leroy, V.: Scalable high-performance architecture for convolutional ternary neural networks on FPGA. In: 2017 27th FPL, pp. 1–7 (2017)
14. Yang, Y., et al.: Synetgy: algorithm-hardware co-design for ConvNet accelerators on embedded FPGAs. In: 2019 ACM/SIGDA FPGA, FPGA 2019, pp. 23–32 (2019)

An Investigation on Vietnamese Credit Scoring Based on Big Data Platform and Ensemble Learning

Quang-Linh Tran[1,2], Binh Van Duong[1,2], Gia-Huy Lam[1,2], Dat Vuong[3], and Trong-Hop Do[1,2($\boxtimes$)]

[1] University of Information Technology, Ho Chi Minh City, Vietnam
hopdt@uit.edu.vn
[2] Vietnam National University, Ho Chi Minh City, Vietnam
[3] Vietnam - Korea University of Information and Communication Technology, Da Nang, Vietnam

Abstract. The credit score is a vital indicator that can affect many aspects of people's lives. However, evaluating credit scores is done manually, so it costs a large amount of money and time. This paper learns from disadvantages of previous research and brings some insights and empirical experiments so as to the advantages of distributed solutions for the problem of credit score in the future. The research compares some feature engineering techniques using a big data platform and ensemble learning methods to find the best solution for predicting the credit score. Since data related to customers' financial activities grows enormously, a big data platform is necessary to handle this amount of data. In this paper, Spark which is a distributed, data processing framework, is used to save and process data. Some experiments are carried out to compare the effectiveness of feature engineering in this problem. Moreover, a comparative study about the performance of ensemble learning models is also given in this paper. A real-world Vietnamese credit scoring data set is used to develop and evaluate models. Four metrics are used to evaluate the performance of credit scoring models, namely F1-score, recall, precision, and accuracy. The results are promising with the highest accuracy of 72.9% in the combination Gradient-boosted Tree and cleaned data set with removing categorical features. This paper is a foundation for using big data platforms to handle financial data and much future research can be carried out to optimize the performance of this paper.

Keywords: Credit scoring · Big data · Ensemble learning · Feature engineering

1 Introduction

Credit score plays an important role in people's lives nowadays. When anyone wants to open a credit card or loan an amount of money, credit institutions will evaluate the credit score of this customer to ensure that he or she can return

N.-T. Nguyen et al. (Eds.): ICIT 2022, LNDECT 148, pp. 289–298, 2022.
https://doi.org/10.1007/978-3-031-15063-0_27

the loan. In addition, the amount of money that credit card holders can spend also depends on their credit score. This score is essential for everyone because if they are assessed that having a low credit score, they cannot loan money to buy a car or a house. This is the reason why a credit score is significant for people today. There is a large amount of data from financial activities such as purchasing a house or loaning a large amount of money. This data is valuable for artificial intelligence to learn and predict the credit score of these people. The amount of data from them are enormous and it increases continuously. When we use this amount of data to build a machine learning model to evaluate credit scores, a huge revolution will happen in the banks' sector. This will create a more trustable, fair, and efficient credit scoring system, giving an accurate credit score to everyone.

In many banks nowadays, evaluating credit scoring procedure depends heavily on manual ways. A large number of people is hired to evaluate the credit score of customers of banks. However, this way shows many disadvantages. Firstly, banks need a lot of people and time to manually evaluate someone's credit score. When an expert receives data about a customer, they need time to read over data and give the score, which is intensively time-consuming. Secondly, these experts can give bias to their acquaintances, and this can result in many serious problems. Therefore, a machine learning model which evaluates the credit score of anyone can replace humans' attendance in this task. Its effectiveness is undeniable from some previous research [6].

With the development of information technology in general and artificial intelligence in particular, many real-world problems have been resolved by computers. Artificial intelligence is applied to all aspects of life, from education to health. Banking is not an exception with a lot of works now can be done by computers such as credit card fraud detection. Credit scoring is a potential area where computers can work well thanks to some state-of-the-art machine learning algorithms. These algorithms can learn valuable insights that hide on customers' data and give an accurate result, which can surpass human performance someday. In this paper, some ensemble learning models will be used to evaluate whether a credit score of a customer is qualified.

In this paper, we carry out several experiments related to many techniques to get the highest result in building credit scoring models. We will mainly focus on feature engineering, which is essential in data mining because data related to customers' financial activities is uncleaned. The improvement in applying feature engineering is significant, giving a considerable improvement in models' performance. In addition, many ensemble learning models can be used for this problem. We would compare the performances of these models in evaluating the credit score.

In the following of this paper, some related works are shown in Sect. 2. Section 3 illustrates some techniques that we use to experiment and information about ensemble learning. The experiments and results are shown in Sect. 4. In Sect. 5, we give the conclusion about this paper and some future works.

2   Related Work

This subject has drawn a lot of attention from scientists, and here are some previous researches that have been carried out in credit scoring evaluation.

In a paper named Machine Learning-Based Empirical Investigation For Credit Scoring In Vietnam's Banking [8], Khanh Quoc Tran and his colleagues did some experiments about using machine learning to investigate the credit score in Vietnam's banking. They used a real-world data set called the Kalapa Credit Score data set, and their results are remarkable. The highest result is 83% F1-score with Random Forest algorithm. Although the results are relatively high, there are still some aspects that we can improve, such as applying feature engineering or using big data platforms to handle data. We will use the same data set as this research to experiment.

An approach proving the effectiveness in resolving the credit scoring problem is ensemble learning. Some ensemble learning methods such as bagging, boosting, and stacking are used to evaluate credit score [9] with the best model is 80.76% accuracy in the research of GangWang et al. In addition, Tounsi and his colleagues [6] prove that Boosting and Credal Decision Tree (CDT) are better than other ensemble algorithms.

With the explosion of data, applying big data platforms for the credit scoring problem was also carried out by Tousil et al. [7]. They use social data instead of traditional financial data to evaluate the credit score, and some surveys on proposed methods are also given to address this problem. Apache Hadoop and Apache Spark are considered to use, but there are no experimental results in their research.

Of all the aforementioned research, there is a main two disadvantage that the scalability of the traditional recommended methods with current data explosion is not fully adequate and some solutions are just on the papers. Therefore, this research sets up experiments on distributed solutions in order to compare the suggested methods with the traditional ones on different aspects (model architectures, data preprocessing).

3   Methodologies

This section introduces the big data platform and the ensemble learning methods employed in this research.

3.1   Big Data Platform

In the era of big data, data is everywhere. Thanks to the development of technological devices, data can be collected from social media, sensors, IoT devices which make the amount of data become enormous that traditional tools and techniques hardly handle. Big data has many applications ranging from banking and securities to communication or education, but how to utilize big data is a problem that traditional tools and techniques cannot process. This is the reason why some frameworks are born to deal with big data. One of the powerful frameworks is Apache Spark which we use for the credit scoring problem in this paper.

Apache Spark [11] initially started in 2009, and it became a part of Apache Software Foundation in 2013. Apache Spark is an open-source cluster computing framework for real-time processing. Spark can be run in many clusters, which utilize the hardware of many devices for resolving a problem. Spark provides an interface for entire programming clusters with implicit data parallelism, and fault-tolerant [10]. Thanks to saving data in multi clusters, when a cluster is broken, data can be retrieved from other clusters, so it provides an efficient fault-tolerant. Spark has some important components that supply users with all necessary tools for collecting data from streaming or other sources to process data and to model data. Spark Core is the base engine for large-scale parallel and distributed data processing. Further, additional libraries which are built atop the core allow diverse workloads for streaming, SQL, and machine learning.

3.2   Ensemble Learning

Decision Tree [4] is a typical tree-based model, which is constructed by many nodes. Decision tree can be used in both classification and regression problems. It has a root and many leaves and decision tree model bases on the rules in these leaves to give predictions. Decision tree uses the Gini index or Entropy to split leaves. The Gini index is computed as:

$$Gini = -\sum_{i=1}^{k} Pi(1 - Pi), \tag{1}$$

The Entropy is computed as follow:

$$Entropy = -\sum_{i=1}^{k} Pi \log_2 Pi, \tag{2}$$

The k is the number of cases and Pi is the probability of case i.

Decision tree is the foundation for Random Forest algorithms and other ensemble learning methods.

Random Forest [2] is a combination of decision trees so it is an ensemble learning method. When a decision tree gives a result, this result will be aggregated with other trees to give the final result. Random Forest helps to reduce the overfitting problem in decision tree because it requires many trees to give the result. Random Forest is an effective ensemble learning method thanks to its mechanism of trees combination.

Gradient-boosted Tree [1] is similar to Random Forest. It also uses a set of decision trees to predict the target label. However, Gradient-boosted tree usually outperforms Random Forest because Random forests use bagging mechanism to build independent decision trees and combine them in parallel. On the other hand, Gradient-boosted Tree uses an algorithm called boosting [5]. It means that Gradient-boosted Tree builds one tree at a time. This additive model works in a forward stage-wise manner to improve the gradient of existing weak learners (tree)[1], sequentially, each new tree corrects the errors of the previous one. In order to evaluate the performance of a tree, the loss function is used. Here is the cross-entropy loss function:

$$L(\hat{y}, y) = -\sum_i y_i \log(\hat{y}), \tag{3}$$

After having the loss of the first tree, the second tree is added so as to lower the loss compared to the first one alone. The loss has to move in direction of lowering its value fastest and the current tree's output is enhanced by that of the previous one. Mathematically, this is given by using the negative derivative of loss with respect to the previous tree's output. This is the reason why this is called Gradient-boosted tree.

$$F(k) = F(k-1) + \eta \times -\frac{\partial L}{\partial F(k-1)} \tag{4}$$

3.3   Important Features Extraction

Important Features Extraction is one of the useful techniques to increase the performance of model. In this paper, important features are extracted from the ensemble learning models on Spark. Some tree-based model calculates the node/feature probability to find out the important features from data set. Normally, the node probability is calculated by the number of samples that reach the node, divided by the total number of samples. When a feature gets a high probability, it has an enormous importance to the model.

Tree-based models extract calculation the importance indicator of a feature by summing the gain, scaled by the number of samples passing through the node. To be more specific, the important feature extraction formula is as in the Eq. 5 below:

$$f_i = \sum_j^k s_j C_j, \tag{5}$$

- f_i: the probability of feature i
- s_j: number of samples reaching node j.
- C_j: the impurity value of node j
- k: nodes j splits on features i.

---

[1] http://www.ccs.neu.edu/home/vip/teach/MLcourse/4_boosting/slides/gradient_boosting.pdf.

4 Experiment

4.1 Data Set and Pre-processing

The raw data set consists of 193 attributes but there are 117 attributes having a missing rate of over 50%. Due to the encrypted attribute names, it is hard to understand the relationship between attributes and their values. The attributes with missing rate grater or equal than 90% are removed. Besides, the complication in datatype needs paying attention to. Therefore, so many investigations have been done to find out the best way to deal with each feature.

The attributes are divided into three groups (based on the datatype and empirical investigation). They are group 1: date and datetime attributes, group 2: unicode ones, and group 3: the remaining features. There are 28 attributes in group 1 (e.g.: *Field_ 1, ngaySinh, A_ startDate, A_ endDate, etc.*). All the values in each one in this group need normalizing to date or datetime form. The group 2 including 30 attributes (e.g.: *Field_ 18, maCv, homeTownCity, brief, etc.*). Values in this group are Vietnamese or English words, sentences, and abbreviations. The concern is the inconsistent expression of the same meaning words or sentences. So, the solution is to lower all characters and remove the unnecessary characters or words from the values (e.g.: *'thủ đô Hà Nội'* or *'tp. Hà nội'* → *'hà nội'*, etc.). The inconsistency of values in each attribute are also normalized (e.g.: *'I'* → 1, *'II'* → 2, *'Ngoài quốc doanh Quận 7'* → *Null*, etc.), then the datatype normalization may be applied if needed. In order to deal with group 3, the numeric datatype correction is used because the attributes in this group mostly have numeric values. The attributes with categorical values are also processed as those in group 2.

The missing values, in categorical and numeric attributes, are treated separately. The missing values in the categorical feature are fulfilled with *'unknown'* value and a very large number is used for imputing missing values in numeric features. Those imputing values are chosen because of their balanced meaning for the real data. By using this method, it is hopefully believed that the data set is fulfilled and the data imputation does not impact too much the original data. After the data set is cleaned, it is believed that the information of each attribute can be made use of as much as possible by generating new attributes. There are 81 new features added and 47 ones deleted. So, after the feature engineering, there are 227 attributes in the processed data set.

4.2 Experimental Procedure

This paper aims to investigate which feature engineering method, and ensemble learning method are the best, so we design the experimental procedure as the following structure. From the Kalapa Credit Score data set, we do some initial pre-processing such as handling missing values and normalizing numeric features in the Spark framework. After that, we divide into 2 data sets, in which one data set is remained categorical features and denoted as Scenario A. The other data set is removed categorical features and denoted as Scenario B. We apply 2 phases of feature engineering into these two data sets to construct new data sets, and then we obtain four new processed data sets corresponding to A and B above and two original data sets. We use three ensemble learning methods as we said in Sect. 3.2 to build the predictive model. Then we use some features which are computed as important of Gradient-boosted Tree in the data set 2 to obtain two new data sets important features. We retrain these two data sets with three models, and finally, we compare the results of different combinations between data sets and ensemble learning methods. In summary, we have two main branches, data having raw categorical features and data removing raw categorical features, denoting as Scenario A and Scenario B, respectively. In each branch, we have four data sets, namely original data set, data set after phase 1, data set after phase 2, and important features data. We use three models to evaluate and compare the results. We randomly divide the dataset to the training set and the testing set with the ratio of 9:1 by using randomSplit in Spark Dataframe.

Important features data is the data set belonging to the best case of extracting important attributes. These features are high correlated and have a significant influence to the target features. A total of six cases are set with the condition that the probability of feature is greater than a threshold - multiple of the probability at which each attribute has the same probability.

$$threshold = \frac{X}{Y} \qquad (6)$$

$$X \in \{0, 1, 2, 3, 4, 5\}$$

- threshold: probability corresponding to x (threshold)
- Y: number of important attributes extracted from the model.

4.3 Experiment Setup

We use MLlib [3] in Spark to build and evaluate models. In the Decision Tree model we use the set of hyperparameter as follow: maxDepth = 5, maxBins = 32, minInstancesPerNode = 1, minInfoGain = 0.0, maxMemoryInMB = 256, cacheNodeIds = False, checkpointInterval = 10, impurity = 'gini', seed = 42.

In the Random Forest, we use 20 trees to build the random forest classifier and the maxDepth is 5, other hyperparameters are as follow: maxBins = 32, minInstancesPerNode = 1, minInfoGain = 0.0, maxMemoryInMB = 256, cacheNodeIds = False, checkpointInterval = 10, impurity = 'gini', featureSubsetStrategy = 'auto', seed = 42, subsamplingRate = 1.0.

In the Gradient-boosted Tree, we using lossType is logistic, the maxDepth is 5, the number of max iteration is 20 and, and other hyperparameters are as follow: maxBins = 32, minInstancesPerNode = 1, minInfoGain = 0.0, maxMemoryInMB = 256, cacheNodeIds = False, checkpointInterval = 10, stepSize = 0.1, seed = 42, subsamplingRate = 1.0, impurity = 'variance', featureSubsetStrategy = 'all', validationTol = 0.01.

4.4   Results and Discussion

Table 1. The result of models on data sets of Scenario A

Dataset	Model	Accuracy	Recall	Precision	F1 score
Data set1	Decision Tree	0.7183	0.5935	0.7068	0.5870
	Random Forest	0.6754	0.5000	0.3377	0.4031
	Gradient Boosted Tree	0.7210	0.5979	0.7114	0.5931
Data set2	Decision Tree	0.7170	0.6029	0.6888	0.6025
	Random Forest	0.6754	0.5000	0.3377	0.4031
	Gradient-boosted Tree	0.7203	0.6020	0.7017	0.5999
Data set3	Decision Tree	0.7170	0.6029	0.6888	0.6025
	Random Forest	0.6754	0.5000	0.3377	0.4031
	Gradient-boosted Tree	0.7192	0.6010	0.6988	0.5988

Table 2. The result of models on data sets of Scenario B

Data set	Model	Accuracy	Recall	Precision	F1 score
Data set1	Decision Tree	0.7254	0.5930	0.7028	0.5891
	Random Forest	0.7202	0.5791	0.7021	0.5677
	Gradient Boosted Tree	0.7292	0.6008	0.7075	0.6001
Data set2	Decision Tree	0.7244	0.5987	0.6921	0.5984
	Random Forest	0.7158	0.5685	0.6997	0.5504
	Gradient Boosted Tree	0.7263	0.6004	0.6968	0.6004
Data set3	Decision Tree	0.7244	0.5987	0.6921	0.5984
	Random Forest	0.7147	0.5575	0.7271	0.5278
	Gradient Boosted Tree	0.7263	0.6004	0.6968	0.6004

As we can see on the Table 1 and 2, the overall performance of these models in classifying credit scoring status of customers is around 70% accuracy and 60% f1-score. In general, Gradient-boosted Tree gives the best performance with 72.1% in the data set 1a and 72.92% accuracy in the data set 1b. This is easily understood because the boosting algorithm in Gradient-boosted Tree is very effective in improving the performance. By contrast, Random Forest shows the worst results in all data sets.

The data set 1, which is the original data set, gives better results than other data sets. The best result in data set 1a is 72.1% accuracy. This proves that the original data set provides sufficiently the data for the model. In contrast, adding more features is not sure to make the performance increasing.

As we can see in the Table 1 and 2, the Scenario B, which is the data set deleting raw categorical features shows higher results than the Scenario A. To be specific, the best result in Scenario B is 72.9% accuracy while the best result in Scenario A is only 72.1%. In addition, the performance of Random Forest in Scenario B is better than that in the Scenario A. This is a huge improvement because the f1-score of Random Forest in Scenario A is about 40% while the f1-score of Random Forest in Scenario B is always over 50%.

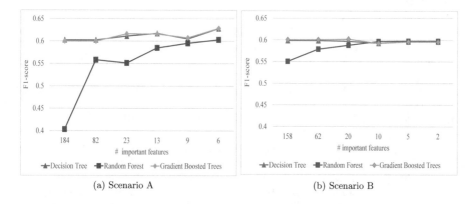

(a) Scenario A   (b) Scenario B

Fig. 1. F1-score on different numbers of important features in data set 2

In Fig. 1, we show the performances of three models on the data set 2 with many important features. We decrease the number of features from all features to the most important features. In Scenario A, there are 184 features in data set 2 and we use the formula in Sect. 3.3 to calculate the importance of each feature and we select a set of important feature to train models. The data for Scenario B is similar. When we decrease the number of features in Scenario A, the f1-score increase in all models, especially in Random Forest, from approximately 40% to about 60%. However, the performances of Decision Tree and Gradient-boosted Tree are fell in Scenario B.

5   Conclusion and Future Work

In this paper, we compare the performances of different feature engineering techniques and various ensemble learning methods for the credit scoring problem. We use the Spark framework, which shows the effectiveness of handling big data. We use a real world data set about Vietnamese customers and the results on this data set are remarkable. Gradient-Boosted tree shows the highest performance at about 0.6% f1-score and 72.92% accuracy. In addition, removing raw categorical features gives a higher result than retaining them. Selecting important features is also an efficient way that improving the model performance that has been proving in this paper.

We have conducted several experiments to improve the performance in classifying customer's credit score status, but many aspects can research in the future. Using streaming data about customers, especially real-time financial activities or social data is a useful way to boost the model's performance. Deep learning models prove their robustness in resolving many problems so apply deep learning models also should be considered to improving performance.

References

1. Friedman, J., Hastie, T., Tibshirani, R.: Special invited paper. Additive logistic regression: a statistical view of boosting. Ann. Stat. 337–374 (2000)
2. Liaw, A., Wiener, M., et al.: Classification and regression by random forest. R news **2**(3), 18–22 (2002)
3. Meng, X., et al.: MLlib: machine learning in apache spark (2015). arXiv: 1505.06807 [cs.LG]
4. Rokach, L., Maimon, O.: Decision trees, vol. 6, pp. 165–192 (2005). https://doi.org/10.1007/0-387-25465-X_9
5. Schapire, R.E.: A brief introduction to boosting. In: Proceedings of the 16th International Joint Conference on Artificial Intelligence, IJCAI 1999, vol. 2, pp. 1401–1406. Morgan Kaufmann Publishers Inc., Stockholm (1999)
6. Tounsi, Y., Hassouni, L., Anoun, H.: An enhanced comparative assessment of ensemble learning for credit scoring. J. Intell. Comput. **10**, 15 (2019). https://doi.org/10.6025/jic/2019/10/1/15-33
7. Tounsi, Y., Hassouni, L., Anoun, H.: Credit scoring in the age of big data -a state-of-the-art. Int. J. Comput. Sci. Inf. Secur. **15**, 134–145 (2017)
8. Tran, K.Q., Duong, B.V., Tran, L.Q., Tran, A.L.-H., Nguyen, A.T., Nguyen, K.V.: Machine learning-based empirical investigation for credit scoring in Vietnam's banking. In: Fujita, H., Selamat, A., Lin, J.C.-W., Ali, M. (eds.) IEA/AIE 2021. LNCS (LNAI), vol. 12799, pp. 564–574. Springer, Cham (2021). https://doi.org/10.1007/978-3-030-79463-7_48. ISBN 978-3-030-79463-7
9. Wang, G., et al.: A comparative assessment of ensemble learning for credit scoring. Expert Syst. Appl. **38**(1), 223–230 (2011). ISSN 0957-4174. https://doi.org/10.1016/j.eswa.2010.06.048, https://www.sciencedirect.com/science/article/pii/S095741741000552Xd
10. Zaharia, M., et al.: Resilient distributed datasets: a fault-tolerant abstraction for in-memory cluster computing. In: Proceedings of the 9th USENIX Conference on Networked Systems Design and Implementation, NSDI 2012, p. 2. USENIX Association, San Jose (2012)
11. Zaharia, M., et al.: Spark: cluster computing with working sets. In: Proceedings of the 2nd USENIX Conference on Hot Topics in Cloud Computing, HotCloud 2010, p. 10. USENIX Association, Boston (2010)

Analysis of the Effects of Stop-word Removal in Hate Speech Detection Problem for Vietnamese Social Network Data

Hanh Hong-Phuc Vo[1,2], Huy Hoang Nguyen[1,2], and Trong-Hop Do[1,2(✉)]

[1] University of Information Technology, Ho Chi Minh City, Vietnam
{18520275,18520842}@gm.uit.edu.vn
[2] Vietnam National University Ho Chi Minh City, Ho Chi Minh City, Vietnam
hopdt@uit.edu.vn

Abstract. The usefulness of removing the stop-word in detecting Vietnamese hate speech on social media has yet to be determined. In this study, statistical hypothesis testing is used to evaluate the effect of stop-word removal in hate speech detection using two datasets, the ViHSD dataset and the HSD dataset of VLSP-HSD, six models, including traditional machine learning models, deep neural learning models, transformer models, and two methods of stop-word selection, word frequency and word form. The results indicate stop-word removal is bad effective in hate speech detection problem on Vietnamese.

Keywords: Stop-word · Hate speech detection · Social network text · Natural language processing · Statistical test · Machine learning

1 Introduction

Stop-word removal is removing words not having meaning in text, as well as be an essential stage in preprocessing of natural language processing. This is a controversial issue due to its effectiveness on different models. Traditional machine learning and statistics-based methods often combine the use of stop-word removal to enhance performance [2]. Recent studies using advanced deep learning methods are frequently not used [3,4].

The effectiveness of stop-word removal is relatively fine depending on the model, language, and the processing problem. Although, stop-word removal is one of the basic preprocessing steps, many people frequently apply this step to every issue in Natural language processing without checking whether it is effective or not. This sometimes takes time and even affects accuracy.

In this paper, the effectiveness of word's stop-word removal for hate speech detection issue on Vietnamese social media data is analysed. Hate speech on online social is any kind of communication in writing that attacks or uses pejorative or discriminatory language with reference to a person or a group on the basis of their characteristics. With rapidly increase in information by second,

N.-T. Nguyen et al. (Eds.): ICIT 2022, LNDECT 148, pp. 299–309, 2022.
https://doi.org/10.1007/978-3-031-15063-0_28

hate speech also increase result in an unfriendly environment in social network. It is essential to detect and handle hate speech on social media to keep a clean environment.

Most comments on social network has abbreviations and idiosyncratic, non-formal phrases, as well as weird characters that result in preprocessing is necessary to achieve high performance. A few methods such as remove special character, normalize abbreviations, tokenize are necessaries for comments on social network but with stop-word removal, its effectiveness remains unanswered.

To analysis the effect of stop-word removal of stop-word removal in hate speech detection problem for Vietnamese social network data, two methods of selecting stop-word, six model includes traditional machine learning, deep neural learning, transformer and two datasets were used in statistical hypothesis testing by two-factor ANOVA with replication and Tukey's HSD.

The contribution of the paper is proving bad effect of stop-word removal to hate speech detection on Vietnamese. Most of the methods of selecting stop-word decrease performance. On the other hand, the statistical hypothesis testing methods indicate that there is a significant different between methods and an insignificant different between models.

2   Fundamental of Stop-Word Removal and Preprocessing in Hate Speech Detection for Social Network Data

The comments and posts made by users on social media sites are known as social media text data. Stranger characters, icons, unaccented letters, abbreviations, slang, teen code, digits and characters mixed, inconsistently breaking marks, and blending foreign words into phrases make up the most of the social network text data.

Based on the characteristics of social media text that mentioned in the previous section, preprocessing is an essential stage that greatly influences the prediction results of models. The techniques combined in the preprocessing are proposed in the following order. In this Fig. 1, (1) data points with no content or duplicated content are removed. (2) changes all letters that are uppercase to lowercase. (3) Convert symbols such as "=)", ":(", ":P", "<3", etc., to emojis. (4) Remove URL, mail, hashtag, mention tag in text data. (5) remove mixed words

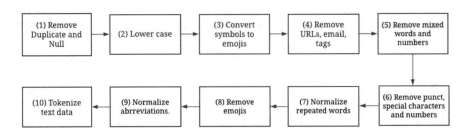

Fig. 1. Preprocessing before stop-word removal

and numbers. (6) words that contain both alphabet and numeric characters, for example "8k", "11tr", "6km", etc., are removed. (7) for example, words that have repeating characters like "hahaha", "okkk", "vclll" are normalized to "ha", "ok", "vcl". (8) remove emojis. (9) in this technique, abbreviations are normalized, i.e. the transformation to its original form, using the Ridge model. This model is trained on a dataset of acronyms for hate speech detection. (10) the text data is tokenized using the Pyvi library.[1]

Stop-word removal is the technique often integrated in the preprocessing phase of most natural language processing tasks. A stop-word list is a collection of words that occur commonly but have little value in supporting predictive models. The words in this list will be removed from the vocabulary and dataset before training the models. A collection of stop-words for the hate speech detection problem is combined with the above preprocessing techniques is able to improve the performance of the models.

3   Analysis of Stop-Word Removal in Hate Speech Detection

3.1   Procedure of Analysis

The experiment in this paper has two main factors: stop-word selection methods and classification model. It is obvious that the accuracy of hate speech detection depends on the stop-word sets, which are obtained from specific selection methods. A method that selects too many words as stop-words will reduce the contextual meaning of the context, leading to lower performances of context-based machine learning models when applying these stop-words. Beside stop-word sets, the accuracy of hate speech detection is determined by the classification machine learning model. To achieve reliability of the conclusions drawn from the experimental result, every treatment, that is a combination of a stop-word selection method and a classification machine learning model, will be applied on two datasets. Analysis of variance (ANOVA) is applied on the experimental results to analyse the effect of stop-word selection methods, classification machine learning model, and the interaction of these two factors on the accuracy of HSD on Vietnamese social data. Tukey's HSD is also used to verify if stop-word removal is useful for HSD problem. The procedure of the analysis is illustrated in Fig. 2.

3.2   Methods for Stop-Word Selection

The methods for selecting stop-words are based on word frequency and word form. Many frequency levels are applied to the datasets and each class to find stop-words in both methods. Only verbs, noun/noun phrases, adj, and adv are selected, and all other words are skipped when selecting by word form. The methods also used as a factor in analysis effect of stop-word removal.

---

[1] https://pypi.org/project/pyvi/.

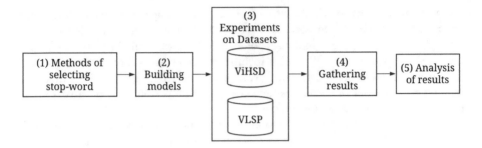

Fig. 2. Procedure for analysing the effectiveness of word removal

Table 1. Results on Stop-word removal and without stop-word removal.

(a) Stop-word removal

	LR	SVM	GRU	BiGRU	CNN	PhoBERT
ViHSD	0.6108	0.6113	0.6293	0.6289	0.6566	0.5704
VLSP	0.6453	0.6613	0.6547	0.6559	0.6763	0.6757

(b) Without stop-word removal

	LR	SVM	GRU	BiGRU	CNN	PhoBERT
ViHSD	0.6286	0.6261	0.6631	0.6716	0.6720	0.6888
VLSP	0.6584	0.6793	0.6787	0.6679	0.6835	0.7084

Moreover, another Table 1(a) and Table 1(b) illustrate the result having stop-word removal and without stop-word removal. A public Vietnamese stop-word list[2] is used to this experiment. The result indicates stop-word removal affect badly to performance on both datasets.

3.3   Models

PhoBERT is a monolingual language models that have been pre-trained for Vietnamese on a big scale. They perform exceptionally well on four Vietnamese NLP tasks.

TF-IDF is used in Logistic Regression and Support Vector Machine to quantify words in a set of documents, which generally compute a score for each word to signify its importance in the document and corpus.

Support Vector Machine (SVM) is a linear model for classification and regression problems. SVM aims to divide the data into classes by drawing a line or hyperplane.

---

[2] https://github.com/stopwords/vietnamese-stopwords.git.

Logistic Regression (LR) is a commonly used statistical method for predicting binomial outcomes. This is one of the most fundamental and well-known classification algorithms approaches.

PhoW2V is a pre-trained Word2Vec word embedding for Vietnamese. It is used to representation for text in which words with the same meaning are represented similarly in GRU, Bi-GRU, Text-CNN.

Gated Recurrent Units (GRU) is an advancement of recurrent neural networks. Compared to Long Short-Term Memory (LSTM), GRU is simpler but faster to train, and its performance is approximately equal to LSTM in several NLP tasks.

Bi-directionalGRU (Bi-GRU) Bi-GRU is a combination of two GRUs, one of which moves forward, starting from the beginning of the data sequence, and the other of which moves backward, starting from the end. Bi-GRU can get information from both the past (ahead) and future (backward) states at the same time thanks to this mechanism.

Convolutional Neural Networks (Text-CNN) is a type of deep learning model that has dominated computer vision problems. Convolution layers, pooling layers, and fully connected layers are the most common building components of CNN.

4   Experiment and Analysis

4.1   Datasets

The datasets used during experiment are ViHSD published by Luu et al. [1] and VLSP-HSD at VLSP shared task 2019 [5]. These two datasets are chosen because they are the popular hate speech detection datasets on Vietnamese in recent years. Both datasets have the same three labels (CLEAN, OFFENSIVE, HATE) and are collected from comments or posts from major social media sites like Facebook or YouTube. In addition to these similarities, the difference between the two datasets mainly comes from the size of the datasets. The Table 2 show some examples in both datasets.

The experiment is performed based on the methods of selecting stop-word, models and two datasets. This process is divided into the following steps, as shown in Fig. 3.

In step (1), preprocessing is conducted on ViHSD and VLSP-HSD datasets. After that, stop-word list is selected and stop-word removal is instantaneously conducted on preprocessed datasets in step (2) and (3). The datasets after step (3) is used to train 6 models, which are LR, SVM, GRU, Bi-GRU, Text-CNN,

Table 2. Examples of datasets

(a) ViHSD dataset

No.	Comments	Label
1	Đố chúng m nhận ra ai *(Guess who)*	0
2	Im mẹ đi thằng mặt lon *(Shut the f*ck up, you're an f*cking c*nt)*	2
3	Lúp lúp như chó *(Hiding like a dog)*	1
4	Hiểu Bùi sủa dơ *(Dirty barking)*	2
5	Quá ngu lồn đi =))) *(Dumb c*nt)*	1

(b) VLSP-HSD dataset

No.	Comments	Label
1	Tại sao cứ đem đam mê của ng khác ra bàn tán soi mói *(Why you pry about other people's passions)*	0
2	Còn dùng nick clone chứ có dám dùng nick thật đéo đâu, dòng thứ đĩ ngu *(Just use clone account, don't f*cking dare use main account, stupid b*tch)*	2
3	ngu quá thành quá ngu.ngu ngu ngu *(Too dumb)*	1
4	CĐV chứ có phải CLB đâu. Thằng add viết ngu vãi *(Fan but not club. The admin writes stupidly)*	2
5	Nứng vl =)) *(Very randy)*	1

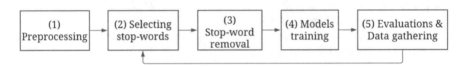

Fig. 3. Experimental process on datasets

PhoBERT in step (4). In step (5), trained models are evaluated and, the evaluation results are gathered for analysis of the effect of stop-word removal. The process from step (2) to step (5) is repeated for each method of selecting stop-word.

4.2 Collecting Experimental Results

In the experiment, there are 204 data points. F1-score macro is used to evaluate result of data points from 2 experiment units, ViHSD and VLSP-HSD datasets. An experiment unit is applied a treatment that be a combination of six models, two methods of selecting stop-word with eight levels in each and baseline that not be stop-word removal.

Figure 4 plots the results obtained from the experiment of stop-word removal on ViHSD dataset and VLSP-HSD dataset. In which, the vertical axis is the F1-score value, the horizontal axis is the methods of selecting stop-word, and the baseline is the result without removing the stop-word. Specifically, the level of selecting stop-word X_Y_Z which

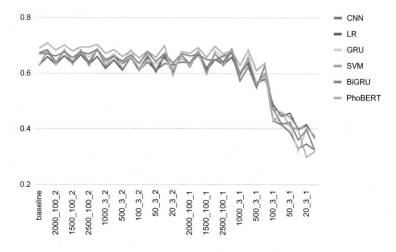

Fig. 4. Experimental stop-word removal result

- Filtering words whose frequency greater than X.
- Filtering words whose frequency in each label is greater than Y.
- If Z equals 2 then only word forms in the Sect. 3.2 are selected. If Z equals 1 then all word forms are selected.

For example, the method 2000_100_2 means filtering words with a frequency greater than 2000, the frequency of occurrence in each label greater than 100, and applying filtering based on the word forms.

Based on Fig. 4, the result of baseline is better than most of stop-word removal levels. However, these differences are not too large, only the last four levels are significantly different.

4.3 Analysis of Experimental Results

Hypothesis Testing of the Effect of Methods of Selecting Stop-Word.
Analysis of variance (ANOVA) is a statistical test developed by Ronald Fisher in 1918 and used until now. Briefly, ANOVA tells us whether there is any significant difference between the means of two or more independent groups. Just like other statistical tests, ANOVA also uses two hypotheses: Null Hypothesis and Alternatives Hypothesis. The null hypothesis in ANOVA is true when all sample means are equal, there are no significant differences between means. On the other hand, if at least one of the sample means is different from the rest, then the Alternative hypothesis is correct.

Two-factor ANOVA with replication is used in this hypothesis testing. This is a variant of ANOVA for tests that have more than one factor or independent variable and more than one sample element for each combination of the levels of factors. In this analysis, methods of selecting stop-word and models are identified as two-factor of the test. Each combination of the levels of these two-factors is called a treatment. To ensure replication principle in design of experiment, two datasets that are ViHSD and VLSP-HSD is used as experiment units.

Verify the Effectivity of Various Methods of Selecting Stop-Word. The limitation of ANOVA is that it only tells us whether there is a difference between the means of the independent groups, it will not specifically indicate which group is different from the others. Therefore, post-hoc analysis is needed in this situation.

Tukey's HSD (honestly significant difference) test is used as a follow-up analysis for the two-factor ANOVA with replications. This is a single-step multiple comparison procedure and statistical test. It can be used to find means that are significantly different from each other.

Table 3. Result of two-factor ANOVA with replication

Source	SS	df	MS	F	P-value	F crit
Stop-word selection method	1.9540	17	0.1149	123.3729	0.0000	1.7181
HSD model	0.0133	5	0.0026	2.8560	0.0183	2.2984
Interaction	0.0457	85	0.0005	0.5771	0.9956	1.3979
Error	0.1006	108	0.0009			
Total	2.1136	215				

Result of the Hypothesis Testing. The Table 3 shows the results of the two-factor ANOVA test on the data collected from the experiment on the ViHSD and VLSP-HSD datasets. Since the P-value of the methods of selecting stop-word (Method) is $0.0000 < 0.05 = \alpha$, it rejects the null hypothesis that the means of the levels of this factor are equal. Therefore, the conclusion drawn with 95% confidence that there are significant differences between the effectiveness of the methods of selecting stop-word. Furthermore, this difference is large, as shown by F value of 133.8991. Similarly, the P-value of the model factor is $0.0063 < 0.05 = \alpha$, it also rejects the Null hypothesis of the model factor (Model) that the means of the levels of this factor are equal and concludes that there are significant differences between the effectiveness of the models. Because the F value is equal to 3.4594, this difference is insignificant. For the Interaction, the P-value, equal to 0.9957, is greater than 0.05 and the F-value is approximately 1, so it can be concluded that there is no interaction between the methods of selecting stop-word and the models.

Post Hoc Analysis of Hypothesis Testing. The results of the ANOVA test show that there are significant differences between the effectiveness of the methods of selecting stop-word, and these differences are large. Tukey's HSD test

Table 4. Result of Tukey's HSD

Method 1	Method 2	Meandiff	P-adj	Lower	Upper	Reject
Baseline	1000_3_1	0.0436	0.0456	0.0004	0.0868	TRUE
Baseline	1000_3_2	0.0156	0.9000	−0.0276	0.0588	FALSE
Baseline	100_3_1	0.2223	0.0010	0.1791	0.2655	TRUE
Baseline	100_3_2	0.0265	0.7377	−0.0167	0.0697	FALSE
Baseline	1500_100_1	0.0233	0.9000	−0.0199	0.0665	FALSE
Baseline	1500_100_2	0.0041	0.9000	−0.0391	0.0473	FALSE
Baseline	2000_100_1	0.0088	0.9000	−0.0344	0.0520	FALSE
Baseline	2000_100_2	0.0054	0.9000	−0.0378	0.0485	FALSE
Baseline	20_3_1	0.3099	0.0010	0.2667	0.3531	FALSE
Baseline	20_3_2	0.0306	0.5266	−0.0126	0.0738	FALSE
Baseline	2500_100_1	0.0066	0.9000	−0.0366	0.0498	FALSE
Baseline	2500_100_2	0.0005	0.9000	−0.0427	0.0437	FALSE
Baseline	500_3_1	0.0822	0.0010	0.0390	0.1253	TRUE
Baseline	500_3_2	0.0182	0.9000	−0.0250	0.0614	FALSE
Baseline	50_3_1	0.2727	0.0010	0.2295	0.3159	TRUE
Baseline	50_3_2	0.0235	0.8991	−0.0197	0.0667	FALSE
Baseline	Available	0.0288	0.6221	−0.0144	0.0719	FALSE

is used to specifically indicate which methods of selecting stop-word is better and whether these differences are statistically significant.

The results of Tukey's HSD are presented in the Table 4. The comparison between methods of selecting stop-word and baseline is focused. It is easy to see that the Meandiff value of the baseline compared to methods of selecting stop-word is entirely positive, which means that not removing stop-word has a positive effect. More specifically, for levels whose value of Reject is TRUE, the effect of removing stop-word is statistically significant that this negatively affects performance of models. For levels where the value of reject is FALSE, although it has a negative effect, we need further consideration because this conclusion is not statistically significant, or the results may be computed by chance. To do this, visualization methods are used.

Visualization Analysis of Hypothesis Testing. Figure 5 depicts the average of the experimental results for models. It is explicit that the results between models are small, the largest distance is approximately 0.025, this supports the result of the ANOVA analysis that there is an insignificant difference between the effectiveness of the models.

Figure 6 illustrate comparison of effectiveness of methods of selecting stop-word on models in ViHSD dataset. Most of the methods decreased F1-score. Figure 7 illustrate comparison of effectiveness of methods of selecting stop-word

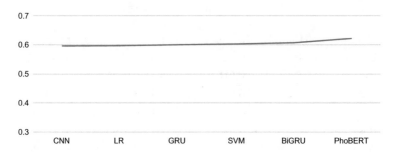

Fig. 5. Experimental stop-word removal result of each model

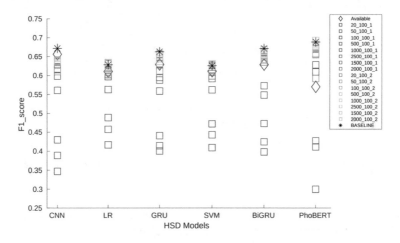

Fig. 6. Experimental stop-word removal result in ViHSD dataset

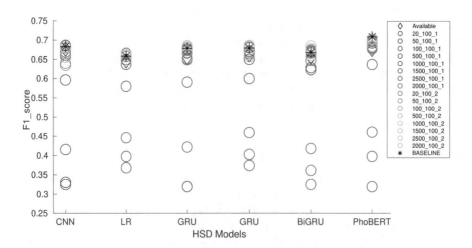

Fig. 7. Experimental stop-word removal result in VLSP dataset

on models in VLSP dataset. A few of methods improves performance. Results from both dataset are suitable to the result of Tukey's HSD that stop-word removal negatively affects the result of hate speech detection.

4.4   Discussion

Through experimental results using two-factor ANOVA and Tukey's HSD, stop-word removal causes decreased performance that lead to stop-word removal should not be done in hate speech detection problem on Vietnamese. The limitation in this experiment is the number of datasets. Because creating datasets take plenty of time and effort, there doesn't have datasets more. Removing stop-word decreases performance, because stop-word is the context of sentences helping model predict correctly. Stop-word removal more, lost context more that result in performance decreased more.

5   Conclusion

In this study, the experiments are conducted to analysis of the effects of stop-word removal in hate speech detection problem for Vietnamese social network data. Two datasets, two methods and six models are used to analysis the effects of stop-word removal by two-factor ANOVA with replication and Tukey's HSD. From results of statistical hypothesis testing methods, stop-word removal should not be done in hate speech detection issue on Vietnamese. Moreover, there are significant differences between methods of selecting stop-word, but the different between models is quite small.

References

1. Luu, S.T., Nguyen, K.V., Nguyen, N.L.-T.: A large-scale dataset for hate speech detection on Vietnamese social media texts. In: Fujita, H., Selamat, A., Lin, J.C.-W., Ali, M. (eds.) IEA/AIE 2021. LNCS (LNAI), vol. 12798, pp. 415–426. Springer, Cham (2021). https://doi.org/10.1007/978-3-030-79457-6_35
2. Nguyen, K.P.Q., Van Nguyen, K.: Exploiting Vietnamese social media characteristics for textual emotion recognition in Vietnamese. In: 2020 International Conference on Asian Language Processing (IALP), pp. 276–281. IEEE (2020)
3. Nguyen, L.T., Van Nguyen, K., Nguyen, N.L.T.: Constructive and toxic speech detection for open-domain social media comments in Vietnamese. arXiv preprint arXiv:2103.10069 (2021)
4. Van Huynh, T., Nguyen, V.D., Van Nguyen, K., Nguyen, N.L.T., Nguyen, A.G.T.: Hate speech detection on Vietnamese social media text using the Bi-GRU-LSTM-CNN model. arXiv preprint arXiv:1911.03644 (2019)
5. Vu, X.S., Vu, T., Tran, M.V., Le-Cong, T., Nguyen, H.: HSD shared task in VLSP campaign 2019: hate speech detection for social good. arXiv preprint arXiv:2007.06493 (2020)

Context Graph Alignment Using Adversarial Learning for Air Pollution Detection on IoT Sensor Systems

Thien Pham[1,3] , Tuan Bui[1,3] , Trung Mai[1,3] , Quan Le[2,3],
Chuong Dang[2,3] , Thuat Nguyen[2,3] , and Tho Quan[1,3(✉)]

[1] Ho Chi Minh City University of Technology (HCMUT),
Ho Chi Minh City, Vietnam
{pcthien.sdh20,tuanbc88,mdtrung,qttho}@hcmut.edu.vn
[2] University of Information Technology, Ho Chi Minh City, Vietnam
{quanlt,chuongdlb,thuatnk}@uit.edu.vn
[3] Vietnam National University Ho Chi Minh City, Ho Chi Minh City, Vietnam

Abstract. Nowadays, with the undeniable development of 5G technologies, Advanced Artificial Intelligence (AAI) systems using IoT sensors have been widely deployed in many different applications. The emerging requirement on these platforms is the identification of real situations (or contexts) based on time-series information given by sensors, which are ideally represented as graphs. However, when dealing with large-scale and abundant big-data continuously generated by IoT sensors, the online graph matching process becomes a high-challenge issue. In this paper, we propose an efficient graph-based context detection to handle this problem. Our proposed approach is based on adversarial learning applied on a low-dimensional space embedded from the nodes of the graphs, making it feasible for a real-time alignment and matching process. Our method is then experimented for detection of context-based air pollution running on a testing IoT system installed in Ho Chi Minh City University of Information Technologies (UIT) which promising results achieved.

Keywords: Adversarial learning · IoT · Deep learning · Context graph

1 Introduction

Along with the evolution of the Industry Revolution 4.0, smart systems that deploy IoT devices on 5G infrastructure combined with the Artificial Intelligence techniques have been increasingly growing lives. When deployed in real life with a fully automated manner, one of the problems with these systems is the *context recognition*. One of the most suitable mechanism to represent such contexts is *graph*. Hence, in order to automatically detect whether a context is occurring, the system must continually perform *graph aligning* between temporary contexts provided by IoT systems with predefined contexts.

© The Author(s), under exclusive license to Springer Nature Switzerland AG 2022
N.-T. Nguyen et al. (Eds.): ICIT 2022, LNDECT 148, pp. 310–319, 2022.
https://doi.org/10.1007/978-3-031-15063-0_29

Assuming that we have an early warning system for environmental pollution built on a network of IoT sensors measuring various environmental parameters such as temperature, humidity, the density of fine particles in the air, wind direction, wind speed, etc. These devices continuously collect and send data to the processing center in the form of time-series data, which are used to from graph describing the current contexts. To know what "pollution" is, one should construct certain *context graphs* depicting pollution and matching it with the collected data from IoT sensors.

In fact, the time series context information transmitted from IoT devices will form a very large network, so the problem of context detection and matching will become a *network alignment* problem. Despite its huge benefits, with the nature of an NP-hard problem [1], network alignment is a challenging task. Many unsupervised approaches have been proposed to solve this problem directly, such as NetAlign [2], UniAlign [3], IsoRank [3], FINAL [3] and REGAL [4]. Unfortunately, these methods fail for dealing with large networks because of exponential computation time.

In this paper, we propose an adversarial learning-based approach to perform the graph alignment problem in real-time. The main contributions of our work are summarized as follows: (i) we propose a context graph model from IoT signals and use this to define the pollution situation in IoT sensor system; (ii) we propose graph embedding method to perform network alignment in real-time; (iii) we propose a mechanism to train our model based on adversarial learning to perform graph alignment in an unsupervised manner. We also show that our alignment process is running in logarithmic complexity, thus it can be applied in real time; and (iv) we conduct extensive tests on real data taken from sensors located at UIT to serve the problem of detecting air pollution. Our framework not only achieves competitive performance but also exhibits robustness to both network structural inconsistency and graph size imbalance.

2   Background

2.1   Homogeneous Graph and Heterogeneous Graph

In graph theory, *homogeneous graph* are graphs whose edges and vertices represent for one type of object only. For example, a vertex serves for a person and edge shows the friendship between two people. In contrast, *heterogeneous graph* are the ones whose edges and vertices serve for different types of objects and relationships. Obviously, heterogeneous graph can be utilized to represent more generic and complex graphs comparing to homogeneous such as knowledge graph. In our work, the context graphs are composed from different types of IoT sensors and represent various types of connection. Thus, those graphs are heterogeneous by nature. It makes us to consider *graph embedding* technique to embed all vertices into a single embedding space for further processing.

2.2   Network Alignment

In network science, network alignment, the task of recognizing node correspondence across different networks, is one of the fundamental problem appearing in many data analysis applications. For example, by detecting accounts from the same user in different social networks, information of that user in one site can be exploited to perform better downstream functions (e.g. friend suggestion or content recommendation) in the other site [3]. In our research, we use network alignment to detect a context graph in a very large graph generated from IoT sensors signals. In theory, this is a NP-hard problem. However, machine learning techniques can help to approximate the solutions. Given the huge size of networks once applied in real applications, supervised learning approaches are not preferable. The state-of-the-art unsupervised graph alignment approaches require some *"anchor links"* to start with, which is not also applicable in our work. To resolve it, we propose an approach based on adversarial learning to handle the network alignment in a totally unsupervised manner without any anchor links.

3   Unsupervised Learning for Air Pollution Detection on IoT Systems

3.1   Case Study

To make it easy for readers to follow, we would like to illustrate with a simple system as follows. Supposed that we have 3 stations where IoT devices are located to monitor the environment as shown in Fig. 1, in which Station $S1$ monitors Monoxide (CO) in the air over time. Meanwhile Station $S2$ and Station $S3$ monitor the Nitrogen Dioxide (amount of NO2) gas over time. We assume that in this situation the signal from station S1 can be received by devices $S2$ and $S3$ and vice versa. This means that there is a physical connection between $S1$ and $S2$, between $S1$ and $S3$. Meanwhile, due to the geographical distance, the signal from $S2$ cannot be received by S3 and vice versa, in other words, $S2$ and $S3$ are not physically interconnected in the system.

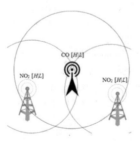

Fig. 1. Simulation of monitoring stations and their connections

In this example, it is supposed that we have only 2 state levels, **H** for high and **L** for low. With a system of three monitoring stations as shown in Fig. 2, data collected over time is stored as illustrated in Table 1, where t_i correspond to time slots. In particular, we do not always get 100% data from IoT sensors, so there are some time slots we do not receive the signal from the sensors, as illustrated with the dark cells in Fig. 2.

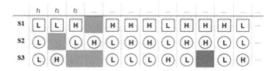

Fig. 2. Collected data from IoT devices over time

With the data obtained as shown in Fig. 2, our system will analyze and construct the graph over time as shown in Fig. 3. Each node of the graph will correspond to one value of one device in a special time slot.

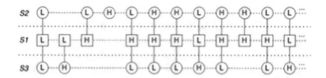

Fig. 3. The graph generated from data collected from IoT devices

Thus, the graph generated from data collected from IoT devices is a heterogeneous graph because there are nodes and edges that belong to different data types. Each node can correspond to the value of different devices and the edge represents different relationships.

Figure 3 illustrates a graph generated from data collected from IoT devices monitoring the environment. In order to detect the context of whether there is pollution or not, we need to have a model graph that defines what is "pollution". We call this graph a *context graph*. Supposed that we define that if the data we obtain there is a situation where the amount of CO is high for three consecutive time slots and (at the same time) the amount of NO2 also is at a high level for at least 2 consecutive time slots, we say that the area (context) is polluted. With such a pollution hypothesis, the pollution will be represented in two context graphs such as Fig. 4a and Fig. 4b. Pollution detection will become the graph detection problem in Fig. 4a or Fig. 4b that built in step 2.

With the data graph as shown in Fig. 3, and the context graph as shown in Fig. 4, we detect the pollution situation by detecting the existence of graphs in Fig. 3 as a sub-graph in Fig. 4. However, in practice, the data graph like Fig. 4 will

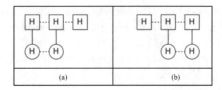

Fig. 4. The graph generated from data collected from IoT devices

be huge, so we will be *slicing* this large graph into sub-graphs over time frames. In the example in Fig. 5, we perform slicing with a time frame of 4 timeslots. As a result, we will detect the existence of the context graph in Fig. 5 with these subgraphs. For example, when slicing the graph as in Fig. 5, we get 2 sub-graphs like those in Fig. 6 and Fig. 7.

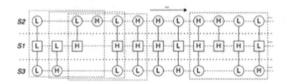

Fig. 5. Slicing data graph

In our method, the matching between two graphs is carried out by finding pairs of closest nodes between two graphs. Thus, when performing matching between the context graph (including 5 nodes) in Fig. 4a with the sub-graph in Fig. 6, we have 5 pairs of matching modes. This is the subgraph that is closest to the graph in Fig. 4a and the graph in Fig. 6. And we easily check that this sub-graph is not a graph showing the concept of pollution.

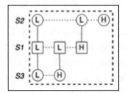   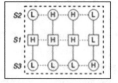

Fig. 6. The first sub-graph   **Fig. 7.** The second sub-graph

Next, when conducting matching between the graph in Fig. 4b with the graph in Fig. 7, we get the subgraph closest to the context graph in Fig. 4b and the graph in Fig. 7. We need to note that the graph can be rotated but does not change the connection properties between nodes and edges. The context graph in Fig. 4b has been rotated (flipped horizontally) (Figs. 8 and 9).

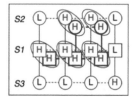
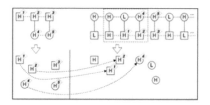
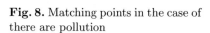

Fig. 8. Matching points in the case of there are pollution

Fig. 9. Sampling the data graph

3.2   Adversarial Learning Training Process

In this section, we build an approach to the unsupervised network alignment problem directly supporting the heterogeneous network representation. This method consists of the following steps

Sampling and Embedding

As shown above, the graphs are embedded in a multidimensional vector space as a set of points. Then the goal of training is to find a rotation matrix that rotates the nodes of the graph such that they close to the respective nodes of the other.

However, in practice, the number of nodes of the graphs is very large and changeable, while the size of the trained rotation matrix is proportional to the square of the number of nodes so we cannot take all of the nodes of the graphs to the training process. Therefore, at each training step, we sample a certain number of nodes in each graph. In our case, the size of the context graph is quite small, so there is no need to sample on the context graph, we just sample on the data graph, as shown in the figure below.

To do that, we maximize the co-occurrence of direct neighbor nodes using log-likelihood function and negative sampling, the popular strategy which helps to speed up the training process. As the networks are embedded independently, for simplicity, we focus on describing the embedding on one network $G = (V, E)$ without distinguishing between the source and the target networks. For every pair of nodes $(u, v) \in E$, the goal is to find the parameters Θ so as to maximize their co-occurrence probability between them:

$$maximize_\Theta \prod_{(u,v)\in E} p(u|v; \Theta) \qquad (1)$$

For network embedding, the parameters Θ are their vector representations (z_u, z_v). Let us denote by $p(E = 1|u, v)$ the probability that (u, v) came from the edge sets. To reduce the computation complexity, the log function is applied to transform the product to sum. The objective function then becomes maximizing the log-likelihood:

$$maximize_\Theta \prod_{(u,v)\in E} \log_{p(E=1|u,v;\Theta)} \qquad (2)$$

Training the Rotation Matrix with a 2-Step GAN Algorithm

We employ the adversarial learning based on GAN for training. Basically, the GAN algorithm has 2 steps: generative and discriminative. In step 1 (GAN phase 1), the GAN network will try to train the parameters of the rotation matrix so that each node in the context graph after rotation is as close as possible to its corresponding node on the data matrix. In step 2 (GAN phase 2), the GAN network tries to train so that each node in the context graph is rotated as far away from the nodes that are not its corresponding nodes on the data graph.

Thus, after a long enough training process, when the GAN network converges, the rotation matrix will bring each node on the context graph to its most logical position. For example, in Fig. 10, the node H of the context graph will rotate to a sufficiently close position which near to the corresponding node H on the data graph. However, among these "close enough" positions, the GAN will choose a position so that it is as far away from the other nodes on the data graph as possible.

In the case that a node on a context graph does not have a corresponding node on the graph sampled from the data graph, the GAN network will focus only on step 2, adjusting the rotation matrix so that this node is far away from other nodes of the graph after rotation as possible.

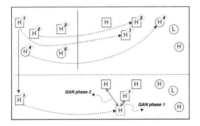
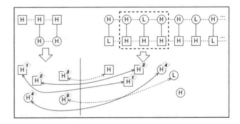

Fig. 10. The 2-steps GAN training process simulation

Fig. 11. The result of matching after the training

Formally, we design our framework as a structured probabilistic model [11] that contains latent variables z_s sampled from Zs and observed variables z_t sampled from Zt. This approach is in line with an unsupervised domain adaption problem as proposed by [8], wherein our case, a domain is represented by a network embedding (source or target).

- The discriminator is presented as a function that takes samples (both real and fake samples) as inputs and Θ_D as parameters.
- The generator is a function that takes z_s as an input and uses W as a parameter. It is worth noting that the learned W of the second player will be used as the mapping function for network alignment. Hence, from now on, we call this process an aligner instead of a generator as described in the original GAN concept.

Both processes have cost functions defined on Θ_D and W, yielding the final cost evaluated as follows.

$$J_W(W|\Theta_D) = -\frac{1}{n}\sum_{i=1}^{n}\log P_D(label = 0|W z_s^i) - \frac{1}{m}\sum_{j=1}^{m}\log P_D(label = 1|z_t^j) \quad (3)$$

where $z_s^i \in Z_s, z_t^j \in Z_t$.

Matching

After getting the reasonable parameters of the rotation matrix, the matching between the two graphs will be carried out by embedding and then by rotating. Next, we will compute the distance matrix between pairs of nodes and perform the pairing as described above. Figure 11 shows the result of pairing, in which the pairs described by solid line association are obvious (labelled) and dashed-linked pairs are pairs formed by the training process.

Theorem 1. *The complexity of our graph matching process is logarithmic.*

Proof. As presented in Fig. 11, the graph matching process include the following steps: (i) Embedding of context graph and sliced graph; (ii) rotation on the coordinates of the embedded vectors; and (iii) find the nearest nodes for each node of the context graph.

Step (i) and (ii) are in fact matrix multiplication actions, thus having constant complexity. Step (iii), once data are properly indexed, runs in is logarithmic time. Thus, the whole process executes in logarithmic complexity.

4   Experiment

To illustrate these steps, we use a real example from an ongoing IoT system, deployed to monitor environmental indicators in the village of the Vietnam National University Ho Chi Minh City. The image of the actual device deployment is shown in Fig. 12 below:

When working, IoT devices connect wirelessly to the router and stores the received data in a relational database. When collecting data into the database, we have a data storage pattern as shown in Fig. 13, with basic data columns, such as Sensor name, Value, and Timestamp.

SENSOR	VALUE	TIMESTAMP
CO	0.22281001508236	2020-11-25T12:22:49.000Z
NO2	0	2020-11-25T12:22:49.000Z
O3	0.24507454037666	2020-11-25T12:22:49.000Z
PLV1	0.00	2020-11-25T12:22:48.000Z
PLV2	0.00	2020-11-25T12:22:48.000Z
PLV3	0.00	2020-11-25T12:22:48.000Z
ANE	0.80	2020-11-25T12:22:48.000Z
WV	SE	2020-11-25T12:22:48.000Z
BAT	90	2020-11-25T12:22:47.000Z
BAT	96	2020-11-25T12:20:00.000Z

of 253273 entries

Fig. 12. The image of the actual device deployment.

Fig. 13. The data storage pattern used to collect data in our relational database.

To evaluate the efficiency and validity of our proposed framework, three real-world networks are adopted for experimental purposes, including protein-protein interaction (PPI) [5], economic network (ECON) [6], and UIT data as above mentioned. These experiments investigate comparative performance between different unsupervised models. Experimental results on the three datasets are presented in Table 1. The best existing unsupervised aligner we evaluate (REGAL) achieves an average accuracy of 32.28% for three datasets at the structural noise level of 0.2. With our method, we further improve it to 89.52%. We achieve an average improvement of 57.24% on node alignment when employing REGAL because while both algorithms use network structure information, REGAL adopts a strict assumption on topology consistency that the two nodes are similar when their neighbor's degree are the same, which makes its model susceptible to a considerable level of structure noise. The mean performance of FINAL and BigAlign on three datasets is 35.57% and 32.28%, respectively.

Table 1. Comparing the performance between existing methods and our method.

Dataset	Algorithm	REGAL	FINAL	BigAlign	Our Method
PPI	ACC	27.02	50.11	39.94	**97.84**
	MAP	36.23	11.98	2.91	**95.90**
	P@5	44.13	18.77	3.49	**99.10**
	P@10	50.06	30.90	5.54	**99.52**
ECON	ACC	35.47	26.02	33.15	**91.03**
	MAP	44.12	16.93	3.75	**86.49**
	P@5	52.44	29.62	3.84	**94.10**
	P@10	59.49	42.51	7.29	**97.10**
UIT	ACC	34.65	41.97	23.79	**79.69**
	MAP	43.31	17.12	3.01	**77.70**
	P@5	50.53	29.57	3.19	**90.56**
	P@10	57.02	39.37	6.55	**94.47**

5   Conclusion

In this paper, we propose a method for automatic context detection for data collected from IoT devices using context networking alignment approach. Especially, our training model is based on adversarial learning and thus enjoying a logarithmic complexity; making it applicable for practical problems such as air pollution detection. Experimenting with a real testing IoT system deployed at UIT, our method demonstrated significant improvement as compared with other state-of-the-art unsupervised network alignment methods.

Acknowledgment. This research is funded under grant number C2021-20-13. We acknowledge the support of time and facilities from Ho Chi Minh City University of Technology (HCMUT), VNU-HCM for this study.

References

1. Bayati, M., Gerritsen, M., Gleich, D.F., Saberi, A., Wang, Y.: Algorithms for large, sparse network alignment problems. In: 9th IEEE International Conference on Data Mining. IEEE, USA (2009)
2. Koutra, D., Tong, H., Lubensky, D.: Big-align: fast bipartite graph alignment. In: IEEE 13th International Conference on Data Mining. IEEE, USA (2013)
3. Zhang, S., Tong, H.: FINAL: fast attributed network alignment. In: Proceedings of the 22nd ACM SIGKDD International Conference on Knowledge Discovery and Data Mining. ACM, USA (2016)
4. Heimann, M., Shen, H., Safavi, T., Koutra, D.: REGAL: representation learning-based graph alignment. In: Proceedings of the 27th ACM International Conference on Information and Knowledge Management. ACM (2018)
5. Hamilton, W., Ying, Z., Leskovec, J.: Inductive representation learning on large graphs. In: Advances in Neural Information Processing Systems. Curran Associates, USA (2017)
6. Rossi, R., Ahmed, M.: The network data repository with interactive graph analytics and visualization. In: Proceedings of the 29th AAAI Conference on Artificial Intelligence, USA (2015)
7. Alexis, C., Guillaume L., Marc'Aurelio, R., Ludovic, D., Hervé, J.: Word translation without parallel data. J. Comput. Res. Repos. (2017)
8. Schonemann, P.H.: A generalized solution of the orthogonal procrustes problem. Psychometrika **31**(1), 1–10 (1966)
9. Smith, S.L., Turban, D.H., Hamblin, S., Hammerla, N.Y.: Offline bilingual word vectors, orthogonal transformations and the inverted softmax. J. Comput. Res. Repos. (2017)
10. Cisse, M., Bojanowski, P., Grave, E., Dauphin, Y., Usunier, N.: Parseval networks: improving robustness to adversarial examples. In: Proceedings of the 34th International Conference on Machine Learning, Australia (2017)
11. Ellis, P.: Extension of phase plane analysis to quantized systems. IRE Trans. Autom. Control **4**, 43–54 (1959)

Deep Convolutional Support Vector Machines for Human Activity Recognition

Phuoc-Hai Huynh[1,2(✉)] and Van Hoa Nguyen[1,2]

[1] Faculty of Information Technology, An Giang University,
An Giang, Vietnam
{hphai,nvhoa}@agu.edu.vn
[2] Viet Nam National University Ho Chi Minh City, Ho Chi Minh City, Vietnam

Abstract. Over the past few decades, the Internet of Things has enabled advancements and innovations to develop modern health care applications. Human Activity Recognition (HAR) is new trend using data generated from the wearable devices to monitor health. The activity monitoring applications require high accuracy of the activity classification that is a challenging time series classification task. In this context, Deep Convolutional Support Vector Machine (DCSVM) model is proposed to perform human activity recognition. This model trains all 1D convolution layers through a support vector machine. The classifying results show that the our proposal improves generalization performance on HAR dataset without preprocessing data.

Keywords: Deep Convolutional Support Vector Machines · Deep convolutional neural network · Support vector machines · Human activity recognition

1 Introduction

In recent years, smartphones and smartwatches are enabled the rapid growth of software in healthcare. Human activity data is collected from wearable sensors. These datasets can be used to research in medical diagnosis and fitness monitoring. In the healthcare system, many applications are developed based on human activity data [21,25]. HAR is the sequences classification of sensor data into clearly defined movements. The first study for HAR dates back to the late '90s [8]. The HAR research community has proposed many methods and techniques during the last 30 years [7]. In data mining, these algorithms are supervised learning tasks to classify a multivariate time series. It plays an important key in the field of medicine and healthcare, especially in detect human diseases and fitness tracking [3]. There are many studies to address this task [3,18,21,25]. However, a major challenge of HAR models is that a feature extraction task usually wastes time and requires experience of researchers [7,24]. These problems are caused by

N.-T. Nguyen et al. (Eds.): ICIT 2022, LNDECT 148, pp. 320–329, 2022.
https://doi.org/10.1007/978-3-031-15063-0_30

the large number of observations collected per second, lead to the data collected is very complex.

Deep convolutional neural network (DCNN) and SVM are two popular learning algorithms. DCNN approaches can automatically learn the features of complex data such as image data [16]. In HAR, some deep learning was used such as [18,22]. For SVM, many studies are used for HAR because it can effectively discriminate features [2,17]. In this work, we implement a Deep Convolutional Support Vector Machines (DCSVM) for accurately classifying complex human activities. The key idea of DCSVM is to train a specialized 1 dimension convolutional network (1DCNN) to automatically learn robust features from human activity data and provide them as input and fed into linear support vector machine classifiers. In our model, a soft-max is replaced by SVM classifier. This approach is beneficial because the classifier is optimized by SVM. The model is trained by using back-propagation and stochastic gradient descent to learn hidden features of human activity data. In addition, we also implement other ten learning algorithms in experiment to compare with proposed model. Our experiment compares the algorithms mentioned above with DCSVM and suggests the most efficient algorithm based on experiment results. The experimental results show that DCSVM model enhances generalization performance on the HAR dataset. Moreover, we also evaluate performance of DCSVM and 1DCNN on subset HAR data. This experiment aims to show the effectiveness of DCSVM on limit training datasets.

This paper consists of 4 parts. Section 2 presents classifying methods. For Sect. 3, the experiment for HAR has been done and evaluated. Section 4 brings things to a close.

2 Methods

In this session, we present about 1DCNN and SVM, then DCSVM is introduced and explained for human activity classification. The Fig. 1 shows the framework of HAR.

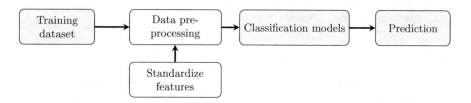

Fig. 1. Experiment pipeline for preprocessing and classification.

Firstly, the training data is normalized. Standardizing a dataset is critical for classification models since individual features that do not like standard normally distributed data may perform poorly.

Secondly, we evaluate the classification performance of our proposal for HAR. Therefore, we construct multiple models to classify human activity data. In experiment, we have constructed a baseline model including linear SVM, k nearest neighbors, decision trees, naive Bayes, and logistics regression (called SVM, kNN, DT, NB, LR). The ensemble models using decision trees include random forests (RF), bagging (BA-DTs), gradient boosting (GB-DTs), and XGboost [5] of decision trees. Moreover, we also implement 1 dimension DCNN (1DCNN). This model has architecture the same as the DCSVM, but it uses soft-max to classify. In summary, our experiments are designed with ten classifiers to compare the predictive of the models.

Finally, these models are trained on training dataset and validated on testing dataset. Accuracy, precision, recall, and F1 scores are four standard categorization metrics that are evaluated. Furthermore, we also use confusion matrix to evaluate models.

2.1 1D Deep Convolutional Neural Network

Convolutional neural networks is most commonly deep learning model, which is widely used by many studies to process images [16]. In these researches, the input of deep neural networks is 2D images. In addition to image classification, it is also applied to speech recognition [20], gene expression data [12], RNA data [11]. However, for human activity data, a sample is 1-dimension vector. Therefore, the classic CNN is not suitable for this kind data. However, a 1-dimension CNN is a special architecture of deep learning that the input of this model is a 1D vector. For HAR, 1DCNN has been used in [9,18,22].

2.2 SVM Algorithm

SVM is popular algorithm for classification task [23]. This algorithm's main goal is to find a hyper-plane in order to maximize the margin and minimize the errors between the two classes. For the non-linear classification problem, SVM uses kernel-based learning technologies [6]. For multi-class issue, there are some method including OVA (one-versus-all) [23] and OVO (one-versus-one) [15]. In HAR, it is usually used because it effectively [1,18].

2.3 Deep Convolutional Support Vector Machines

In this work, we implement a DCSVM model for HAR. The model is designed from convolution, batch-normalize, max-pooling, drop-out, fully-connected layers, then joined with SVM layer. The architecture of DCSVM is displayed in the Fig. 2. The layers consist of a input, two 1D-Convolution (C1 and C2), two 1D-Max-Pooling (M1 and M2), a fully-connected (FC), and an output layer using a SVM.

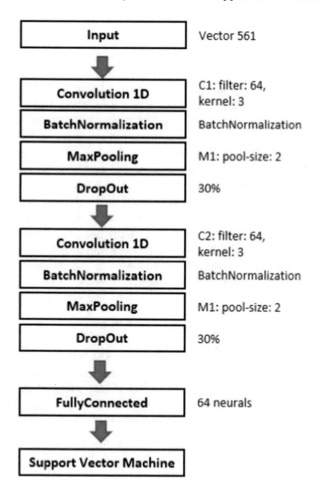

Fig. 2. The architecture of Deep Convolutional Support Vector Machine

In this architecture, the nodes of layer n are linked to the nodes of the layer $(n+1)$. This process helps to decrease the number of parameters of model. Moreover, it also makes one dimension CNN an effective model for HAR. Sixty-four feature maps are used and a kernel size of 3 on both convolution layers. The input of max-pooling layers are received from information of the convolutional layer. This layer aims to reduce the learned features. Besides, we also use Batch-Normalize in this model, and it helps improve the effective training of the network. Moreover, the Drop-Out method is also used, which prevents all nodes in a layer from synchronously optimizing their weights. In process training, the convolution layers learn features from human activity data. An advantage of DCSVM is that we don't use any feature selection technology in pre-processing data. This model can learn latent features of data. The dropout layers is a effective way to handle over-fitting issue of model. Finally, the learnt features are transformed into a vector and delivered to the SVM layer.

The key idea of the DCSVM model is used 1D convolution for human activity data. This model is suitable to extract latent features of sequences data. Therefore, the main advantage of using 1DCNN is that it can learn from the raw time series data directly. Moreover, the local dependency as well as scale in-variance of DCSVM can be bring benefit for predicting. Furthermore, the scale-invariant for different paces or frequencies is the second additional benefit.

For the traditional 1DCNN, it uses the soft-max or 1-*of*-*k* encoding at top network. For DCSVM, the soft-max layer is replaced by SVM classifier. We use one-vs-rest approach [23] to extend SVM in this work. Prediction using SVM is similar to using a soft-max. However, the soft-max layer aims to minimize cross entropy or maximizes the log likelihood. For DCSVM, SVM classifier in top network tries to find the maximum margin between different classes. We propose using the multi-class SVM objective to train robust features learned by 1DCNN for classification tasks for our model. The weights of lower layers are learned by back-propagating from the SVM.

3   Experimental Details and Results

In section, the combine model of 1DCNN and SVM is used to predict activities of human. The models are implemented in Python using Scikit library (www. scikit-learn.org), linear SVM [4]. The DCSVM and 1DCNN are implemented based on TensorFlow (www.tensorflow.org) and Keras (www.keras.io) libraries. The server used was specified as two Intel Xeon E5 CPU 22-core 2.7 GHz. It has 32 GB RAM.

3.1   Datasets of Human Activities

In experiment, the dataset is downloaded from UCI (University of California at Irvine) database [10]. It has six classes. The dimension of data is that 561 features are extracted from sensor signals present in a smartphone. There are 10299 samples out of which 70% (7352 samples) is trained and 30% (2947 samples) was tested. The description of dataset are fully available in [2].

3.2   Parameter Evaluation

For baseline classification models, SVM and LR use $C=1$, kNN uses 7 neighbors. 100 trees are used to build RF, BA-DTs, GB-DTs, ET-DTs, XGBoost. Gini impurity is used for splitting algorithms in decision trees. The Adam [14] is used to tune DCSVM. The model is fit for a fixed number of 100 epochs. A good value for batch size is 32 samples.

3.3   Classification Results

This session displays the results of the experiment. The classifying results of DCSVM are compared to 10 models including NB, DT, BA-DTs, kNN, RF, GB-DTs, ET-DTs, LR, XGB, SVM and 1DCNN. Table 1 presents resutls of models. The maximum values are bolded, while the second values are italicized. The DCSVM results are presented in the last column. From columns 2 to 11, the classifying results of ten models are presented. Figure 3 shown the comparison accuracy results of using 11 models. The results highlight the DCSVM model that achieved best classification performance on HAR dataset [2] with all classification metrics including accuracy, precision, recall and F1 scores. Moreover, from Fig. 4, we can observe that it is not biased towards any particular label.

Table 1. Classification results of models on HAR dataset

Models	NB	DT	BA DTs	kNN	RF	GB DTs	ET DTs	XGB	LR	SVM	1DCNN	DCSVM
Accuracy	57.14	80.96	82.15	89.58	91.92	92.74	93.82	93.82	96.23	95.76	*97.62*	**98.30**
Precision	57.3	82.1	85.937	89.6	92.8	93.23	94.79	93.9	96.3	95.8	*97.65*	**98.30**
Recall	57.1	80.2	92.13	89.1	91.9	92.70	92.70	93.8	96.2	95.8	*97.62*	**98.30**
F1	48.8	79.5	92.12	89.1	91.9	92.68	92.68	93.8	96.2	95.7	*97.63*	**98.31**

As we can see in Table 1, For the HAR dataset, the DCSVM model produces the best results. The DCSVM achieves an overall accuracy of 98.3%, outperforming other methods such as kNN (89.58%), DT (80.96%), NB (57.14%), LR (96.23%), and SVM (95.76%). Compare with ensemble of decision trees models, Table 1 and Fig. 3 show that it improves the accuracy mean of 6.38, 4.48, 16.15, 5.56, 4.48% points obtained by RF, XGB, BA-DTs, GB-DTs and ET-DTs, respectively.

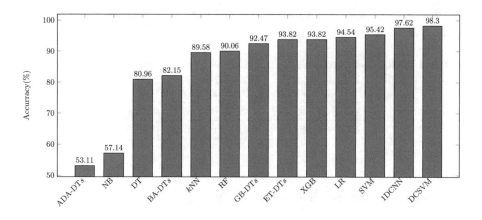

Fig. 3. Comparison accuracy of models

The performance of SVM layer and soft-max layer is compared and evaluated. In order to evaluate, a confusion matrix is used to evaluate DCSVM and 1DCNN. In Table 1, using soft-max layer, the transitional 1DCNN has the accuracy (97.62%), precision (97.65%), recall (97.62%) and F1 score (97.63%). The Fig. 3 is compared result of models. Using the SVM classifier, the DCSVM improves by 0.68% points obtained by 1DCNN. In the confusion matrix of the 1DCNN and DCSVM models 4, it is clear that our model has the least false positives.

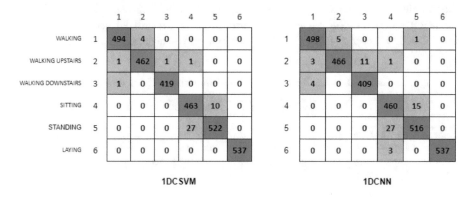

Fig. 4. Confusion matrix of the 1DCNN and DCSVM models

In addition, we are also interested in the performance of DCSVM on a limited training dataset. Therefore, the proposed model also tests on a subset of the HAR dataset with a limited number of samples. We trained DCSVM and 1DCNN on 50% of samples (3676 samples). The performance two models are presented in Table 2. It is clear that DCSVM has better performance than 1DCNN on all classification indices. For a small subset of the HAR dataset, the accuracy score of our model decreased by only 1.05%, while 1DCNN decreased by 3.73%. As result classification, DCSVM can effectively classify on limit dataset and predicts better 1DCNN.

Table 2. Classification results of 1DCNN and DCSVM on subset HAR dataset

	Accuracy	Precision	Recall	F1
1DCNN	93.89	94.05	93.89	93.91
DCSVM	97.25	97.27	97.25	97.25

4   Comparison with Related Works

Over the years, many HAR techniques have been implemented and various methods have been proposed to handle a classification process that provides reliable results. In this session, we discuss and compare with these other studies. We focus the results of studies on HAR dataset [2].

Table 3. Classification results of related works

Study	Model proposed	Accuracy	Precision	Recall
Proposed model	**DCSVM**	*98.3*	**98.3**	**98.3**
[18]	SVM and 1DCNN	96.6, 96.13		
[19]	Markov model	91.3		
[13]	Markov and CRF models		91.42	91.76
[1]	HF-SVM		89	
[21]	Decision tree	88.02		
[25]	Random forest	90		
[17]	Bayes, MLP, kNN, SVM-RBF	95.2; 78.42; 96.94; **99.26**		

In study of [18], a feed-forward DCNN is used. It was found that the SVM (96.6%) and the proposed 1DCNN (96.13%) were the best-performing models. The Markov algorithm is suggested for HAR in [19]. In the following paper [13], the authors are used the hidden markov and conditional random fields models. The modified multi-class SVM model is proposed for the benchmark HAR dataset in [2]. Decision tree is used in [21]. A random forest is proposed in this paper [25]. Bayes, kNN, MLP, SVM based on HAR tasks was discussed in [17].

Moreover, we also compare performance of the DCSVM and related works (Table 3). The comparison result show that DCSVM provides remarkable accuracy and performs well for HAR. It is show that our model classifies well with time series data as well as this approach take advantages of 1DCNN and SVM. However, the it's disadvantage is a long time to train.

5   Conclusions

In conclusion, a new DCSVM architecture is presented for HAR tasks. According to experimental results, our model outperforms other approaches. These improvements are due to our models' ability to extract more discriminative features by combining 1 dimension convolutions layers with support vector machines. We plan to provide further experiments, more HAR datasets, as well as comparisons with the other methodologies. In addition, the network architecture will update to find best architecture for this kind data.

References

1. Anguita, D., Ghio, A., Oneto, L., Parra, X., Reyes-Ortiz, J.L.: Energy efficient smartphone-based activity recognition using fixed-point arithmetic. J. Univers. Comput. Sci. **19**(9), 1295–1314 (2013)

2. Anguita, D., Ghio, A., Oneto, L., Parra, X., Reyes-Ortiz, J.L., et al.: A public domain dataset for human activity recognition using smartphones. In: Esann, vol. 3, p. 3 (2013)
3. Avci, A., Bosch, S., Marin-Perianu, M., Marin-Perianu, R., Havinga, P.: Activity recognition using inertial sensing for healthcare, wellbeing and sports applications: a survey. In: 23th International Conference on Architecture of Computing Systems 2010, pp. 1–10. VDE (2010)
4. Chang, C.C., Lin, C.J.: Libsvm: a library for support vector machines. ACM Trans. Intell. Syst. Technol. (TIST) **2**(3), 1–27 (2011)
5. Chen, T., He, T.: Xgboost: extreme gradient boosting (2021)
6. Cristianini, N., Shawe-Taylor, J., et al.: An Introduction to Support Vector Machines and Other Kernel-Based Learning Methods. Cambridge university press (2000)
7. Ferrari, A., Micucci, D., Mobilio, M., Napoletano, P.: Trends in human activity recognition using smartphones. J. Reliable Intell. Environ. **7**(3), 189–213 (2021). https://doi.org/10.1007/s40860-021-00147-0
8. Foerster, F., Smeja, M., Fahrenberg, J.: Detection of posture and motion by accelerometry: a validation study in ambulatory monitoring. Comput. Hum. Behav. **15**(5), 571–583 (1999)
9. Goh, J.X., Lim, K.M., Lee, C.P.: 1D convolutional neural network with long short-term memory for human activity recognition. In: 2021 IICAIET, pp. 1–6 (2021)
10. Guvenir, H.A., Acar, B., Demiroz, G., Cekin, A.: A supervised machine learning algorithm for arrhythmia analysis. In: Computers in Cardiology 1997, pp. 433–436. IEEE (1997)
11. Huynh, P.H., Nguyen, V.H., Do, T.N.: Novel hybrid DCNN-SVM model for classifying RNA-sequencing gene expression data. J. Inf. Telecommun. **3**(4), 533–547 (2019)
12. Huynh, P.H., Nguyen, V.H., Do, T.N.: Improvements in the large p, small n classification issue. SN Comput. Sci. **1**(4), 1–19 (2020). https://doi.org/10.1007/s42979-020-00210-2
13. Kim, E., Helal, S., Cook, D.: Human activity recognition and pattern discovery. IEEE Pervasive Comput. **9**(1), 48–53 (2009)
14. Kingma, D.P., Ba, J.: Adam: a method for stochastic opoimization (2017)
15. Krebel, U.G.: Pairwise classification and support vector machines. Advances in kernel methods: support vector learning, pp. 255–268 (1999)
16. Krizhevsky, A., Sutskever, I., Hinton, G.E.: Imagenet classification with dcnns. Adv. Neural. Inf. Process. Syst. **25**, 1097–1105 (2012)
17. Marinho, L.B., de Souza Junior, A.H., Rebouças Filho, P.P.: A new approach to human activity recognition using machine learning techniques. In: Madureira, A.M., Abraham, A., Gamboa, D., Novais, P. (eds.) ISDA 2016. AISC, vol. 557, pp. 529–538. Springer, Cham (2017). https://doi.org/10.1007/978-3-319-53480-0_52
18. Muralidharan, K., Ramesh, A., Rithvik, G., Prem, S., Reghunaath, A., Gopinath, M.: 1D convolution approach to human activity recognition using sensor data and comparison with machine learning algorithms. Int. J. Cogn. Comput. Eng. **2**, 130–143 (2021)
19. Ronao, C.A., Cho, S.B.: Har using smartphone sensors with two-stage continuous hidden markov models. In: 2014 10th International Conference on Natural Computation (ICNC), pp. 681–686. IEEE (2014)
20. Sainath, T.N., Mohamed, A.R., Kingsbury, B., Ramabhadran, B.: Deep convolutional neural networks for lvcsr. In: Speech and Signal Processing, pp. 8614–8618 (2013)

21. Sánchez, V.G., Skeie, N.O.: Decision trees for human activity recognition modelling in smart house environments. Simul. Notes Eur. **28**(4), 177–184 (2018)
22. Tsokov, S., Lazarova, M., Aleksieva-Petrova, A.: Accelerometer-based human activity recognition using 1D-CNN. In: IOP Conference Series: Materials Science and Engineering, vol. 1031, p. 012062. IOP Publishing (2021)
23. Vapnik, V.: The support vector method of function estimation. In: Nonlinear Modeling, pp. 55–85. Springer (1998). https://doi.org/10.1007/978-1-4615-5703-6_3
24. Weiss, G.M., Yoneda, K., Hayajneh, T.: Smartphone and smartwatch-based biometrics using activities of daily living. IEEE Access **7**, 133190–133202 (2019)
25. Xu, L., Yang, W., Cao, Y., Li, Q.: Human activity recognition based on random forests. In: 13th ICNC-FSKD, pp. 548–553 (2017)

Distributed Deep Learning for Big Data Time Series Analysis

Dinh-Quang Hoang[1,2], Dang-Khoa Tran[1,2], Viet-Thang Le[1,2],
The-Manh Nguyen[1,2], and Trong-Hop Do[1,2(✉)] (ID)

[1] University of Information Technology, Ho Chi Minh City, Vietnam
{18521294,18520936,18520356,18521084}@gm.uit.edu.vn, hopdt@uit.edu.vn
[2] Vietnam National University, Ho Chi Minh City, Vietnam

Abstract. This paper deals with the problem of traffic flow prediction in
large scale, which has three major challenges. First, the prediction must
be performed in large scale that requires lots of computation. Second,
achieving good accuracy of time series prediction is challenging for most
traditional machine learning algorithms. Third, the amount of training
data is so big that makes it very difficult to train on a single machine.
This paper solves all these problem by proposing a big data time series
analysis system using distributed deep learning. The entire system runs
on Apache Spark big data framework, which allows the prediction to be
performed in large scale. A Temporal Convolutional Networks (TCN)
model is proposed and applied to achieve high accuracy of traffic flow
prediction. The model are trained on a huge amount of training data
using a distributed deep learning framework namely BigDL. The experimental results show that the system can predict the traffic flow in large
scale at accuracy levels that are much higher than that of traditional
machine learning models.

Keywords: Big data · Time series analysis · Distributed deep learning

1 Introduction

The achievements and contributions of Artificial Intelligence (AI) to our world
are increasing day by day. One of the important factors contributing to the success of AI is deep learning. Applications of deep learning have supported people
with many things from work to life. In the healthcare field, disease diagnosis
systems can diagnose many times faster and more accurately than humans. Sentiment analysis or fraud detection systems used by Google, Twitter, Facebook,
etc. have brought them huge profits. In transportation, many traffic problems
such as traffic jams, traffic accidents, etc. are also significantly solved thanks to
intelligent traffic management systems, smart cars or self-driving cars are also
one of the outstanding achievements of AI today. In order to get quality deep
learning models that can be applied in practice as above, a good algorithm is
not enough, data is the core to create a good model. In recent years, the amount

N.-T. Nguyen et al. (Eds.): ICIT 2022, LNDECT 148, pp. 330–338, 2022.
https://doi.org/10.1007/978-3-031-15063-0_31

of data has increased rapidly due to the development of science and technology, especially social networks, e-commerce, sensors, etc. led to a data explosion and the emergence of Big Data. This large and variety amount of data is both an advantage and a challenge for current deep learning. Some fields in Big Data such as Computer Vision, Voice Recognition have applied Deep Learning to improve model results. Deep learning algorithms extract meaningful abstract representations of the raw data through the use of an hierarchical multi-level learning approach, where in a higher-level more abstract and complex representations are learnt based on the less abstract concepts and representations in the lower levels of the learning hierarchy [1]. However, in addition to the advantages, the characteristics of big data also created many challenges for deep learning algorithms such as dealing with streaming input data, huge amount of data, high-dimensional data, large-scale models, etc.

Time series data is one of the most widely used data types today, even real time series data. Time series data analysis is widely used in forecasting tasks for management, monitoring, etc. Algorithms or models for time series problems are often more complex than other problems because they do not only exploit relationships and correlations between attributes, but also need to exploit relationships among time points to get the best performance. Current deep learning algorithms such as Recurrent Neural Network (RNN), Long-Short Term Memory (LSTM), etc. have given high performance for real-time problems. But to do that, the training time for these models is quite long, it can be many days, many weeks, even many months. Not only that, the cost of system resources consumed for training is very expensive. And when the volume of data is constantly increasing, up to hundreds of millions, billions of data points, models with billions of parameters, the current server and machine resources may not be enough to train the model.

To contribute to solving the above problems, in this paper, a method for big data time series analysis by deep learning is proposed. This method based on distributed computing principles to build deep learning models for big data time series analysis on Apache Spark platform. This deep learning model uses the BigDL framework, which is a distributed deep learning framework for big data. BigDL implements distributed, data parallel training directly on top of the functional compute model of Spark [2]. The experiment was conducted on the prestigious large traffic time series dataset published on IEEE DataPort. The experimental results is the comparison of the performance of the deep learning model trained by the traditional method with the deep learning model trained by the proposed method and between the models trained by the proposed method and the proposed method through evaluation measures and training time. Experimental results will be presented in detail in the following sections of the paper.

2   Fundamental of Big Time Series Analysis and Distributed Deep Learning

2.1   Big Data Processing

Big Data is a term that refers to very large and complex data sets including structured, semi-structured or unstructured data, which makes data processing software system's inability to collect, manage and process data within a reasonable period of time. This data comes from many different sources such as from sensors that collect traffic information, weather forecasts, updates, images, videos on social networking sites, transactions on exchanges. e-commerce, etc. Currently, there is still no exact answer to the question of what data is called Big Data. People often judge whether data is Big Data or not based on 5V's: Volume, Velocity, Variety, Veracity and Value Volume is a very large volume of data, measured in Terabytes, Exabytes, and more. Velocity is the growth of the data in terms of speed. This growth rate is extremely fast. Variety is the variety of data. Traditionally we used to be organized as structured data, today most of the stored data is created unstructured (video, images, audio, data from sensors, etc.). Veracity is the reliability or accuracy of data and Value is the value of the data. Data is only really useful when it is collected from a reliable data source and the data must bring some value to the individual, business or organization that owns it. Big data processing includes all tasks from collection, storage, preprocessing, feature extraction, model training, etc. For example, big data must use cloud technology to store because the volume is very large. Big data analysis also needs to reach real-time speeds and handle a variety of data types.

To serve tasks related to big data processing, Apache Spark was born and became an extremely useful platform. Apache Spark is an open source framework used mainly for big data analysis, machine learning and real-time processing based on distributed processing. Apache Spark is faster, easier to use, and more flexible than traditional platforms like Hadoop. By performing computations on the cluster that enables high-speed data analysis when reading and writing data. This processing speed is due to the fact that Spark's computations are performed at the same time on many different machines and are performed either in-memory or entirely in RAM. Apache includes many libraries for various tasks from SQL to machine learning, real-time data processing, and support for many programming languages like Scala, Python, Java, and more.

2.2   Time Series Analysis

Time Series Analysis or Sequence Analysis is the next prediction in a given input sequence based on what was previously observed. Predictions can be anything that might happen next: a symbol, a number, the next day's weather, the next term in the speech, and so on. Time series analysis can be very useful in applications such as stock market analysis, weather forecasting, and product recommendations. There are many approaches to the problem of time series analysis such as based on probabilistic statistical models, machine learning, deep learning, etc. Auto Regressive Integrated Moving Average (ARIMA) is a traditional time series forecasting model based on the hypothesis of stationary series and constant error variance. The model will represent multiple linear regression equations of the input variables (dependent variables) as 3 components, Auto Regression (AR), Intergrated (I) and Moving Average (MA).

Prophet is also a powerful model for time series forecasting. Prophet is a procedure for forecasting time series data based on an additive model where non-linear trends are fit with yearly, weekly, and daily seasonality, plus holiday effects. It works best with time series that have strong seasonal effects and several seasons of historical data. Prophet is robust to missing data and shifts in the trend, and typically handles outliers well [3]. Prophet model is built based on 3 main components: the trend function g(t), seasonality function s(t) and holidays function h(t). Prophet's output is calculated according to the formula shown in (1).

$$y(t) = g(t) + s(t) + h(t) + \epsilon_l. \tag{1}$$

In addition, in deep learning, Long-Short Term Memory (LSTM) is a commonly used model for time series analysis. LSTM is designed as a special form of RNN to avoid long-term dependency. The predictions of the LSTM network ar:e always adjusted based on past experience. The core components of the LSTM model are the cell states and the task of controlling information flows in the network is performed by gates, the LSTM model includes update gate controls the retrieval of information of the previous class (or unit), output gate decides how much information to output and forget gate decides what information to keep and which information is close to being ignored. Suppose x_t is the input at time t, h_t is the output of which cell state C_t. The LSTM algorithm is shown in the order of equations from (2) to (6).

$$f_t = \sigma \left(W_f \cdot [h_{t-1}, x_t] + b_f \right) i_t = \sigma \left(W_i \cdot [h_{t-1}, x_t] + b_i \right) \tag{2}$$

$$\tilde{C}_t = \tanh \left(W_C \cdot [h_{t-1}, x_t] + b_C \right) \tag{3}$$

$$C_t = f_t * C_{t-1} + i_t * \tilde{C}_t \tag{4}$$

$$"o_t = \sigma \left(W_o [h_{t-1}, x_t] + b_o \right) \tag{5}$$

$$h_t = o_t * \tanh \left(C_t \right) \tag{6}$$

2.3   Distributed Deep Learning

Distributed machine learning refers to multi-node machine learning algorithms and systems designed to improve performance, increase accuracy, and scale to larger input data sizes [4]. Distributed deep learning is a field in distributed machine learning and is applied in many different fields because of its effectiveness. Distributed deep learning includes two main types, data parallelism and model parallelism. In data parallelism, the input data is divided into partitions equal to the number of worker nodes (machines) and processed in parallel.

Model weights are replicated for all worker nodes and trained on the partitioned data for that node. The resulting weights or gradients at the nodes after each iteration will be aggregated at the Parameter Server and shared back to the nodes to calculate for the next iteration. The results at Parameter Server can be aggregated in many different ways, the simplest example is parameter averaging. Instead of partitioning the data for worker nodes, in model parallelism, the model itself is divided into parts that are trained simultaneously across different worker nodes. All worker nodes use the same data set and they also send the resulting model weights to the Parameter Server for aggregation and the Parameter Server returns the latest model. Model parallelism is much more difficult to implement and only works well in models with naturally parallel architectures. Therefore, the model proposed in this paper uses data parallelism.

BigDL is one of the distributed deep learning frameworks for Apache Spark, which has been used by a variety of users in the industry for building deep learning applications on production big data platforms released in 2019 [5]. Recently, BigDL released version 2.0 with the combination of the original BigDL and Analytics Zoo. With powerful features like DLlib, Orca, RayOnSpark, Chronos, etc. BigDL is really powerful for Big Data analysis tasks.

3   Proposed System

3.1   Proposed TCN Model

The model used in the experiment is a basic TCN model (Temporal Convolutional Network) with an architecture of 3 temporal blocks. Each temporal block consists of two 1D Causal Convolutional layers and alternately Normalization, Dropout and ReLU activation layers as described in the architecture in Fig. 1. The model is designed with an input sequence length of 288 corresponding to one day's speed information. The number of input features is 5, each of which contains speed information for a road in the dataset. The number of outputs equal to the input is 5, containing the forecasting results for the next time. Dropout ratio used is 0.1.

3.2   Distributed Deep Learning Training

From the model architectures that have been designed, the paper trains deep learning models by distributed method, in the form of data parallelism. Distributed training process is carried out according to the procedure Fig. 2. The entire process is done on the Spark platform First, historical data on traffic speed is collected from the data source. Then, the entire data is passed through preprocessing steps on the Apache Spark platform to clean and convert the data back to the correct model training format. The cleaned data is divided into three parts: Training Data, Validation Data and Test Data as shown in Fig. 2. After the data preparation step, proceed to use two libraries, Orca and RayOnSpark in BigDL to set up a distributed environment. This distributed

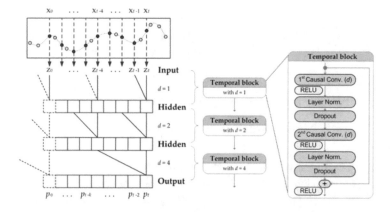

Fig. 1. Proposed TCN model

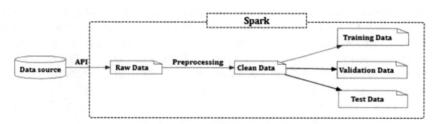

Fig. 2. Data collection and splitting procedure

environment is a cluster of distributed machines running on the Apache Spark platform. The training and validation data was fed into the newly established distributed environment and converted into the standard time series data format in BigDL TSDataset using the Chronos library. Chronos also supports the implementation of the designed deep learning model on a distributed environment to facilitate the training process. According to the form of data parallelism introduced above, data will be partitioned into equal parts and distributed among machines in Spark's distributed cluster (worker node). The installed deep learning model is also replicated to all machines in the distributed cluster and trained on separate pieces of data on those machines. The details of the training process are shown in Fig. 3.

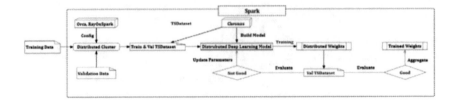

Fig. 3. Training procedure

During distributed training, the model weights are updated after the forward-backward propagation has been completed at the machines in the distributed cluster. After the forward-backward process, the machines will calculate the local gradients and update the weights for the model according to the parameter synchronization method. Specifically, the main model weight is divided into n partitions for calculation, the local gradients are also divided into n parts corresponding to the partitions, the partitions will perform the synthesis of the corresponding local gradients. for that partition by summing and performing a weight update for that partition. Each partition after completing the weight update will send the results to the Parameter Server to synthesize into a complete set of weights for the model, preparing for the next forward-backward step. This process is repeated until the training is complete. After completing the distributed training for the model, the model is tested on the test set according to the procedure of Fig. 4. If the test results are not good, the model will be edited, updated the parameters and re-trained. On the contrary, if the test results are good, the model will be kept and the training phase will end, moving to the evaluation and error analysis phase. The process of evaluating and analyzing model errors on the test dataset is also distributed similarly to the training process, the Chronos library provides the ability to reload the model from the trained weight set to progress. Testing, evaluation and fault analysis. The test and evaluation results are presented in detail in the experimental part of the paper.

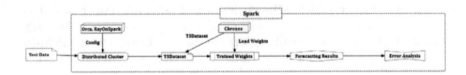

Fig. 4. Test procedure

4 Experiment

4.1 Experimental Environment

The whole experiment is done on Ubuntu 20.04 environment with 400 GB RAM and 24 CPUs Intel(R) Xeon(R) E5-2670 0 @ 2.60 GHz, 24 cores. The training data goes through several preprocessing steps such as deleting unnecessary columns, group by, pivot, interpolation to convert to the correct format of the model input. This entire data preprocessing is done on the Spark platform, written in the Python programming language supported by PySpark.

4.2 Experiment Results

After the experimental process, the distributed TCN model shows very good predictive ability on many evaluation levels including several hours, 1 day and 1 week. The model's forecast results are shown in Figs. 5, 6 and 7. The resulting graphs show that the deviation between the model's forecast results and the actual speed is very small. There are times when my prediction is almost completely accurate. The model training time is also relatively short at 6939.01 s. From all the above results, it can be seen that the distributed trained deep learning model gives the performance not only not inferior to the traditional training methods, but also can scale the model on big data.

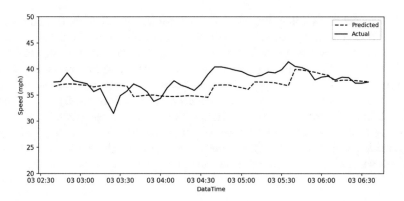

Fig. 5. Forecast results of the model in a few hours

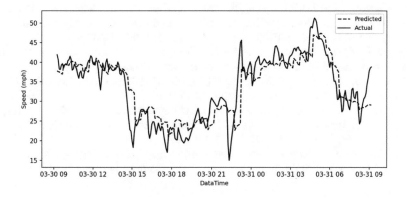

Fig. 6. Forecast results of the model in 1 day

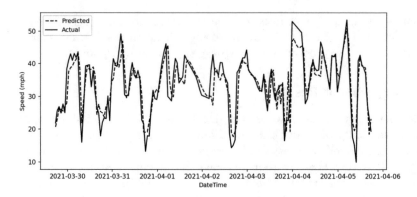

Fig. 7. Forecast results of the model in 1 week

5   Conclusion

Traffic flow prediction is a very challenging problem in practical due to both the required huge computation and the difficulties of time series prediction problem. This paper proposed a traffic flow prediction system that run on Apache Spark big data framework to allow predictions to be performed in large scale. A deep learning model namely Temporal Convolutional Networks is proposed for traffic flow prediction and achieve very high performance compared to that achieved with traditional machine learning model. Due to the huge amount of training data, the model is trained on a computer cluster using a distributed deep learning framework namely BigDL. The experimental results show that the proposed system can predict the traffic flow at very high accuracy in large scale.

References

1. Najafabadi, M.M., Villanustre, F., Khoshgoftaar, T.M., Seliya, N., Wald, R., Muharemagic, E.: Deep learning applications and challenges in big data analytics. J. Big Data **2**(1), 1–21 (2015). https://doi.org/10.1186/s40537-014-0007-7
2. Dai, J.J., et al.: Bigdl: a distributed deep learning framework for big data. In:Proceedings of the ACM Symposium on Cloud Computing (2019)
3. Taylor, S.J., Letham, B.: Forecasting at scale. PeerJ Preprints 5:e3190v2 (2017). https://doi.org/10.7287/peerj.preprints.3190v2
4. Galakatos, A., Crotty, A., Kraska, T.: Distributed Machine Learning. In: Liu, L., Özsu, M.T. (eds.) Encyclopedia of Database Systems. Springer, New York, NY (2018). https://doi.org/10.1007/978-1-4614-8265-9
5. Dai, J.J., et al.: Bigdl: a distributed deep learning framework for big data. In: Proceedings of the ACM Symposium on Cloud Computing, pp. 50–60 (2019)

Safety Helmet Detection at Construction Sites Using YOLOv5 and YOLOR

Van Than Tran[1] [iD], Thanh Sang To[1] [iD], Tan-No Nguyen[2] [iD],
and Thanh Danh Tran[1(✉)] [iD]

[1] Faculty of Civil Engineering, Ho Chi Minh City Open
University, Ho Chi Minh City 700000, Vietnam
`danh.tt@ou.edu.vn`
[2] Department of Civil Engineering, Kyungpook National
University, Daegu 41566, Republic of Korea

Abstract. Wearing a helmet is mandatory for workers at construction sites. It is very important for the safety of workers during work. In many scenarios, detecting workers not wearing helmets can prevent possible occupational accidents in time. Recently, with the rapid development of deep learning, convolutional neural networks (CNNs) have been widely applied in many problems including object detection. The constantly evolving object detection technology has resulted in a series of YOLO algorithms with very high accuracy and speed being used in various scene detection tasks. This paper presents a deep learning approach to solve the above problems. We propose a helmet detection method based on two models, namely YOLOv5 and YOLOR, using a dataset of 900 collected images. The two models are compared and analyzed. The experimental results show that the mAP@0.5 of YOLOR reached 87.3%, significantly larger than that of the YOLOv5 model with mAP@0.5 of only 77.6%, proving the effectiveness of helmet detection using the YOLOR model.

Keywords: Helmet detection · Object detection · YOLOv5 · YOLOR

1 Introduction

Currently, wearing a helmet is mandatory for workers at construction sites. However, due to ineffective on-site supervision and low safety awareness of workers, occupational accidents related to not wearing helmets still occur frequently. To limit accidents, it is necessary to detect construction workers not wearing helmets in real time, thereby preventing possible occupational accidents in time.

Recently, with the rapid development of deep learning, convolutional neural networks (CNNs) have been widely applied in many problems, such as recognizing speech [1, 2], identifying images [3–7], or making predictions [8–10].

Target detection is to find all objects of interest in an image. It consists of two sub-tasks: the determination of target category and the positioning of the target. Object detection methods based on CNNs can be categorized into two categories: one is a two-stage method (e.g., the RCNN series [11–14]), and the other is a one-stage method (e.g.,

N.-T. Nguyen et al. (Eds.): ICIT 2022, LNDECT 148, pp. 339–347, 2022.
https://doi.org/10.1007/978-3-031-15063-0_32

YOLO [15, 16], SSD [17]). In theory, the two-stage method usually has better detection performance while the one-stage method is more efficient. The one-stage target detection algorithm really achieves end-to-end training.

In this paper, we focus on the one-stage detector. Specifically, we compare the performance of the YOLOv5 and YOLOR [18] models for helmet detection.

The rest of this paper is organized as follows: Sect. 2 introduces the method of YOLOv5 and YOLOR [18]. In Sect. 3, we present the experimental results and discussion. The paper closes with a conclusion in Sect. 4.

2   Methods

2.1   YOLOv5

In 2020, the fifth version of YOLO proposed by Utralytics named as YOLOv5 outperforms all previous versions in both speed and accuracy. It is the product of continuous integration and innovation based on YOLOv3 [15] and YOLOv4 [16].

The overall network architecture of YOLOv5 is similar to that of other YOLO algorithm series divided into three main architectural blocks, namely the Backbone network for feature extraction from images, the PANet network to generate a feature pyramids network, and the Output network to complete target detection, presented in Fig. 1.

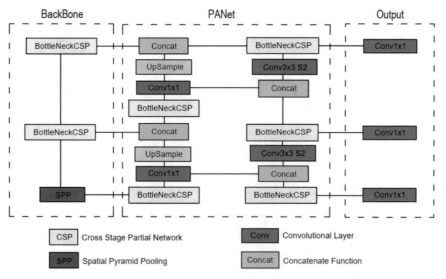

Fig. 1. YOLOv5 architecture.

2.2   YOLOR

You Only Learn One Representation (YOLOR) [18] is a state-of-the-art object detection model, different from YOLOv1-YOLOv5.

YOLOR [18] is a unified network to integrate implicit knowledge and explicit knowledge. Figure 2 illustrates the unified network architecture.

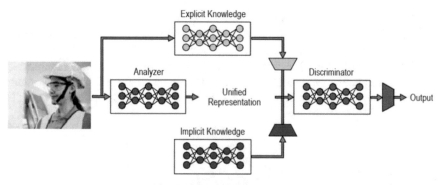

Fig. 2. YOLOR architecture.

2.3 Dataset

Dataset is very important for deep learning-based object detection algorithms. The dataset used in this paper is mainly collected through Internet crawlers. It includes two-class, helmet and no-helmet. In total, 900 images have been collected to create training set, validation set, and testing set according to the ratio of 7:2:1, which is detailed in Table 1.

Table 1. The number of samples used for training, validation, and testing.

Sub-dataset	YOLOv5	YOLOR
Training	630	630
Validation	180	180
Testing	90	90
Total	900	900

The dataset contains various scenes and helmets of different scales, which are shown in Fig. 3.

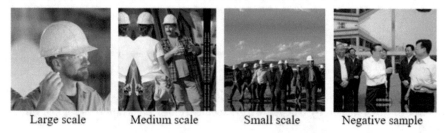

| Large scale | Medium scale | Small scale | Negative sample |

Fig. 3. Representative samples in the dataset.

The statistics on the number of targets in the two dataset classes are shown in Fig. 4(a), in which helmets account for the majority. Figure 4(b) is a normalized map of the target location and it shows the positions of the targets scattered in the coordinate system, more concentrated in the vertical direction and scattered in the horizontal direction. Figure 4(c) is a map of the standardized target size. It can be seen that the size distribution of target is relatively concentrated, and the width and height are mainly distributed at 0–0.2. The targets are mainly small in size.

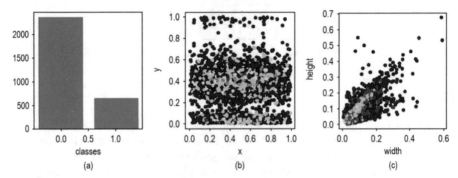

Fig. 4. Statistical results of the dataset: (a) the number of targets in each class; (b) normalized target location; (c) normalized target size.

2.4 Metrics

Precision (P) and recall (R) are often used to evaluate the performance of a model, defined by Eqs. (1) and (2).

$$P = \frac{TP}{FP + TP} \tag{1}$$

$$R = \frac{TP}{FN + TP} \tag{2}$$

where TP is a true positive, FP is a false positive, FN is a false negative.

Precision (P) and recall (R) interact with each other. In order to combine these two metrics, another metric for measuring detection accuracy was introduced, called average precision (AP). This metric (AP) is defined by Eq. (3).

$$AP = \int_0^1 P(R)dR \tag{3}$$

In addition, for multi-class object detection problems, the detection accuracy of the model will be evaluated by calculating the mean average precision (mAP), determined by Eq. (4).

$$mAP = \frac{1}{C} \sum_{i=1}^{C} AP_i \tag{4}$$

where C is the number of the target category.

3 Results and Discussion

3.1 The Training Process of the Models

First, to satisfy the input required for the selected architectures, we convert the training images to 416×416 pixels. The programming language used for the algorithm in this paper is Python. PyTorch deep learning framework was used. The two YOLO models (YOLOv5, YOLOR) are trained to determine which model is more suitable for the helmet detection problem.

In the YOLO family, there is a combination of three loss functions calculated using bounding box regression score, objectness score, and classification score, which can be expressed as in Eq. (5).

$$
\begin{aligned}
loss = {}& 1 - IoU + \frac{\rho^2(b,b^{gt})}{c^2} + \frac{16}{\pi^4} \frac{(\arctan \frac{w^{gt}}{h^{gt}} - \arctan \frac{w}{h})^4}{1 - IoU + \frac{4}{\pi^4}(\arctan \frac{w^{gt}}{h^{gt}} - \arctan \frac{w}{h})^2} \\
& - \sum_{i=0}^{S^2} \sum_{j=0}^{B} W_{ij}^{obj} [\hat{C_i^j} \log C_i^j + (1-\hat{C_i^j}) \log(1 - C_i^j)] - \lambda_{noobj} \sum_{i=0}^{S^2} \sum_{j=0}^{B} (1 - W_{ij}^{obj})[\hat{C_i^j} \log C_i^j + (1-\hat{C_i^j}) \log(1 - C_i^j)] \\
& - \sum_{i=0}^{S^2} \sum_{j=0}^{B} W_{ij}^{obj} \sum_{c=1}^{C} [\hat{p_i^j}(c) \log(p_i^j(c)) - (1 - \hat{p_i^j}(c)) \log(1 - p_i^j(c))]
\end{aligned} \quad (5)
$$

where S^2 is the number of a grid in the input image, B is the number of bounding box in the grid.

The training epoch setup for YOLOv5 and YOLOR was to 500, 100, respectively. The validation loss is used to monitor the training process after each epoch, the loss results on the training and validation sets are shown in Fig. 5. These results show that as the number of iterations increases, the three different types of losses decrease. In other words, the network model can correctly identify the target in the training set.

3.2 Performance of the Models

In this section, the performance of the two models is compared using metrics of precision, recall, mAP@0.5:0.95, and mAP@0.5.

The curves describing the performance during the training of the two models are shown in Fig. 6. It is noteworthy that the performance metrics of the YOLOR model are significantly higher than that of the YOLOv5 model, except for the precision metric. Here, the key performance metric is mAP. This proves that the YOLOR model is more suitable for the helmet detection problem.

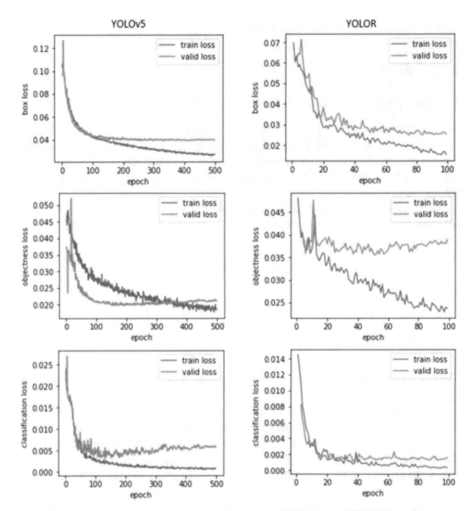

Fig. 5. Loss on the training set and validation set of YOLOv5 and YOLOR models.

The comparison results of four metrics of the two models are presented in Table 2. It can be seen that the mAP@0.5 of the YOLOR model (87.3% mAP@0.5) exceeds that of YOLOv5 (77.6% mAP@0.5) by 9.7%. The results indicated that YOLOR is significantly superior in helmet detection.

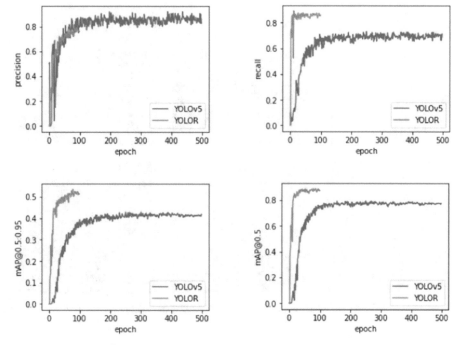

Fig. 6. Performance of YOLOv5 and YOLOR models.

Table 2. The performance of YOLOv5 and YOLOR models.

Metric	YOLOv5	YOLOR
Precision	0.826	0.762
Recall	0.715	0.858
mAP@0.5:0.95	0.420	0.513
mAP@0.5	0.776	0.873

In order to better show the experimental results, we found some images in the testing set to visualize the results. As shown in Fig. 7, YOLOR has a good ability to detect helmets of different colors and sizes. For those objects containing multiple targets, YOLOR can also detect them. In addition, YOLOR also detects people who are not wearing helmets.

4 Conclusion

In this paper, we compare the helmet detection results of two YOLOv5 and YOLOR models through experiment. As a result, the helmet detection result of the YOLOR model outperforms YOLOv5 in our dataset. The mAP@0.5 value in the YOLOR model

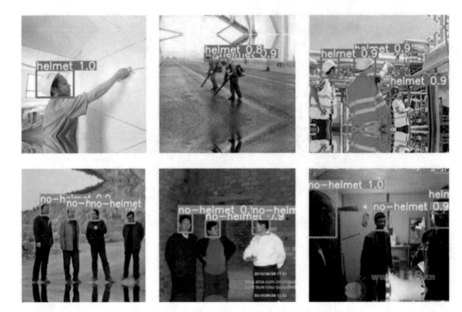

Fig. 7. Visualization results of YOLOR from testing set.

is 87.3%, while in the YOLOv5 model is only 77.6%. In deep learning, data plays a very important role. The larger the amount of data, the better the performance and generalizability of the model. Therefore, a larger number of real scene samples have to be collected in the future studies. On the other hand, no one model is best for all tasks. In order to find the best one, further experiments need to be conducted with several other models such as YOLOX, Scaled-YOLOv4, etc.

References

1. Narayanan, A., Prabhavalkar, R., Chiu, C.C., Rybach, D., Sainath, T.N., Strohman, T.: Recognizing long-form speech using streaming end-to-end models. In: 2019 IEEE Automatic Speech Recognition and Understanding Workshop (ASRU), pp. 920–927. IEEE, December 2019
2. Shaqra, F.A., Duwairi, R., Al-Ayyoub, M.: Recognizing emotion from speech based on age and gender using hierarchical models. Procedia Comput. Sci. **151**, 37–44 (2019)
3. Ho, T.T., Kim, G.T., Kim, T., Choi, S., Park, E.K.: Classification of rotator cuff tears in ultrasound images using deep learning models. Med. Biol. Eng. Comput. **60**, 1269–1278 (2022). https://doi.org/10.1007/s11517-022-02502-6
4. Ho, T.T., et al.: A 3D-CNN model with CT-based parametric response mapping for classifying COPD subjects. Sci. Rep. **11**(1), 1–12 (2021)
5. Park, S.S., Tran, V.T., Doan, N.P., Hwang, K.B.: Evaluation of damage level for ground settlement using the convolutional neural network. In: Ha-Minh, C., Tang, A.M., Bui, T.Q., Vu, X.H., Huynh, D.V.K. (eds.) CIGOS 2021. LNCE, vol. 203, pp. 1261–1268. Springer, Singapore (2022). https://doi.org/10.1007/978-981-16-7160-9_128
6. Park, S.S., Tran, V.T., Lee, D.E.: Application of various YOLO models for computer vision-based real-time pothole detection. Appl. Sci. **11**(23), 11229 (2021)

7. Dinh, V.Q., Munir, F., Azam, S., Yow, K.C., Jeon, M.: Transfer learning for vehicle detection using two cameras with different focal lengths. Inf. Sci. **514**, 71–87 (2020)
8. Ho, T.T., et al.: Deep learning models for predicting severe progression in COVID-19-infected patients: retrospective study. JMIR Med. Inform. **9**(1), e24973 (2021)
9. Do, D.T., Lee, J., Nguyen-Xuan, H.: Fast evaluation of crack growth path using time series forecasting. Eng. Fract. Mech. **218**, 106567 (2019)
10. Nguyen, D.L.H., Do, D.T.T., Lee, J., Rabczuk, T., Nguyen-Xuan, H.: Forecasting damage mechanics by deep learning. CMC-Comput. Mater. Continua **61**(3), 951–977 (2019)
11. Girshick, R., Donahue, J., Darrell, T., Malik, J.: Rich feature hierarchies for accurate object detection and semantic segmentation. In: Proceedings of the IEEE Conference on Computer Vision and Pattern Recognition, pp. 580–587 (2014)
12. Girshick, R.: Fast R-CNN. In: Proceedings of the IEEE International Conference on Computer Vision, pp. 1440–1448 (2015)
13. Ren, S., He, K., Girshick, R., Sun, J.: Faster R-CNN: towards real-time object detection with region proposal networks. In: Advances in Neural Information Processing Systems, vol. 28 (2015)
14. He, K., Gkioxari, G., Dollár, P., Girshick, R.: Mask R-CNN. In: Proceedings of the IEEE International Conference on Computer Vision, pp. 2961–2969 (2017)
15. Redmon, J., Farhadi, A.: YOLOv3: an incremental improvement. arXiv preprint arXiv:1804.02767 (2018)
16. Bochkovskiy, A., Wang, C.Y., Liao, H.Y.M.: YOLOv4: optimal speed and accuracy of object detection. arXiv preprint arXiv:2004.10934 (2020)
17. Liu, W., Anguelov, D., Erhan, D., Szegedy, C., Reed, S., Fu, C.-Y., Berg, A.C.: SSD: single shot multibox detector. In: Leibe, B., Matas, J., Sebe, N., Welling, M. (eds.) ECCV 2016. LNCS, vol. 9905, pp. 21–37. Springer, Cham (2016). https://doi.org/10.1007/978-3-319-46448-0_2
18. Wang, C.Y., Yeh, I.H., Liao, H.Y.M.: You only learn one representation: unified network for multiple tasks. arXiv preprint arXiv:2105.04206 (2021)

Image Segmentation of Concrete Cracks Using SegNet

Tan-No Nguyen[1] , Van-Than Tran[2] , Seung-Wook Woo[1] ,
and Sung-Sik Park[1(✉)]

[1] Department of Civil Engineering, Kyungpook National
University, Daegu 41566, Republic of Korea
sungpark@knu.ac.kr
[2] Faculty of Civil Engineering, Ho Chi Minh City Open University, Ho Chi Minh City 700000,
Vietnam

Abstract. Inspecting flaws in a structure are vital for engineering applications, especially in concrete projects. The goal of this paper was to employ semantic segmentation model named as SegNet to identify concrete cracks for the continuously and automatically structural health monitoring. The commonly used Adaptive Moment Estimation algorithm and Stochastic Gradient Descent algorithm were applied for optimization. Various recently objective loss functions were served as the evaluation function for image segmentation. Different raw input images of concrete cracks under various conditions such as the shape of cracks, width of cracks, rough or smooth surfaces of backgrounds, were divided for training and validation subsets. The findings revealed that both optimizers performed the similar accuracy by using the intersection over union for concrete crack inspections. In addition, dice, tversky, and focal tversky losses showed better than binary cross-entropy and lovasz losses in terms of the overall accuracy of image classification problems.

Keywords: Image segmentation · Crack detection · SegNet · Optimization

1 Introduction

In the past few decades, a growing number of structural health monitoring (SHM) techniques of buildings have been introduced to maintain the health states of structure as well as to reduce human safety [1]. Among them, non-destructive testing methods are widely and effectively applied for monitoring the structural conditions in comparison with the human visual inspection in terms of detecting hidden damages.

Deep learning method has been known as an advanced non-destructive technique for identifying damages from images relating to the applications of object detection, classification, and semantic segmentation issues [2–4]. Recently, semantic image segmentation models have been introduced as a superior performance in detecting cracks of structures [5, 6]. A semantic segmentation, termed as SegNet has been shown as the top model for semantic segmentation [5, 7, 8]. Since SegNet has shown successfully

N.-T. Nguyen et al. (Eds.): ICIT 2022, LNDECT 148, pp. 348–355, 2022.
https://doi.org/10.1007/978-3-031-15063-0_33

in semantic problems, it is used for detecting the concrete cracks in this study. To train the SegNet model, an effective optimization approach should be taken into account. For better decision making, two solvers namely Adaptive moment estimation (Adam) and Stochastic gradient descent (SGD) are employed [10]. In addition, applying sophisticated loss functions are essential for achieving reliable results. Five recent loss functions, namely binary cross-entropy, dice, tversky, focal tversky, and lovasz losses are utilized. Finally, the assessment of the reliability of crack detection is a difficult task. To identify the overall accuracy of SegNet model, a common performance metric for the object detection named as intersection-over-union (IoU) is implemented.

The contributions of this paper are to: (1) investigate Adam algorithm and SGD algorithm to optimize the implementation of SegNet; (2) compare the performance of crack detection using different loss functions.

Section 2 describes SegNet, initial parameters of Adam and SGD and performance evaluation methods. The findings are presented in Sect. 3. Lastly, some important concluding remarks and subsequent studies are mentioned in Sect. 4.

2 Methodology

2.1 SegNet Network Architecture

SegNet has been known as an encoder-decoder architecture for pixel-wise semantic segmentation [7]. This architecture is demonstrated in Fig. 1. Each encoder layer corresponds to the decoder layers. The decoder upsamples the input feature map by max-pooling to produce a sparse map. The last mapping is connected to a softmax procedure to identify each pixel. The encoder-decoder structure is trained with the supervised learning task in comparison with the unsupervised feature training of other approaches [11, 12].

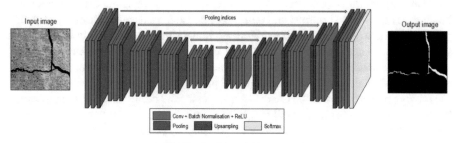

Fig. 1. SegNet architecture.

2.2 Dataset

The training process of deep convolutional encoder-decoder structure model is a type of supervised learning task. SegNet network requires a variety of labeled images of concrete cracks. A publicly available dataset from [6] is chosen for training SegNet model. The dataset consists of 100 RGB color images of concrete cracks with manually annotated segmentations presenting the subject with a binary image. In this study, totally 100 RGB color images of concrete cracks with a resolution of 544 × 384 were divided into 70% for training, 30% for validation subsets. For better generalization performance of the training model, these images are different conditions such as the shape of cracks, width of cracks, rough or smooth surfaces of backgrounds. Some representative samples and their segmentations are given in Fig. 2.

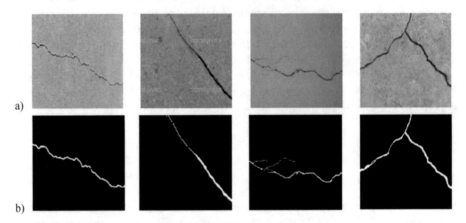

Fig. 2. Representatives of database: a) Source samples, b) Ground truth.

2.3   Training Parameter and Evaluation Metric

This paper conducts the SegNet architecture implementation in Python. The recent and efficient deep-learning framework PyTorch is applied to implement neural networks flexibly. In addition, to improve and optimize the results of datasets during the training process of a network, a reasonable optimization algorithm and a suitable loss function should be considered. For optimization algorithm, adaptive moment estimation (Adam) and stochastic gradient descent (SGD) is adopted in this study. Adam is a combination of momentum algorithm and root mean squared propagation algorithm to converge faster. This study applies for the 0.0001 learning rate and the 0.9 decay rate. SGD optimization has been shown as an efficient weight update technique to train networks. 0.01 learning rate of and 0.9 momentum are implemented for SGD solver. The size of batch setup was 16 and optimization was carried out up to 200 epochs for both optimizers. All initially physical parameters of optimizers can be found in Table 1. Additionally, in to monitor the learning process after each epoch, activation functions, namely binary cross-entropy (bce), dice (dice), tversky (tversky), focal tversky (focal_tversky), and lovasz (lovasz) looses are used.

For assessment metric, Intersection-Over-Union (IoU) metric is increasingly employed for object detection. In general, the IoU metric is defined as follows:

$$IoU = \frac{\text{Target} \cap \text{Prediction}}{\text{Target} \cup \text{Prediction}}$$

where, intersection (Target ∩ Prediction) consists of pixels obtained from images of prediction and ground truth, whereas union (Target ∪ Prediction) is composed of all pixels gained from either the prediction image or the target image.

Table 1. Initial parameters of optimization algorithms.

Parameter	Optimizer	
	Adam	SGD
Leaning rate	0.0001	0.01
Momentum	0.90	0.90
Batch	16	16
Iteration	200	200

3   Results and Discussion

For the qualitative evaluation of SegNet model, the output of SegNet predictions at different epochs based on Adam optimization algorithm under the training test presented in Fig. 3.

The qualitative results showed a good performance of SegNet predictions with various rough, smooth, clear, or dark background surfaces of concrete cracks or various concrete crack conditions. The model indicated a slight improvement of performance with the training iteration of up to 10, whereas it achieved a high improvement of performance after the 50th iteration. The output increased consistently for the epochs between 100 and 200.

The optimized model after finishing the training process was evaluated by new input images. The visualization of findings was found in Fig. 4. It is clear that the predicted images reflected well the real cracks in the raw samples.

In terms of quantitative assessment of SegNet, different optimization algorithms and different loss functions were applied. The changes in loss function values on training set and changes in IoU values on validation set using Adam algorithm and SGD algorithm were showed in Fig. 5 and Fig. 6, respectively.

For applying Adam algorithm, it is noteworthy that the lovasz loss function had the lowest convergence rate in comparison with the bce, dice, tversky, and focal_tversky losses. These losses demonstrated the same convergence rate for the epochs between 100 and 200.

For applying SGD algorithm, the experimental results illustrated the same trend of convergence speed of various loss functions as using Adam algorithm. While the highest convergence speed can be found in the dice loss, the lowest can be obtained from lovasz loss. However, all loss functions demonstrated the convergence speed of less than 0.5 at the 200th epoch.

While the lovasz loss revealed the lowest convergence rate in Adam and SGD optimizations, it performed better than the bce loss function in enhancing the performance of the model. In addition, the best performance can be observed from the dice loss for both optimizers.

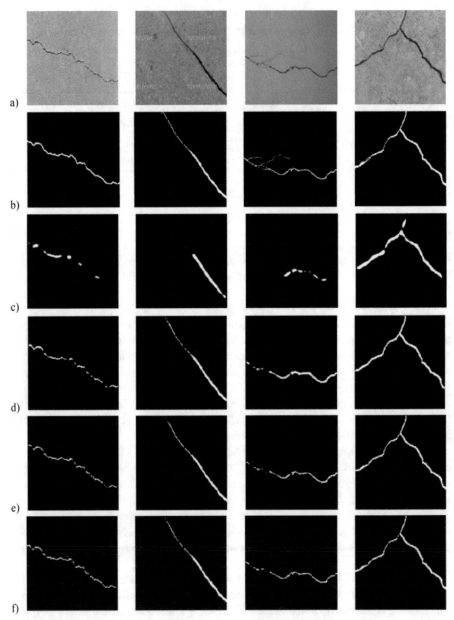

Fig. 3. Qualitative assessment for SegNet model predictions of concrete crack detection dataset using Adam algorithm and Focal_Tversky loss function under the training set: (a) Source samples, (b) Ground truth, (c) 10th epoch, (d) 50th epoch, (e) 100th epoch, (f) 200th epoch.

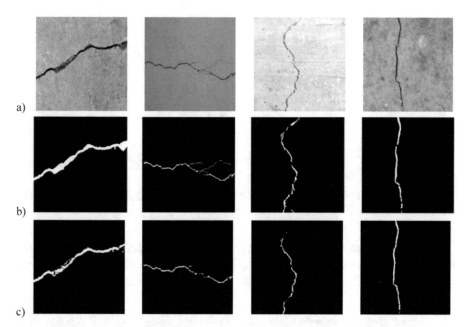

Fig. 4. Optimized SegNet model for crack detections using Adam algorithm and Focal_Tversky loss function under the testing test: (a) Test samples, (b) Ground truth, (c) Outputs

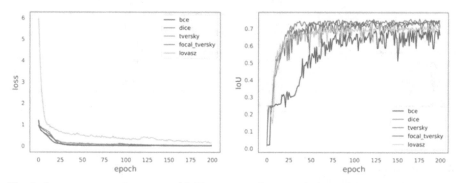

Fig. 5. Loss versus epoch curves and IoU versus epoch curves during training set and validation set, respectively, using Adam algorithm.

The quantitative comparisons of using two optimizers with different loss functions were shown in Table 2. It can be seen that a slight difference in the detection accuracy of concrete cracks was found in five loss functions at the 200th epoch. The lowest precisions of 0.70 and 0.68 were observed in the bce loss using for Adam optimization and SGD optimization, respectively, whereas the highest precision of 0.76 and 0.75 were obtained from the dice loss using for Adam and SGD, respectively. The accuracy of approximately 0.74 was found in other loss functions. In other words, the loss functions used for Adam and SGD optimizations in this study performed successfully in enhancing the performance of SegNet.

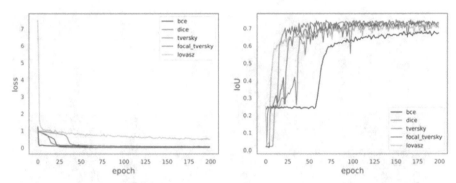

Fig. 6. Loss versus epoch curves and IoU versus epoch curves during training set and validation set, respectively, using SGD algorithm.

From the validation result, it is highlighted that the accuracy of Adam was similar to that of the SGD using for optimizing the feature layers in the SegNet. In other words, choosing the loss functions is essential for image segmentation problems.

Table 2. The quantitative comparison of validation subset in the 200th epoch.

Loss function	Optimizer	
	Adam	SGD
bce	0.70	0.68
dice	0.76	0.75
tversky	0.75	0.74
focal_tversky	0.75	0.74
lovasz	0.72	0.73

4   Conclusion

The goal of this study is that of detecting the cracks of concrete structure using SegNet model for image segmentation. The experimental results illustrated that concrete crack detection methods using Adam and SGD optimization algorithms are similar accuracies when applying different loss functions. In other words, choosing a suitable loss function was essential for detecting, classifying, and localizing an object. In this paper, the dice, tversky, and focal Tversky losses outperformed the bce and lovasz on concrete crack inspections. As for future work, this approach will be extended to enhance the result of experimental tests applied for structural health monitoring.

Acknowledgment. This study was supported by a National Research Foundation of Korea (NRF) grant funded by the Korean government (MSIT) (No. NRF-2018R1A5A1025137).

References

1. Dixit, S., Sharma, K.: A review of studies in structural health monitoring (SHM). In: Proceedings of the Creative Construction Conference 2019, CCC 2019, pp. 84–88. Budapest University of Technology and Economics, Hungary (2019)
2. Yuan, X., Shi, J., Gu, L.: A review of deep learning methods for semantic segmentation of remote sensing imagery. Expert Syst. Appl. **169**, 114417 (2021)
3. Pathirage, C.S.N., Li, J., Li, L., Hao, H., Liu, W., Ni, P.: Structural damage identification based on autoencoder neural networks and deep learning. Eng. Struct. **172**, 13–28 (2018)
4. Park, S.S., Tran, V.T., Lee, D.E.: Application of various YOLO models for computer vision-based real-time pothole detection. Appl. Sci. **11**(23), 11229 (2021)
5. Chen, T., et al.: Pavement crack detection and recognition using the architecture of segNet. J. Ind. Inf. Integr. **18**, 100144 (2020)
6. Liu, Y., Yao, J., Lu, X., Xie, R., Li, L.: DeepCrack: a deep hierarchical feature learning architecture for crack segmentation. Neurocomputing **338**, 139–153 (2019)
7. Badrinarayanan, V., Handa, A., Cipolla, R.: SegNet: a deep convolutional encoder-decoder architecture for image segmentation. arXiv preprint arXiv:151100561 (2015)
8. Song, C., et al.: Pixel-level crack detection in images using SegNet. In: Chamchong, R., Wong, K.W. (eds.) MIWAI 2019. LNCS (LNAI), vol. 11909, pp. 247–254. Springer, Cham (2019). https://doi.org/10.1007/978-3-030-33709-4_22
9. Kingma, D.P., Ba, J.: Adam: a method for stochastic optimization. arXiv preprint arXiv:1412. 6980 (2014)
10. Zhang, T.: Solving large scale linear prediction problems using stochastic gradient descent algorithms. In: Twenty-First International Conference on Machine Learning, ICML 2004, Banff, Alberta, Canada, p. 116 (2004)
11. Kavukcuoglu, K., Sermanet, P., Boureau, Y.L., Gregor, K., Mathieu, M., Cun, Y.: Learning convolutional feature hierarchies for visual recognition. In: 24th Annual Conference on Neural Information Processing Systems 2010, NIPS 2010, Vancouver, Canada, p. 10 (2010)
12. Ranzato, M., Huang, F.J., Boureau, Y.L., LeCun, Y.: Unsupervised learning of invariant feature hierarchies with applications to object recognition. In: 2007 IEEE Conference on Computer Vision and Pattern Recognition, CVPR 2007, Minneapolis, MN, USA, pp. 1–8 (2007)

Intelligent Scene Text Recognition in Streaming Videos

Minh-Quan Ha[1], Vy-Hao Phan[1,2], Bao D. Q. Nguyen[1,2], Hoai-Bao Nguyen[1,2], Trong-Hop Do[1,2(✉)] (ID), Quang-Dung Pham[3], and Nhu-Ngoc Dao[4] (ID)

[1] University of Information Technology, Ho Chi Minh City, Vietnam
{19522076,19520524,19520402,19520405}@gm.uit.edu.vn
[2] Vietnam National University, Ho Chi Minh City, Vietnam
hopdt@uit.edu.vn
[3] Faculty of Information Technology, Vietnam National University of Agriculture,
Hanoi, Vietnam
pqdung@vnua.edu.vn
[4] Department of Computer Science and Engineering, Sejong University,
Seoul, South Korea
nndao@sejong.ac.kr

Abstract. The Scene text Recognition is a problem of extracting textual information from images and videos and plays an increasingly important roles in the era of consuming visual media through social networks. Besides, The execution speed of the models is also shown based on fbs to compare the performance between methods in real-time services occasion. We believe that by conducting a detailed examination of using many modern Deep Learning architectures including SAST, EAST, Rosetta on the Vietnamese scene text dataset also called Vintext developed by the VinAI organization, we may contribute considerable knowledge improving the quality of Vietnamese deep learning field.

Keywords: Streaming videos · Scene text recognition · Deep learning

1 Introduction

Nowadays, there is an increasing number of people using the social platforms (such as Facebook, Instagram, Twitter, etc.) for posting their memories and their real-life images. Those images could be every aspects of the world and contain words in any style and shape. It is becoming the primary way of consuming information while the volume of photos being uploaded has grown exponentially in the last many years, which poses a tough challenge of textual information retrieval, also called *Scene text Recognition*.

The last decade has witnessed the explosion of many algorithms and deep learning architectures which help overcome the obstacles of English language problem. However, nobody can give a clear answer to which one works the most effective on the Vietnamese language dataset. To solve the issue, we conduct this project in order to examine many state-of-the-art models of the Scene

© The Author(s), under exclusive license to Springer Nature Switzerland AG 2022
N.-T. Nguyen et al. (Eds.): ICIT 2022, LNDECT 148, pp. 356–365, 2022.
https://doi.org/10.1007/978-3-031-15063-0_34

Fig. 1. Many samples from the Vintext dataset for scene text recognition.

Text Recognition issue including SAST [6], EAST (Efficient Accurate Scene Text Detector) [7], Rosetta [1] and CRNN [5]. We adopt the VinAI's Vintext dataset which is a new benchmark dataset for Vietnamese scene text recognition [4].

The Sect. 2 illustrates the details of the Vintext dataset used in our work. The next Sect. 3 gives the brief information about the Scene text Recognition architectures' mechanism in this study. The Sect. 4 describes how we set up the experiment and the way we evaluate those architectures. The final important Sect. 5 is the process of analyzing the errors raised by the models experimented in the hope of deep understanding and improving the deep learning networks used for the Vietnamese Scene text problem.

2   Dataset

In this study, we conduct experiments on Vintext dataset [4], The dataset contains 2,000 completely annotated photos with 56,084 text instances, making it the largest Vietnamese scene text dataset. Each text instance had a quadrilateral bounding box that was coupled with the ground truth character sequence. We divided the data into three groups at random. :

- Train: 1200 images
- Val: 300 images
- Test: 500 images

In addition to Vietnamese dataset, the reasons we choose this dataset to conduct evaluation on Deep Learning model is that, All images are taken in a normal life situation so it contain casual scenes with many shop signs, billboards, and propaganda panels. These things make the dataset become more challenging and suitable to evaluate the performance of complex Deep Learning architecture.

3 Method

3.1 SAST

The segmentation-based **S**ingle-shot **A**rbitrarily-**S**haped **T**ext detector (SAST) proposed by Pengfei Wang and his collaborators has introduced a number of new techniques for the task. SAST [6] is a multi-task learning framework based on a *Fully Convolutional Network* (FCN) [3] and and has three parts combined.

The first component is the stem network, which is constituted by ResNet-50, FPN and *Context Attention Block* (CAB). FPN helps blend a feature map X of $1/4$ size of the input images. When dealing with more complex text instances, such as extended text, two CABs are stacked to capture rich circumstantial information and help ease the challenges created by the narrow receptive field. By linking two CABs that combine pixel-wise contextual information both horizontally and vertically, long-range dependencies from pixels can be obtained.

The TCL (text center line), TBO (text border offset), TVO (text vertex offset), and TCO (text center offset) maps are projected as a multi-task issue in the second component.

The final part which is so-called post-processing part adopts the notion of *Point-to-Quad Assignment* to achieve text instance segmentation. This stage entails(1) recognizing text quadrangles using TCL and TVO maps,(2) binarizing the TCL map using a defined threshold (as in EAST[7]), and(3) grouping the outputs of bounding boxes in the binarized TCL map into text instances.

3.2 EAST

EAST (Efficient Accurate Scene Text Detector) [7] take a role of text detecton model. This model consists of two parts which are a fully convolutional network (FCN) [3] for getting the results of features from the images and Non-Maximum Suppression part to generate the correct box for the final results.

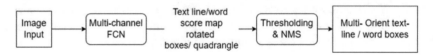

Fig. 2. The pipeline of EAST

The main idea of the method is getting trained of the model to predict where the text was placed in images.

The Structure of Text Detection FCN. The model consists of three components : Feature Extractor system (PVANet) [2], Feature Merging Branch and Ouput layer (Fig. 5)

- *Feature Extractor (PVANet)* [2]: Consists of many convolutional neural network with convolutional layer and pooled layer with interleaving convolution and pooling layers, pre-trained on the ImageNet dataset.
- *Feature Merging Branch:*
  The operator $[;]$ denotes concatenation along the channel axis, while g_i is the merge base and h_i is the merged feature map (Eq. 1).

$$g_i = \begin{cases} unpool(h_i) & if \ i \leq 3 \\ conv_{3\times3}(h_i) & if \ i = 4 \end{cases} \tag{1}$$

The feature map from the previous stage is fed to an unpooling layer to double its size before being concatenated with the current feature map in each merging stage. Following that, a $conv_{1\times1}$ bottleneck [8] reduces the number of channels and reduces computation, followed by a $conv_{3\times3}$ that fuses the data to produce the final output of this merging stage. Following the last merging step, a $conv_{3\times3}$ layer generates the merging branch's final feature map and feeds it to the output layer. This process is briefly illustrated in the Eq. 1 and 2.

$$h_i = \begin{cases} f_i & if \ i = 1 \\ conv_{3\times3}(conv_{1\times1}(g_{i-1}, f_i)) & otherwise \end{cases} \tag{2}$$

- Output layer: The final output get from Feature merging branch session is sent to the $conv_{1\times1}$ layer with 1 channel, giving a map score ranging from [0–1]. The final result is also passed through RBOX or QUAD Geometry which provides a multichannel geometry map (Fig. 5).

3.3   CRNN

In this section, we introduced the method using for scene text recognition which is CRNN [5]. This model can adapt the role of sequence to sequence model, especially for image recognition tasks. Although CRNN is the old method of image recognition tasks, we can take advantage of this by scoping each word in bunch of images to use for training model rather than use the images with many words.

The Proposed Network Architecture. The network architecture design of CRNN include three parts, which are the convolutional layers, the recurrent layers, and a transcription layer, from bottom to top.

Firstly, the model uses a bunch of CNN layers to extract features from each input images. At the next step, the recurrent network take a role of writing the predicted context from the extracted features from previous layers. Lastly, the transcription part lying on the top of model structure used for transforming the predictions from RNN layers to label sequence.

3.4   Rosetta

Rosetta [1] which is the Facebook's scalable OCR system has been deployed in production and Facebook's applications. In this paper, we adopted the Text recognition model called *"CTC model"* because it uses a sequence-to-sequence CTC loss.

Dynamically programmed, this loss aims to compute the conditional probability of a label predicted. Rooted in the ResNet-18 architecture, the convolutional body has a last layer which doesn't use any recurrent network and calculates directly the most likely character at every position of the input word.

As regards the image pre-processing stage, the word images are resized to 32×128 pixels if they are bigger. Therefore, for most words, no distortion is produced.

4   Experiment

We conduct an experiment on scene text problems including detection and recognition separately stages, especially we use EAST and SAST for detection architecture and apply CRNN and Rosetta for recognition stage. At the detection stage, we compare the productivity of each model throughout four metrics which are precision, recall, hmean and fps. In terms of recognition, we just use accuracy for evaluation each model. The results of the experiment are described in the Table 1.

In addition, we adjusted slightly the original data from Vintext to adapt the model in recognition task. Each image in Vintext dataset contains many words inside, so we decided to crop each word in the image into individual image to apply in recognition model. This makes the models able to focus on each word and recognize it better.

Table 1. Experiments on Detection and recognition stage in different metrics (**bold** indicates best performance in a specific metric)

	Detection			Recognition	
	EAST	SAST		CRNN	Rosetta
precision	0.57019	**0.71692**	acc	0.20352	**0.48547**
recall	0.61251	**0.62799**	fbs	3762.1	**4364.3**
hmean	0.59059	**0.66951**			
fps	**27.61172**	3.54279			

As shown in Table 1, we can see the results of each model from each stage. At the detection stage, experimental results from SAST model show superiority over the results from EAST in three metric which are precision, recall and hmean. However, EAST achieves faster execution times with 27.61 fps. Regarding to recognition results, Rosetta method gives the better results than CRNN in both accuracy and execution times.

5   Error Analysis

5.1   Detection

Table 2. IoU

Model name	IoU	Number of boxes	Number of pictures	Average boxes per picture	Percentage
EAST	<0.5	4072	273	13.945	48.21%
	>= 0.5	4375	283	14.983	51.79%
SAST	<0.5	4024	274	13.78	47.32%
	>= 0.5	4479	281	15.34	52.68%

The Table 2 show the statistical results on predicted samples between two models which are EAST and SAST.Based on IOU, The results printed out that SAST had the better results than EAST. In term of detection, we divide the errors into 4 main types: Tilted text error, Text size error, Blurred text error, and Special fonts error.

(a) SAST   (b) EAST

Fig. 3. Tilted text error

The detection result depends on how much the text is tilted. For example, in the Fig. 3, on the outer left sign, texts are excessively distorted so EAST and SAST cannot recognize them. On the other hand, the right and top sign is pretty clear. While EAST still cannot point out, SAST seems to be better than EAST because it recognize almost all the text on these signs. However, the drawback of SAST is that there are spare boxes and the bounding box area is not accurate. There should be one bounding box corresponding "GPP" instead of two bounding boxes saying "GP" and "PP".

Small text is also a challenge to any detection model and even to human's eyes. They are usually missed or confused by small non-text surrounding details.

(a) SAST   (b) EAST

Fig. 4. Text size error

(a) SAST   (b) EAST

Fig. 5. Blurred text error

In Fig. 4, SAST can detect more small texts than EAST with less noise boxes but getting all of them is impossible.

Text that has color almost the same as background is considered blurry. For example, "AXO", Ử , "HÀNG", "TẠP", "HÓA" are the 5 blurred words in this case, they cant be recognized by EAST nor SAST. In general, we can see that SAST is the better detection model compare to EAST.

5.2   Recognition

In this section, after the experiment with model recognition is completed. We recognize that there are a bunch of error in the results of prediction, so we divide the errors into 2 main types:

1. Inability to recognize: Cases where model cant recognize text and the result usually is "#", "##", "###" or ""(empty)
2. Incorrect word: Cases where the extracted text has no meaning, completely/partly wrong or a character or a punctuation is usually mistake for another.

The Fig. 6 illustrates four typical cases where the textual information in the images cannot be recognized by the models. Those are when the words are of bizarre shape, too small to read, blurred, or blended in with their surroundings.

Table 3. Frequency of error letters

word	frequency	word	frequency	word	frequency	word	frequency
#	1887	á	98	ở	29	ắ	11
n	1360	3	95	/	26	ữ	10
h	976	5	85	ủ	26	õ	8
t	749	8	84	ò	25	!	8
c	662	ô	78	ổ	24	)	8
g	565	4	78	ợ	24	z	8
i	557	w	68	ứ	24	ù	8
a	543	7	66	ã	23	%	7
o	430	:	65	í	22	è	6
u	340	6	63	ẽ	21	ỗ	6
e	290	ố	62	ŷ	21	ự	6
m	289	ơ	61	ụ	20	*	5
s	286	f	60	ừ	20	ĩ	4
p	259	ạ	59	ề	20	j	4
r	256	q	53	ẻ	19	ằ	4
l	237	ọ	53	ú	19	ỳ	3
0	234	â	53	ậ	19	ẫ	3
b	227	x	52	-	18	'	3
đ	199	ầ	51	ẳ	18	+	2
v	189	ệ	48	ẩ	18	@	2
k	168	ế	48	ũ	17	"	2
d	166	ớ	44	ọ	17	ỹ	2
à	163	ó	39	(	17	<	1
y	157	ể	37	ỏ	15	ẳ	1
.	132	ê	37	ẹ	14	?	1
9	132	ờ	34	ử	14	]	1
å	127	ị	34	õ	14	ẵ	1
á	127	ă	32	é	14	_	1
ư	127	ồ	31	ý	13		
1	124	ỉ	31	ẹ	12		
2	112	ì	30				

(a) Strange shape.

(b) Small

(c) Blur

(d) Text's color is as same as background

Fig. 6. Not Recognition error

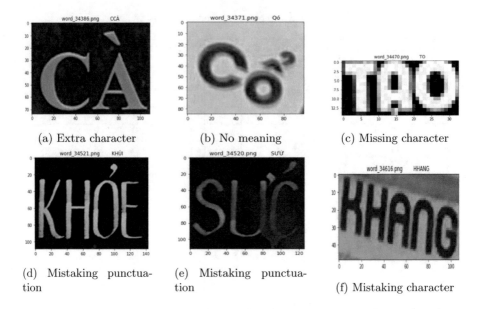

(a) Extra character   (b) No meaning   (c) Missing character

(d) Mistaking punctua-tion   (e) Mistaking punctua-tion   (f) Mistaking character

Fig. 7. Incorrect word error

The text showed in Fig. 7 are clear. However, the prediction for these cases are incorrect. This is because the recognition model was not trained with sufficient Vietnamese text data. As Fig. 7f show, some fonts or handwriting has characters look almost the same. In this case, "K" and "H" looks alike which lead to mistaken, instead of the right word "KHANG" it turn out to be "HHANG".

Acknowledgement. This work was supported by the National Research Foundation of Korea (NRF) grant funded by the Korea government (MSIT) (No. 2021R1G1A1008105).

References

1. Borisyuk, F., Gordo, A., Sivakumar, V.: Rosetta: large scale system for text detection and recognition in images. In: Proceedings of the 24th ACM SIGKDD International Conference on Knowledge Discovery & Data Mining, pp. 71–79 (2018)
2. Kim, K.H., Hong, S., Roh, B., Cheon, Y., Park, M.: Pvanet: deep but lightweight neural networks for real-time object detection. arXiv preprint arXiv:1608.08021 (2016)
3. Lin, T.Y., Dollár, P., Girshick, R., He, K., Hariharan, B., Belongie, S.: Feature pyramid networks for object detection. In: Proceedings of the IEEE Conference on Computer Vision and Pattern Recognition, pp. 2117–2125 (2017)
4. Nguyen, N., et al.: Dictionary-guided scene text recognition. In: Proceedings of the IEEE Conference on Computer Vision and Pattern Recognition (CVPR) (2021)
5. Shi, B., Bai, X., Yao, C.: An end-to-end trainable neural network for image-based sequence recognition and its application to scene text recognition. IEEE Trans. Pattern Anal. Mach. Intell. **39**(11), 2298–2304 (2016)

6. Wang, P., et al.: A single-shot arbitrarily-shaped text detector based on context attended multi-task learning. In: Proceedings of the 27th ACM International Conference on Multimedia, pp. 1277–1285 (2019)
7. Zhou, X., et al.: East: an efficient and accurate scene text detector. In: Proceedings of the IEEE Conference on Computer Vision and Pattern Recognition, pp. 5551–5560 (2017)

Mangosteen Fruit Detection Using Improved Faster R-CNN

Trung Hai Trinh[1]([✉]) [iD], Xuan Thien Bui[1], Thu Huong Tran[2],
Ha Huy Cuong Nguyen[1], and Khanh Duy Ninh[3]

[1] Vietnam – Korea University of Information and Comunication Technology, No. 470 Tran Dai Nghia Street, Da Nang, Vietnam
{tthai,bxthien.18it3,nhhcuong}@vku.udn.vn
[2] Thu Dau Mot University, No. 6 Tran Van Van Street, Thu Dau Mot, Binh Duong, Vietnam
[3] The University of Danang – University of Science and Technology, No. 54 Nguyen Luong Bang Street, Da Nang, Vietnam
nkduy@dut.udn.vn

Abstract. Agriculture is progressively getting access to scientific and technological breakthroughs in the digital era. In order to achieve smart agriculture, huge farms must be monitored and managed using advanced technologies. Anticipating and grading ripe fruit at harvest may aid in lowering storage costs and capturing market demand. Monitoring the ripening phase of the fruit also helps in the management of input and output criteria, which has practical implications in the harvesting process such as calculating the proper amount of water and nutrients at the end of the harvest, reducing traditional labor force, resulting in financial and human resource savings. In this research, we present a technique for identifying and classifying the ripening stage of mangosteen in agricultural fields. The study employs a two-stage approach based on a Faster R-CNN, a deep neural network which uses Region Proposal Network to extract the image region containing the item for the classification and identification of the mangosteen's location, and an improved RoI (Region of Interest) Pooling algorithm by adding a RoI Align layer to optimize feature data during training. We enhanced both of the speed and accuracy while processing huge and complicated data sets using the suggested methodology. When employing a dataset of 10,000 photos of ripe mangosteen, our model outperforms the one-stage approach in terms of accuracy while maintaining real-time speed.

Keywords: Faster R-CNN · Region proposal network · RoI align layer · Image classifier · Object detection

1 Introduction

The four representative pillars of the 4.0 technological revolution are Artificial Intelligence, Internet of Things, Big Data, and Cloud Computing. They have produced remarkable scientific accomplishments and have a significant effect on the economy, politics, as well as society. The adoption of advanced technology has aided the agricultural industry

© The Author(s), under exclusive license to Springer Nature Switzerland AG 2022
N.-T. Nguyen et al. (Eds.): ICIT 2022, LNDECT 148, pp. 366–375, 2022.
https://doi.org/10.1007/978-3-031-15063-0_35

in transitioning from a subsistence to an export model. Smart farming (slow fertilization and one-time spraying of biological products, use of sensors to evaluate water level, humidity, nutrients, and so on) has helped Vietnam increase output beyond prior harvests, reduce environmental pollution, and save labor. Intelligent monitoring and control systems are also used to monitor and regulate temperature, humidity, light, and other environmental conditions to aid plant development. However, the aforementioned models have shortcomings in several areas, such as identifying and classifying ripe fruit at harvest or identifying pests and illnesses that harm leaves and fruits.

Computer vision and image processing are widely used in a wide range of fields. Ripe fruit identification via pictures captured by digital cameras is being researched and used in practice. Image processing and object detection require a substantial amount of computer power as well as large data sets. In object identification and classification, color space is utilized to distinguish things from the background. Mangosteen fruit identification is a challenging problem since mangosteen fruits have many similarities and may be obscured by surrounding vegetation. To address the aforementioned issues, we decided to develop an automatic system for mangosteen fruit recognition by inheriting a multi-layer convolutional neural network (CNN) and employing a two-stage Faster R-CNN algorithm for mangosteen fruit identification and classification.

In this paper, we discuss the following topics. The second part highlights important related studies. The third section examines the collected data and the data enrichment and annotation tools. The fourth part provides our proposed technique, and the fifth section summarizes the experimental findings. Finally, we will focus on section six and analyze development and the anticipated result in future.

2 Related Work

The paper offers an object identification strategy for distinguishing ripe mangosteen from images in a training data set. The output required to determine the position of the mangosteens on the image and whether they are ripe or not. The input can be a photo or a video with one or more objects from which the features can be retrieved. By processing image units with numerous convolutional layers and pooling layers, convolutional neural networks (CNNs) can extract information from convolutional layers. Fast R-CNN has been created from this, included two major phases. Selective search is utilized first to find the best-suited bounding box (called as Region of Interest or RoI) positions, and then CNNs are employed to extract those bounding boxes [1]. To find the suggested areas, R-CNN is combined with a selective search strategy, which serves as the foundation for Fast R-CNN [2, 3]. Another method is the You Only Look Once (YOLO) [4] detector, which uses a single neural network to analyze the entire image, divides it into regions, and predicts bounding boxes Inkyu Sa et al. [6] suggested using the convolutional neural network approach to train the fruit identification model in "Fruit Detection System Using Deep Neural Networks", Byoungjun Kim, You-Kyoung Han [7], propose an improved vision-based method of detecting strawberry diseases using a deep neural network (DNN) capable of being incorporated into an automated robot system. Jose Luis Rojas-Aranda... [5, 8] presented an image classification method, based on lightweight Convolutional Neural Networks (CNN) to increase the classification accuracy, different input features are added into the CNN architecture, Changqing Cao; Bo Wang; Wen-rui Zhang et al. [9] propose an improved algorithm based on faster region-based CNN (Faster R-CNN) for small object detection. Using the two-stage detection idea, in the positioning stage, we propose an improved loss function based on intersection over Union (IoU) for bounding box regression, and use bilinear interpolation to improve the regions of interest (RoI) pooling operation to solve the problem of positioning deviation, in the recognition stage. Tang, Y., Chen, M et al. [10] suggested a deep neural network-based training method to detect fruits from photographs. Muresan, H., and Oltean, M. [11] suggested utilizing Convolutional Neural Networks (CNN) with computer vision to create a fruit recognition system. A mathematical model was created and implemented in Python, including the use of TensorFlow.

Advances in agricultural technology improve productivity, efficiency, and product quality, thereby fostering long-term sustainability. Detecting crops by identifying or classifying similar ones is a traditional method. Based on images of tomatoes and other plants, an analysis of fruit shape variation has been developed. Dark backgrounds were used to extract size information from sliced tomatoes. The paper on detecting and local-izing fruit pickers with vision-based methods [12, 21] proposes a technical solution for fruit classification using deep learning. Automatic fruit identification by Based on the features extracted, CNN and SVM to classify the presence of the object and the paper on fruit recognition and object detection with deep learning use multi-layer neural net-works to detect and classify post-harvest fruits. And there are many studies using CNN recognition model and the improved models are proposed to be applied in digital image detection and recognition [13–20] in the reference.

3   Data Collection and Preprocessing

In this work, we collect data through firsthand surveys on agricultural fields and a variety of publicly available datasets from around Vietnam, allowing us to obtain photos of man-gosteen from a variety of orchards. When the mangosteen was ripe, we used specialized cameras and the weather was relatively fine to capture the best images. This collection of 4000 mangosteen images is divided into three sub-datasets: the training dataset is used as the main image data for model training. Validation dataset used in the process of estimating the trained model is good and adjusting the parameters accordingly and finally the test dataset used for evaluating the fit of the final model same on the training dataset. The first 3100 shots were used as the training set, the next 800 as the validation set, and the last 100 as the testing set. Here are a few images from the dataset (Fig. 1).

We use the Roboflow Annotate tool to automatically assign data labels that simplify the processing of raw photos into a computer vision model for training deployment, will be used to name the gathered images. With this tool, you may create a rectangle around the visible fruit. We acquired not just the bounding boxes, but also the pixels corresponding to each fruit. The annotation job is to extract the feature areas of pineapple based on morphological properties such as eyes, color, shape, and so on. The end output will be an annotated dataset saved in Microsoft COCO format from Roboflow Annotate.

We have labeled over 4000 images and have collected over 10000 characteristic coordinates of ripe mangosteen with a single label of "Ripe". The result of this process is 4000 image files and 4000 text files with corresponding coordinates (Fig. 2).

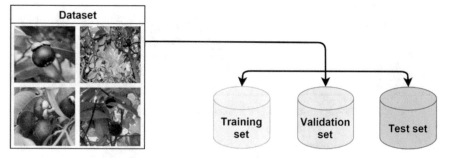

Fig. 1. Mangosteen fruits images dataset

Fig. 2. Annotate images using the Roboflow tool

A preprocessing step is required for the photographs. In this stage, we filter out photos that are fuzzy, noisy, or lack objects. During this phase, photos will also be returned to the same 416×416 resolution. To shorten training time and increase model performance, a clean and synchronous dataset is required. Because the number of photos in the present dataset is insufficient for the training model to handle the features, we do data enrichment. Image processing procedures such as transform, rotate, expand, and transform color are used, and the cutoff approach is used to augment the data set without overfitting. With the image processing technique introduced along with data enhancement, the initial data set is increased to 10000 images, which helps the training model to be able to predict the unknown, lost outcomes. 1 piece of data or different size (Fig. 3).

4 Proposed Method

The purpose of this study is to present a technical method for classifying and localization ripe mangosteen. The suggested technique employs an enhanced Faster R-CNN algorithm to speed up real-time for object detection (Fig. 4).

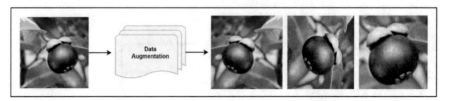

Fig. 3. Data Augmentation and transformations for mangosteen fruits images

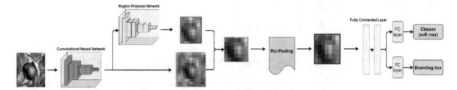

Fig. 4. Faster R-CNN system architecture

4.1 Faster R-CNN

Region Proposal Network (RPN). Faster R-CNN [2] is an improvement over two close relatives, R-CNN and Fast R-CNN. It avoids the Selective search (the algorithm will generate roughly 2000 areas for each image), which are zones that can contain a range of things and sizes. This approach analyzes around 2000 suggestion regions for each image, which is time consuming, and it is produced from another model, which is costly to train. After completing the CNN pre-train for image feature extraction, Faster R-CNN will use a fine-tune RPN (region proposal network) model for the proposal region task, which is initialized by the pre-train image classifier. It is in charge of classifying the positives (the area containing the object) and the negatives (the background). Following work in the training model is done in the same way as Fast R-CNN, employing the RPN's proposal areas. The RPN and the detection network are now exchanging convolution layers. Faster R-CNN combines the RPN and Fast R-CNN recommendation algorithms to speed up feature processing [2] (Fig. 5).

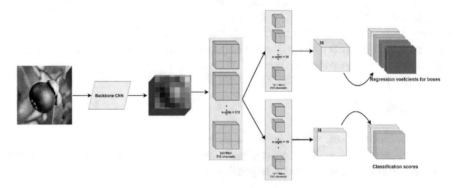

Fig. 5. Mangosteen image extracted region proposed through Region Proposal Network

RoI Align. In this stage, we address various concerns that have arisen as a result of RoI Pooling. Too much data is lost as a result of RoI Pooling. The distinction between RoI Pooling and RoI Align is quantization. RoI Align does not aggregate data via quantization. It splits the original RoI into 9 equal-sized boxes, with the size of each box defined by the mapped RoI and the pooling layer. To calculate the data value, four sampling points must be extracted from each box. The location of the pooling layer will be represented by a total score of four points (the average score can be used). The end output has 512 layers.

This shows that RoI Pooling's data processing loses too much important information. Red corresponds to the pixel data lost after entering the RoI Pooling layer, and yellow corresponds to the remaining pixel data. However, with RoI Align, feature data will not be lost, instead will be offset by contiguous pixel data. The corresponding green color for pixel values is added to the highlight value of the image. Ensure full features and no loss of information (Fig. 6).

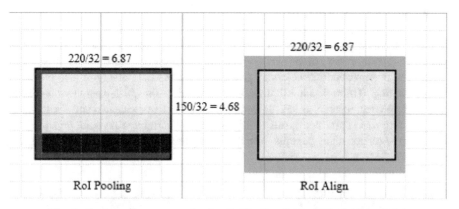

Fig. 6. Compare data extracted using RoIPooling (left) and RoI Align (right)

Output of the Training Model. The RoI Align layer generates a fixed-size feature vector that represents the proposed area. Data sharing is optimized since many area suggestions for the same image are particularly likely to overlap. The feature is divided in to two brances by the bounding box regression to localize and softmax estimator $K + 1$ (+1 background) to classify after passing through the fully connected layer.

4.2 Mangosteen Images Classifier

The identification and categorization of ripe mangosteen in real time is a tough challenge. As a result, the technological solutions in this work are separated into two sections. The input picture is first received and processed before being sent into the Faster R-CNN training model. The weight value will be returned by the stage outcome. The second step is constructing the classifier. Ripe or unripe mangosteen may be distinguished at this stage. Images will be transmitted to the Faster R-CNN classifier in real-world circumstances, such as while performing tests on agricultural fields, then the classifier procedure will be carried out on the trained classifier to find the mangosteen (Fig. 7).

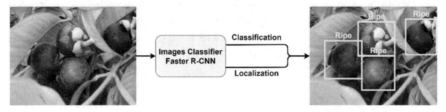

Fig. 7. Classifier of mangosteen by Faster R-CNN model with 2 branches of classification and localization

5   Experimental Results

We created a system capable of analyzing and assessing photos based on the trained model. The input photos comprise ripe and unripe mangosteens recognized using Faster R-CNN, with improved classifiers to guarantee that the data does not lose too much during training. Through actual studies, the classifier can aid farmers in classifying photos of ripe mangosteen, as well as finding mangosteen in concealed and small places, and detecting pests that impair fruit quality. The study's findings are quite practical, and farmers have reaped many benefits from delivering items for domestic and international export (Fig. 8).

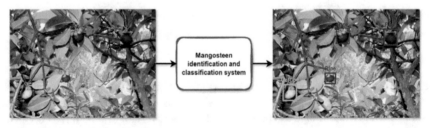

Fig. 8. Input frame (passes the detection and classification system); Output frame recognized

We trained for 6000 epochs in order to examine the reported Mean Average Precision completely (mAP). The predicted values are derived using two metrics, Precision and Recall, which are characterized by four parameters: True Positive (TP), True Negative (TN), False Positive (FP), and False Negative (FN) (FN). Figure 9 is a 2D chart based on the above two metrics that shows the greatest accuracy of 95.23%, the average variation of roughly 94%, and the false negative is clearly decreased from 16% to 11%. This is required in order to determine if the training model achieves a high accuracy or not. We also evaluate the loss of the bounding box regression and classification outputs to confirm that they are indeed more efficient than Fast-CNN. When all of the experiment's

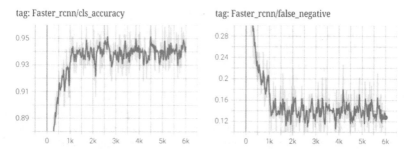

Fig. 9. Diagram to evaluate the accuracy and loss of the training model

situations are considered, the problem of data loss is fairly minor, and the technique obtains extremely good validation findings.

We compared the model trained with YOLOv4 with data of 10000 images along with 6000 iterations corresponding to the Faster R-CNN model. The results showed a marked difference in accuracy and data loss. With Faster R-CNN (~95%) the accuracy is significantly improved compared to YOLOv4 (~78%). Because the difference in the input of the different training process of the two models has an influence on the accuracy. Therefore, it is extremely reasonable to use Faster R-CNN in the identification process (Fig. 10).

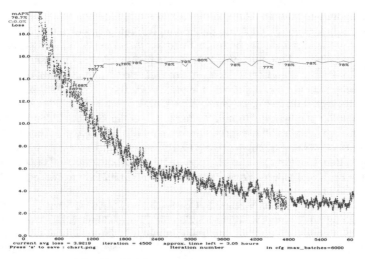

Fig. 10. YOLOv4 model evaluation diagram

6   Conclusion

We have described an efficient method to identify mangosteen fruits on fields by finding and categorizing them using an improved Faster R-CNN algorithm in this paper. The study presents a new technique to reduce data loss during the feature synthesis process. Furthermore, we suggest an RPN model with the goal of speeding up the feature extraction and information exchange processes as well as minimizing the computational time of model training stage. In addition, improving the RoI Pooling layer with the added RoI Align layer has improved the data representation, helping the region proposal not to lose information during feature extraction. The obtained test results on a real environment are quite high, showing the suitability of our proposed technique for practical applications in assisting farmers with the fruit monitoring and caring procedures. The findings of the study are applicable to large-scale farms. In terms of real-time capability, it can be ignored because the process of mining proposed areas of the two-stage Faster R-CNN model slows down the recognition process, but in return, the recognition accuracy is improved even better than the YOLOv4 one-stage model.

References

1. Uijlings, J.R.R., van de Sande, K.E.A., Gevers, T., Smeulders, A.W.M.: Selective search for object recognition. Int. J. Comput. Vis. **104**, 154–171 (2013). https://doi.org/10.1007/s11263-013-0620-5
2. Girshick, R.: Fast R-CNN. In: Proceedings of the IEEE International Conference on Computer Vision (ICCV), pp. 1440–1448 (2015)
3. Ren, S., He, K., Girshick, R., Sun, J.: Faster R-CNN towards real-time object detection with region proposal networks. In: Proceedings of the IEEE International Conference on Computer Vision, Santiago, pp. 1440–1448 (2016)
4. Redmon, J., Divvala, S., Girshick, R., Farhadi, A.: You only look once unified, real-time object detection. In: IEEE Conference on Computer Vision and Pattern Recognition (CVPR), Las Vegas, NV, pp. 779–788 (2016). https://doi.org/10.48550/arXiv.1506.02640
5. Song, Y., Glasbey, C.A., Horgan, G.W., Polder, G., Dieleman, J.A., Heijden, G.V.D.: Tomatic fruit recognition and counting from multiple images. Biosyst. Eng. **118**, 203–215 (2014). https://doi.org/10.1016/j.biosystemseng.2013.12.008
6. Sa, I., Ge, Z.Y., Dayoub, F., Upcroft, B., Perez, T., McCool, C.: Deep fruits: a fruit detection system using deep neural networks. Sensors **16**, 1222, 1–23 (2016). Autonomous Sweet Pepper Harvester Project. https://doi.org/10.3390/s16081222
7. Kim, B., Han, Y.K., Park, J.H., Lee, J.: Improved vision-based detection of strawberry diseases using a deep neural network. Front. Plant Sci. **11**, 559172 (2021). https://doi.org/10.3389/fpls.2020.559172
8. Rojas-Aranda, J.L., Nunez-Varela, J.I., Cuevas-Tello, J.C., Rangel-Ramirez, G.: Fruit classification for retail stores using deep learning. In: Figueroa Mora, K.M., Anzurez Marín, J., Cerda, J., Carrasco-Ochoa, J.A., Martínez-Trinidad, J.F., Olvera-López, J.A. (eds.) MCPR 2020. LNCS, vol. 12088, pp. 3–13. Springer, Cham (2020). https://doi.org/10.1007/978-3-030-49076-8_1
9. Cao, C., Wang, B., Zhang, W.: An improved faster R-CNN for small object detection (2019). INSPEC accession number: 18897667. Electronic ISSN: 2169-3536. https://doi.org/10.1109/ACCESS.2019.2932731
10. Tang, Y., Chen, M., Wang, C., Luo, L., Li, J., Lian, G., et al.: Recognition and localization methods for vision-based fruit picking robots: a review. Front. Plant Sci. **11**, 510 (2020). https://doi.org/10.3389/fpls.2020.00510
11. Muresan, H., Oltean, M.: Fruit recognition from images using deep learning. Acta Univ. Sapientiae Inform. **10**, 26–42 (2018). https://doi.org/10.2478/ausi-2018-0002

12. Rajeshwari, P., Abhishek, P., Srikanth, P., Vinod, T.: Object detection: an overview. Int. J. Trend Sci. Res. Dev. (IJTSRD) **3**, 1663–1665 (2019)
13. Zhang, Y., Sohn, K., Villegas, R., Pan, G., Lee, H.: Improving object detection with deep convolutional networks via Bayesian optimization and structured prediction. In: Computer Vision and Pattern Recognition (2016). arXiv:1504.03293 [cs.CV]
14. Simonyan, K., Zisserman, A.: Very deep convolutional networks for large-scale image recognition. In: Computer Vision and Pattern Recognition, pp. 1–14 (2015). arXiv:1409. 1556v6
15. Erhan, D., Szegedy, Ch., Toshev, A., Anguelov, D.: Scalable object detection using deep neural networks. In: The IEEE Conference on Computer Vision and Pattern Recognition (CVPR), pp. 2147–2154 (2014)
16. Lu, Y., Javidi, T., Lazebnik, S.: Adaptive object detection using adjacency and zoom prediction. In: Computer Vision and Pattern Recognition (2016). arXiv:1512.07711 [cs.CV]
17. Zhao, Z.Q., Zheng, P., Xu, S., Wu, X.: Object detection with deep learning: a review. IEEE Trans. Neural Netw. Learn. Syst. Publ. (2019). arXiv:1807.05511 [cs.CV]
18. Munera, S., Amigo, L.M., Blasco, J., Cubero, S., Talens, P., Alexios, N.: Ripeness monitoring of two cultivars of nectarine using VIS-NIR hyperspectral reflectance imaging. J. Food Eng. **214**(3), 29–39 (2017)
19. Kim, S., Ji, Y., Lee, K.: An effective sign language learning with object detection based ROI segmentation. In: Second IEEE International Conference on Robotic Computing (IRC), Laguna Hills, CA, pp. 330–333 (2018)
20. Nguyen, H.H.C., Nguyen, D.H., Nguyen, V.L., Nguyen, T.T.: Smart solution to detect images in limited visibility conditions based convolutional neural networks. In: Hernes, M., Wojtkiewicz, K., Szczerbicki, E. (eds.) ICCCI 2020. CCIS, vol. 1287, pp. 641–650. Springer, Cham (2020). https://doi.org/10.1007/978-3-030-63119-2_52
21. Nguyen, H.H.C., Luong, A.T., Trinh, T.H., Ho, P.H., Meesad, P., Nguyen, T.T.: Intelligent fruit recognition system using deep learning. In: Meesad, P., Sodsee, D.S., Jitsakul, W., Tangwannawit, S. (eds.) IC2IT 2021. LNNS, vol. 251, pp. 13–22. Springer, Cham (2021). https://doi.org/10.1007/978-3-030-79757-7_2

Naturalness Improvement of Vietnamese Text-to-Speech System Using Diffusion Probabilistic Modelling and Unsupervised Data Enrichment

Tung Tran[1,2,3(✉)] [iD], Tuan Nguyen[1,2] [iD], Hung Bui[1,2] [iD], Khuong Nguyen[3] [iD],
Nghia Gia Vo[4], Tran Vu Pham[1,2] [iD], and Tho Quan[1,2] [iD]

[1] Ho Chi Minh City University of Technology (HCMUT),
268 Ly Thuong Kiet Street, District 10, Ho Chi Minh City, Vietnam
{tung.tran.1805,tuan.nguyen991,hung.bui0089,ptvu,qttho}@hcmut.edu.vn
[2] Vietnam National University Ho Chi Minh City, Linh Trung Ward,
Thu Duc District, Ho Chi Minh City, Vietnam
[3] FPT Software Company Limited, D1 street, District 9,
Ho Chi Minh 700000, Vietnam
khuongnd6@fsoft.com.vn
[4] Department of Information and Communications of Binh Dinh Province,
Quy Nhon City, Binh Dinh Province, Vietnam

Abstract. Speech synthesis, which aims to generate natural and comprehensible speech from input text, is a popular research topic with a wide range of industrial applications. However, it appears to be a difficult problem due to its strong dependency on data, particularly for accent-sensitive and multi-dialect languages, e.g. Vietnamese. Perhaps the most common model applied in this area is Tacotron 2, using *Recurrent Neural Network* (RNN) and *Convolutional Neural Network* (CNN) architectures. Still, Tacotron 2 has not yet achieved the expected naturalness, possibly because it was not sophisticated enough to capture the natural expression of human voice. Moreover, with a low-resource language like Vietnamese, to collect a sufficient training dataset for this task is also a non-trivial problem. Hence, in this paper we propose an end-to-end framework with Grad-TTS, a denoising diffusion probabilistic model, as an acoustic model in the *Text-to-speech* (TTS) system instead of the traditional approach employed by Tacotron 2. The proposed approach helps us achieved a more natural synthesized speech, as depicted in the experiments. Furthermore, we also introduce an unsupervised approach to collect Vietnamese data from the Internet resource as well as to pre-process the data before putting it into training. This helps solve the problem of lacking Vietnamese data, and enhance our outcome. We released the dataset for further development of TTS system for Vietnamese at: https://bit.ly/3rnNsFi.

Keywords: Speech synthesis · Text-to-speech · Generative modelling · Grad-TTS · Low-resource language · Vietnamese · HMI

© The Author(s), under exclusive license to Springer Nature Switzerland AG 2022
N.-T. Nguyen et al. (Eds.): ICIT 2022, LNDECT 148, pp. 376–387, 2022.
https://doi.org/10.1007/978-3-031-15063-0_36

1 Introduction

Text-to-speech (TTS) [2] is a technology that converts text into speech. In interactive, speech-based systems such as *human-machine interaction* (HMI), the issue of voice synthesis hence is very crucial. Currently, there are roughly 6500 languages in the world. Among them, English and Chinese are the most popular with several deep investigations. However, there are still some other low-resource languages, e.g. Vietnamese, which have not been well exploited. Regardless of whether the language has rich or low resources, a large amount of data is required to construct and train such a system. The typical process is to hire a professional native speaker to read dozens of hours of materials in a clear environment. Moreover, the environment must be accompanied with a high-quality microphone, maintaining consistency in the actor's voice quality. Up till recently, there are no carefully recorded and annotated corpora for TTS systems in so-called low-resource languages like Vietnamese. Mostly, one resource which can be collected is from the shared task TTS challenge of the VLSP Campaign. In VLSP Campaign 2019[1], the group "ZALO" won the competition with the best performance. In their design, Tacotron 2 [15], a deep learning model combining CNN [1] and RNN [10], was chosen as the acoustic model. Furthermore, Tacotron 2 is typically used in several studies. Intuitively, though, the voice is not as natural as expected.

Besides the Tacotron 2, many other techniques are applied to the TTS problem. Recently, some innovative models based on *Generative Adversarial Networks* (GAN) [4] have been proven to outperform traditional models (Tacotron 2). The *Diffusion Probabilistic Model* (DPM) [6] is the most recent addition to the generative modelling methods and one example is Grad-TTS [13]. Hence, we integrate Grad-TTS as an acoustic model into our proposed approach with the strong expectation of outperforming Tacotron 2. Although the generative network performs well in English, how well it performs in Vietnamese remains a mystery. In the result section, we respond to that question.

There are currently few documents on how to develop a TTS system for Vietnamese, a low-resource language. As a result, we plan to create a comprehensive document on how to design a Vietnamese TTS system. Furthermore, the Vietnamese dataset that is publicly available is limited. Therefore, we propose an unsupervised data enrichment approach for collecting and pre-processing Vietnamese data from publicly accessible data sources.

Our main contributions are summarized as follows.

- We develop an unsupervised data enrichment approach in which data is gathered from publicly available sources via crowdsourcing.
- We introduce an end-to-end framework with Grad-TTS for a more natural Vietnamese TTS.

This paper is organized as follows. Section 1 describes our motivation and our main contribution. Section 2 gives background on the problem. Section 3 discusses some related works that have been carried out for Vietnamese. Section 4

---

[1] https://vlsp.org.vn/vlsp2019/eval.

shows the process of Unsupervised Data Enrichment approach. Section 5 discusses the Grad-TTS-based model approach to the TTS problem. Section 6 demonstrates the results and experimental setting. Lastly, Sect. 7 draws a conclusion and makes recommendations for future work.

2   Background

There are 3 main components in the speech synthesizer, also known as a TTS model: Text Analysis, Acoustic Model, and Vocoder, as illustrated in Fig. 1. Those components can be briefly described as follows.

Fig. 1. 3 main component in a TTS model

Text Analysis, also known as "front-end", converts input text into linguistic features that provide comprehensive information on pronunciation and prosody.

Acoustic model generates acoustic features based on the input linguistic features or directly from a character or phoneme. Then, these acoustic features will be transformed into waveforms by using vocoders.

Vocoder receives linguistic or acoustic features to generate waveforms.

3   Related Works

3.1   TTS for Vietnamese Language

In [5], VOS, a concatenative-based speech synthesis system, was presented. They use syllables as the basic synthesis for Vietnamese unit selection TTS systems. The major disadvantage of this method is that unit selection synthesis necessitates a considerable amount of speech database storage. Furthermore, the intonation and rhythm are poor.

VTed [16], a Vietnamese TTS system developed under the Mary TTS platform, used the *Statistical Parametric Speech Synthesis* (SPSS) [18] approach with a *Hidden Markov Model* (HMM) [14] based architecture. The result demonstrates the need for more specific work on the tonal level to improve the automatic synthesis of Vietnamese.

In [9], "ZALO" group used deep learning to improve their TTS system. Their core system was based on the Tacotron 2 and the WaveGlow neural vocoder. When compared to the SPSS approach, the results obtained with this approach were substantially superior. The dataset they used to train the model, however, was not made public.

In VLSP 2020[2], the best team's [11] synthetic voice was created with Tacotron 2 and Hifigan vocoder, achieving 89.3% compared to the human voice in terms of naturalness. However, they still had big issues with long sentences, particularly improper prosodic phrasing and bad pronunciation of loan words.

The TTS system with the Tacotron 2 acoustic model is considered a classic deep learning approach. DCA-Tacotron 2 [3] is an improved version of Tacotron 2 that uses a new location-relative attention mechanism called Dynamic Convolution Attention from the additive energy-based family to tackle text alignment problems (DCA). Despite the improvements, DCA-Tacotron 2 is still unable to solve long texts generation. Therefore, the latest generative models are predicted to perform better. According to [13], the TTS system using Grad-TTS as an acoustic model can solve all of the problems described above (unnatural voice, need for a very large database, lack of emotion, and inability to handle long sentences). On the other hand, Grad-TTS has recently performed exceptionally well on English datasets. Grad-TTS's ability to perform well on a Vietnamese dataset remains a mystery, as Vietnamese is regarded as a challenging language due to distinct traits that differ significantly from English. In this work, we build a TTS system using the Grad-TTS model, as well as illustrate how to pre-process data to improve a model dramatically.

3.2 Generative Modelling in TTS

In [17], the authors proposes a flow-based generative network for speech synthesis with control over speech variation and style transfer named Flowtron. Flowtron is optimized by maximizing the likelihood and allows for control of speech variation and style transfer. Flowtron also learns a latent space that stores non-textual information. In terms of speech quality, *mean opinion scores* (MOS) reveal that Flowtron matches state-of-the-art TTS models.

This paper [8] suggests Glow-TTS as a flow-based generative model for parallel TTS which does not require an external aligner. The suggested approach searches for the most likely monotonic alignment between text and the latent representation of speech on its way by integrating the properties of flows and dynamic programming. Not only does it achieve a comparable speech quality, Glow-TTS also achieves an order-of-magnitude speedup over the Tacotron 2.

The research [7] introduced the TTS system that integrates the denoising diffusion probabilistic model (DDPM) into the core architecture. Diff-TTS allows a controlled trade-off between quality and inference speed. It can be trained with half the number of parameters of the Tacotron 2 and Glow-TTS while generating high-fidelity audio. Moreover, Diff-TTS also effectively controls speech prosody.

As shown in the studies above, the acoustic generative model produces excellent outcomes in comparison to previous methods. All of this, however, is only available in English. In this research, we built a TTS system for Vietnamese using the generative model technique (specifically, Grad-TTS). According to

---

[2] https://vlsp.org.vn/vlsp2020/eval.

MOS, Grad-TTS's outcome in English is also superior to the previous models, which is why we chose Grad-TTS for our TTS system.

4   Unsupervised Data Enrichment

In the field of deep learning, data are just as crucial as algorithms when it comes to producing the best possible results. Even if you have a fantastic architecture, if it is trained on low-quality data, the outcomes are doubtful to be satisfactory.

The Vietnamese TTS dataset is currently limited, and if exists, is not made public. In this paper, we propose an approach for unsupervised data enrichment, hoping this approach will help the TTS community collect more data, especially for the Vietnamese community who is suffering from Vietnamese being a low-resource language. As a result, we also publish our dataset for use in Vietnamese TTS Research Community. This dataset is named ViSpeech.

Perhaps, the InfoRe dataset[3] used in TTS Shared Task in VLSP Campaign 2019 [12] is the only freely available Vietnamese dataset. However, InfoRe has a number of drawbacks, including the fact that it was captured in a variety of settings, the dataset is not consistent, and there is a lot of noise. We show in below sections that our ViSpeech dataset can overcome the disadvantages that InfoRe currently has.

4.1   Data Description

First, we describe the ViSpeech dataset used to train the model. The data is collected from a single speaker's audiobooks, which lasted roughly 20 h. There are around 10,000 utterances, each last 3-7 s with corresponding phonemes. There are two fields in the metadata file.

- ID is the name of the corresponding .wav file.
- Phonemes are analyzed from corresponding transcripts.

  Audiobooks were chosen because they have the following qualities.

- The reader's voice is clear, loud, and inspirational.
- There is little noise because it was recorded in a professional environment.
- There is almost no variation in expression or voice because the same person reads and writes from the same book.
- The audio book library is extensive and diversified.

For these reasons, audio books can be used by the TTS community to train their models as a very high-quality data source. Since data on the internet does not have careful labels, that is why we call it Unsupervised Data.

The data collection and unsupervised enrichment process to build the ViSpeech dataset is shown in Fig. 2. Section 4.2 will go over this procedure in depth.

---

[3] https://bit.ly/33C4dUW.

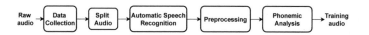

Fig. 2. Data Processing Schema. Raw audio is fed into the system, which is then processed through data collection, split audio, automatic speech recognition, pre-processing, and phonemic analysis. The training audio is the final output.

4.2 Data Collection and Unsupervised Data Enrichment

Data Collection First, we'll choose audio audiobooks spoken by a speaker that last about 20 h. Then use pydub library's export function to convert to wav format.

Split Audio Second, use pydub[4] library's spilt_on_silence function to split audio with silence. There are two important parameters:

- minimum_silence_length (in ms) is the minimum silence length used to separate audio.
- silence_thresh (in dBFS) sounds below this number will be considered silence, default is −16 dBFS.

A lot of noisy data will be created if the values for these two parameters are not chosen correctly. The right value was determined by looking at the waveform image of an audio clip to determine the speaker's interruption rule. We acquired utterances with a duration of 3-7 s after this step.

Automatic Speech Recognition Third, we will generate transcript for each utterances using azure.cognitiveservices.speech[5] function of Azure services.

Pre-processing,

- Filter out outliers by plotting a scatter plot of the full dataset's duration. We need to remove such outliers from the data since they are the noisy points.
- We apply a linear regression with two variables: duration and the number of words in each file, since the amount of words in a phrase is proportional to the duration of the audio file. The graph's outliers will then be removed.
- Normalized Transcription with numbers, ordinals, and monetary units expanded into full words (UTF-8).
- Finally, the transcript will be stripped of all punctuation so that the model may learn to interrupt on its own.

Phonemic Analysis Finally, we will create a phonemic analysis for each transcript. It is the dataset's second attribute that we will utilize to train the model. We use vPhon[6]'s transformation rules to analyze the phoneme for each transcript. The machine will learn from the phoneme better than the original transcript because the phoneme is a form of representation closer to the sound.

---

[4] https://github.com/jiaaro/pydub.

[5] https://bit.ly/33fuOHV.

[6] https://github.com/kirbyj/vPhon.

5   Grad-TTS-Based Synthetic Model Approach

Compared to the traditional approach, Tacotron 2, we recommend the Grad-TTS-based synthetic model approach, which can give much more natural results. The overall architecture of our end-to-end TTS system is shown in Fig. 3. It includes vPhon as a Text Analysis, Grad-TTS as an Acoustic Model and HiFi-GAN as a Vocoder. The Vocoder's output is then included in the loss.

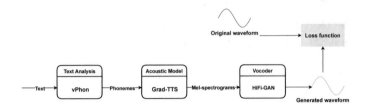

Fig. 3. Our TTS system architecture. Text Analysis is vPhon. Acoustic Model is Grad-TTS. Vocoder is HiFi-GAN. The input is text and the final output is waveform.

Here, we employ vPhon as the text analysis module, which is a Vietnamese phonetizer, based on the *International Phonetic Alphabet* (IPA).

In the architecture, we proposed HiFi-GAN for our vocoder. Although HiFi-GAN was trained on English, it also worked for Vietnamese without the need for finetuning on Vietnamese. Therefore, we focus on the acoustic model, Grad-TTS.

Grad-TTS, with a score-based decoder, generates mel-spectrograms by gradually transforming the noise data predicted by the encoder and aligned by the duration predictor. With the help of stochastic differential equations, Grad-TTS is able to make the reconstructing process flexible by explicitly controlling the trade-off between sound quality and time inference.

Figure 4 shows us the overall framework, consisting of 3 modules: encoder, duration predictor and decoder. As for the encoder and duration predictor, we use exactly the same architectures as in Glow-TTS. The Diffusion Probabilistic Model appears in the decoder.

Diffusion Probabilistic Model has 2 steps:

- The forward diffusion step:

$$dX_t = \frac{1}{2}\Sigma^{-1}(\mu - X_t)\beta_t dt + \sqrt{\beta_t}dW_t, t \in [0; T] \tag{1}$$

where W_t is the standard Brownian motion.

We can find immediately X_T from X_0 by Terminal distribution:

$$Law(X_T|X_0) \rightarrow \mathcal{N}(\mu, \Sigma), T \rightarrow \infty \tag{2}$$

- The reverse diffusion step:

$$dX_t = \frac{1}{2}(\Sigma^{-1}(\mu - X_t) - \nabla\mathrm{log}p_t(X_t))\beta_t dt \tag{3}$$

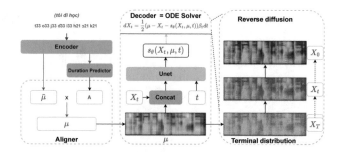

Fig. 4. Grad-TTS architecture includes encoder, duration predictor, decoder. The output of encoder, aligned by duration predictor, is used to generate waveform by decoder.

5.1 Inference

Input: an input text sequence $x_{1:L}$ of length L consists of characters or phonemes.
Output: mel-spectrograms $y_{1:F}$, where F is the number of acoustic frames.

First, the input $x_{1:L}$ is converted into $\tilde{\mu}_{1:L}$, which is used by the duration predictor module to produce hard alignment A between $\tilde{\mu}_{1:L}$ and $\mu_{1:L}$.

Second, $\mu_{1:L}$ are passed into the decoder. In the formula (3), the expression $\nabla \log p_t(X_t)$ is hard to calculate so it is approximated by $s_\theta(X_t, \mu, t)$ from Unet model, defining an ordinary differential equation (ODE):

$$dX_t = \frac{1}{2}(\mu - X_t - s_\theta(X_t, \mu, t))\beta_t dt \tag{4}$$

which can be solved by the first-order Euler scheme. Here, we chose $\Sigma = I$ to simplify the whole feature generation pipeline.

5.2 Training

One training objective of Grad-TTS is to minimize the distance between aligned encoder output μ and target mel-spectrogram y.

$$\mathcal{L}_{enc} = -\sum_{j=1}^{F} \log\varphi(y_j, \tilde{\mu}_{A(j)}, I) \tag{5}$$

where $\varphi(., \tilde{\mu}_i, I)$ is a probability density function of $\mathcal{N}(\tilde{\mu}_i, I)$

The duration predictor DP is trained with *Mean Square Error* (MSE) criterion in logarithmic domain:

$$d_i = \log\sum_{j=1}^{F} \mathcal{I}_{A^*(j)=i}, i = 1, ..., L \tag{6}$$

$$\mathcal{L}_{dp} = MSE(DP(sg[\tilde{\mu}]), d) \tag{7}$$

where $\mathcal{I}$ is an indicator function, $\tilde{\mu} = \tilde{\mu}_{1:L}$, $d = d_{1:L}$ and stop gradient operator $sg[.]$ is applied to prevent $\mathcal{L}_{dp}$ from affecting encoder parameters.

As for the loss related to the DPM (Diffusion Probabibistic Model), it is calculated:

$$\mathcal{L}_{diff} = E_{X_0,t} \left[\lambda_t E_{\xi_t} \left[\left\| s_\theta(X_t, \mu, t) + \frac{\xi_t}{\sqrt{\lambda_t}} \right\|_2^2 \right] \right] \tag{8}$$

where $\lambda_t = 1 - e^{-\int_0^t \beta_s ds}$.

6   Experiments

In this section, we trained a DCA-Tacotron 2 [3] model on the publicly available data (InfoRe dataset) as our *Baseline* model, with the same parameters and configuration as in the original paper. We then evaluated the effectiveness of our proposed data enrichment and framework on both the InfoRe dataset and our ViSpeech dataset, compare to the baseline model. With each of the dataset, we used 90% of the samples as training data, and the 10% left for validation. Before that, we select and leave-out 40 sentences for testing.

6.1   Experimental Details

For baseline model, we used the Pytorch implementation[7]. For the proposed method, we based on the source code of the original paper [13]. Each model was trained for 2500 epochs on a single GPU (NVIDIA RTX 3090 with 16 GB memory).

All acoustic model are paired with the same HiFi-GAN vocoder, as stated in Sect. 5.

To make subjective evaluation of TTS models, a MOS test was conducted to compare our system with others. Each sample is rated by at least 8 raters on a scale from 1 ("Bad") to 5 ("Excellent") with 0.5 points increments.

6.2   Results

MOS results are presented in Table 1. It shows that, our model, trained on 2 other datasets, were able to achieved a better performance in the naturalness of the synthesized speech, with a large gap compare to previous approach. Furthermore, the model achieves almost natural synthesis with a MOS value of 4.0046.

Table 1. Mean Opinion Score (MOS) evaluations with 95% confidence intervals computed from the t-distribution for various systems.

System	MOS
Ground truth	4.4245 ± 0.0526
OURS trained on ViSpeech	4.0046 ± 0.0660
OURS trained on InfoRe	3.5423 ± 0.0750
DCA-Tacotron 2	2.6535 ± 0.1005

---

[7] https://github.com/bshall/Tacotron.

There is a notable differences between the MOS values of the model, trained on ViSpeech, compare to InfoRe. This result show the advantages of our data enrichment method.

Ablation study. For the ablation study, we are interested in the performance of the proposed generative modelling approach, especially without the reverse diffusion process. Table 2 presented the MOS values of the ablated system. The score is notably lower than that of the original model. Upon further investigation, we realized the reasoning behind the performance drop is the spectrogram generated by encoder, contains a lot of noisy sounds. However, the main vocal voice is as natural and expressive as that of the decoder output. An example can be seen in Fig. 5. We believed this can be beneficial for building expressive and controllable TTS, since the hidden state of the model, the encoder output, consists of information that us, human, can understand and further exploit.

Table 2. MOS evaluations between Encoder and Decoder outputs of our model.

System	MOS
OURS encoder output	2.5551 ± 0.0762
OURS decoder output	4.0046 ± 0.0660

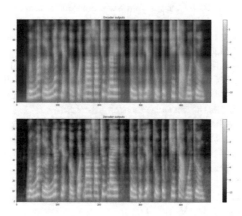

Fig. 5. Differences between the spectrogram of the encoder and decoder of the model. The decoder output spectrogram is much more details than that of the encoder.

7   Conclusion

In this paper, we introduced a novel method for building and training TTS system for Vietnamese, through generative modelling, which we hope to open a new direction of research for Vietnamese Speech synthesis system. We also proposed and illustrated a data processing technique for collecting and utilizing available resources from the Internet, rather than exclusively and expensively data collection methods. We shown that, through our methods, our system trained on processed data achieved a SOTA MOS result, which confirm the benefits of using our methods. In future works, we will investigate other choices for drift and diffusion coefficient in SDE. Concurrently, we want to intergrate timbre information from ground-truth speech to encoder output for more expressive and controllable TTS.

Acknowledgments. This research is funded by Binh Dinh Foundation for Science and Technology Development under grant number 02-02-2020. We acknowledge the support of time and facilities from Ho Chi Minh City University of Technology (HCMUT), VNU-HCM for this study.

References

1. Albawi, S., Mohammed, T.A., Al-Zawi, S.: Understanding of a convolutional neural network. In: 2017 International Conference on Engineering and Technology (ICET), pp. 1–6 (2017). https://doi.org/10.1109/ICEngTechnol.2017.8308186
2. Baart, J.L., van Heuven, V.J.: From text to speech; the mitalk system: Jonathan Allen, M. Sharon Hunnicutt and Dennis Klatt (with Robert C. Armstrong and David Pisoni): Cambridge University Press, cambridge, 1987, xii+216, & #x00A3;25.00. Lingua **81**(2), 265–270 (1990). https://doi.org/10.1016/0024-3841(90)90014-C,https://www.sciencedirect.com/science/article/pii/002438419090014C
3. Battenberg, E., et al.: Location-relative attention mechanisms for robust long-form speech synthesis (2020)
4. Goodfellow, I.J., et al.: Generative adversarial networks (2014)
5. Ha, V.Q.D., Tuan, N.M., Nam, C.X., Nhut, P.M., Quan, V.H.: VOS: the corpus-based etnamese text-to-speech system. J. Res. Dev. Inf. Commun. Technol (2010). https://doi.org/10.32913/mic-ict-research.v3.n7.285
6. Ho, J., Jain, A., Abbeel, P.: Denoising diffusion probabilistic models (2020)
7. Jeong, M., Kim, H., Cheon, S.J., Choi, B.J., Kim, N.S.: Diff-tts: a denoising diffusion model for text-to-speech (2021)
8. Kim, J., Kim, S., Kong, J., Yoon, S.: Glow-tts: a generative flow for text-to-speech via monotonic alignment search (2020)
9. Lam, V., Phan, K., Dinh, T., Khuong, D., Nguyen, Q.: Development of zalo vietnamese text-to-speech for VLSP (2019)
10. Lipton, Z.C., Berkowitz, J., Elkan, C.: A critical review of recurrent neural networks for sequence learning (2015)
11. Nguyen, T.T.T., Nguyen, H.K., Pham, Q.M., Vu, D.M.: Vietnamese text-to-speech shared task VLSP 2020: remaining problems with state-of-the-art techniques. In: Proceedings of the 7th International Workshop on Vietnamese Language and Speech Processing, pp. 35–39. Association for Computational Lingustics, Hanoi, Vietnam (2020). https://aclanthology.org/2020.vlsp-1.7
12. Nguyen, T.T.T., Nguyen, X.T.: Text-to-speech shared task in VLSP campaign 2019: evaluating vietnamese speech synthesis on common datasets (2019)
13. Popov, V., Vovk, I., Gogoryan, V., Sadekova, T., Kudinov, M.: Grad-tts: a diffusion probabilistic model for text-to-speech (2021)

14. Rabiner, L., Juang, B.: An introduction to hidden markov models. IEEE ASSP Mag. **3**(1), 4–16 (1986). https://doi.org/10.1109/MASSP.1986.1165342
15. Shen, J., et al.: Natural tts synthesis by conditioning wavenet on mel spectrogram predictions (2018)
16. Trang, N., d'Alessandro, C., Rilliard, A., Tran, D.D.: Hmm-based tts for hanoi vietnamese: issues in design and evaluation. In: Proceedings of the Annual Conference of the International Speech Communication Association, INTERSPEECH, pp. 2311–2315 (2013)
17. Valle, R., Shih, K., Prenger, R., Catanzaro, B.: Flowtron: an autoregressive flow-based generative network for text-to-speech synthesis (2020)
18. Zen, H., Tokuda, K., Black, A.: Statistical parametric speech synthesis. Speech Commun. **51**, 1039–1064 (2007). https://doi.org/10.1016/j.specom.2009.04.004

Real-Time Traffic Congestion Forecasting Using Prophet and Spark Streaming

Ngan-Linh Nguyen[1,2], Hoang-Thong Vo[1,2], Gia-Huy Lam[1,2],
Thanh-Binh Nguyen[1,2], and Trong-Hop Do[1,2(✉)] (iD)

[1] University of Information Technology, Ho Chi Minh City, Vietnam
{18520989,18521462,18520832}@gm.uit.edu.vn,
{binhnt,hopdt}@uit.edu.vn
[2] Vietnam National University, Ho Chi Minh City, Vietnam

Abstract. Traffic prediction system is one of the principal components of an intelligent traffic system (ITS). This system relies on data collected through vehicle-to-vehicle communication, probe vehicle monitoring, speed estimation, and vehicle counting based on vehicle tracking to predict the state of traffic. One of the most demanding tasks in traffic prediction is traffic congestion prediction. Upon predicted traffic congestion, traffic flow control and other intervention can be performed to prevent or at least reduce future congestion, which eases potentially disastrous effects of traffic congestion on the environment, society, and the economy. Traffic congestion prediction is a challenging task in several aspects. First, the prediction needs to be precise. Second, it needs to be made promptly so that any intervention can be meaningful. Third, the system needs the capacity to process a huge amount of data to provide the result for tens of thousands of locations in the map of a city simultaneously. This study proposes a traffic prediction system using Prophet and Spark Streaming. The entire system is built on Apache Spark, which is a Big data processing framework that can be scaled to process a huge amount of data. Spark Streaming is applied to process the streaming data and make real-time forecasting of the traffic flow. The Prophet model, which can capture long-range temporal sequences of data is used to predict traffic flow. The proposed system is shown to achieve good performance based on experimental results with the PEMS-BAY public transport dataset.

Keywords: Traffic analysis · Real-time · Big data · Time series · Deep learning

1 Introduction

Transportation is one of the essential needs of people all over the world. According to data from the Union Internationale des Transports Publics, an average of 168 million passengers use metros every day to travel in 178 cities in 56 countries. Thanks to transportation, the supply chain of goods and people's travel needs

© The Author(s), under exclusive license to Springer Nature Switzerland AG 2022
N.-T. Nguyen et al. (Eds.): ICIT 2022, LNDECT 148, pp. 388–397, 2022.
https://doi.org/10.1007/978-3-031-15063-0_37

are continuously met and circulated everywhere and at any time. The explosion in the number of personal vehicles, the increase in greenhouse gas emissions, the traffic jams, and the number of deaths due to traffic accidents are among the issues that deserve great attention in recent years. Intelligent transportation systems (ITS) [7] is the solution proposed throughout the years to solve the above problems. Nevertheless, ITS only solves a small part of the problems that occur, such as reducing environmental pollution, improving community health, reducing traffic congestion, and increasing the number of private vehicles. In large cities, high population densities and large populations lead to overloaded transport systems that are unable to respond to rapid population growth. Policymakers and experts from different fields have come together to discuss and propose outstanding solutions among them the implementation of ITS with the core technology of artificial intelligence.

Traffic prediction is a complex problem, also an obstacle that needs to be solved to build ITS and a research topic that receives a lot of attention from researchers. The challenges of this problem are processing large amounts of data in real time with high accuracy and infrastructure to store large amounts of data to query in the fastest time for prediction. In major cities around the world, responding to real-time data processing with high accuracy can instantly reduce traffic congestion at intersections, ensure continuous traffic flow in the city. Currently, topics in traffic prediction tasks, such as flow, speed, demand, travel time, occupancy [11] are the fundamental problems discussed. In this paper, we present a traffic prediction system using Prophet and Spark Streaming which is used to predict the average speed of vehicles at different times based on historical data. We base the performance of the model evaluation on the measures for the time-series regression problem achieved top results with the characteristic properties of time-series data, such as trend, seasonal, and stationary. In addition, the system can also process large amounts of data in real time. We stored this amount of data distributed in databases in the overall architecture of the system. When large amounts of data are needed, we will query it from databases at real-time speed for prediction.

2   Related Work

Deep Learning Methods. In 2017, diffusion convolutional recurrent neural network, a model for spatiotemporal forecasting tasks, achieved top results on the traffic prediction benchmark dataset [6]. In the same year, a decentralized deep learning-based method based on the congestion state of the neighboring stations was presented by Fouladgar et al. [4]. This limitation inhibits deep learning methods from accurate traffic forecasting.

Apache Spark. A comprehensive and flexible architecture based on distributed computing platform for real-time traffic control was proposed by Amini et al. in which the architecture was based on a systematic analysis of the requirements of the existing traffic control system [1]. Saraswathi et al. presented a system used to predict the number of vehicles on different roads based on the processing technique of streaming data to reduce traffic congestion and visualize the traffic conditions analysis in real-time [9]. Anveshrithaa et al. introduced the proposed model is aspire to predict traffic flow information by integrating Spark and Kafka along with deep neural networks [2]. In the study of data surveys and big data tools, author Jiang et al. have detailed the use of the above tools for traffic estimation and prediction problems [5]. The limitation of these methods is only propose real-time processing systems or predict traffic jams based on the number of vehicles distributed on the road.

3   Proposed Architecture

In this section, we present an overview of the system's architecture, from the input data collected from the traffic sensors to the output data, which is the predicted speed of traffic on road networks and an understanding of a traffic situation from traffic data.

Figure 1 represents the overall procedure of our work in this paper. Distributed data across traffic sensors are collected and streamed via API [12,13]. Spark Streaming [3,8] is an extension of spark core API that allows developers to process real-time big data and ensure fault tolerance. As a feature of streaming, it helps to scale data from one node to thousands of nodes for execution. Each data node includes many tasks to execute queries, so the processing is quite fast. In addition, it also integrates with DStream, which represents the series of RDDs in Spark, so any function in Apache Spark can process data. A more convenient way for real-time data analysis, it also integrated the streaming feature on Spark SQL, MLib, GraphX to make it easy to execute queries as well as apply predictive models. The data are then either written into storage for retraining new models and interpreting data trends or used as inputs for near-future graph traffic predictions. We updated the prediction model [10] very week for the best performance and resource efficiency. Note that it also written newly streamed data into storage and contributes to expand the available dataset.

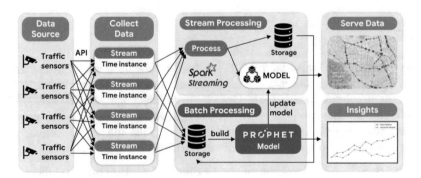

Fig. 1. Overall procedure of traffic prediction system

4   Experiments

4.1   Dataset

PEMS-BAY [6] traffic dataset is collected by California Transportation Agencies (CalTrans) Performance Measurement System (PeMS). 325 sensors were selected in the Bay Area and 6 months of data collected ranging from Jan 1st 2017 to May 31st 2017 for the experiment. With the time window of 5 min, the total number of observed traffic data points is 16,937,179. Details about the dataset split will be better demonstrated based on circumstances in the experiment settings section.

4.2   Experiment Procedure and Settings

We divided our approaches into forecasts on a single sensor and on multiple sensors. We use PEMS-BAY first three months data (from Jan 1st to March 31st) as training data for both approaches. Our approaches are described in Fig. 2.

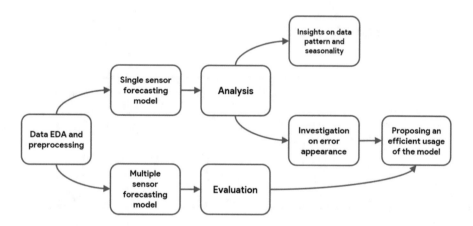

Fig. 2. Detailed experimental procedure of single and multiple sensor forecasting

Talking about Prophet training parameters, since our training data spreads over 3 months, which does not cover a whole year, we set year_seasonality = False, weekly_seasonality = True and daily_seasonality = True.

In single sensor forecasting, after training on first 3 months data, we obtain weekly, daily seasonality, and also absolute error throughout the last three months test data, next we investigate the reasons for error appearance.

In multiple sensors forecasting, we aim to compare weekly MAEs, and see how strong MAE increases if we use a fixed model trained only on the first three months versus retraining the model every week using previous 3 month data for training.

Finally we propose a correct usage of the framework and present a real time forecasting framework using Spark streaming to eagerly take current time as input and produce predictions for the next five minutes, thirty minutes and ninety minutes as well as show the prediction on an interactive world map.

4.3   Metrics

Suppose $y = y_1, y_2, y_3,..., y_n$ represents the ground truth, $\hat{y} = \hat{y}_1, \hat{y}_2, \hat{y}_3,...,\hat{y}_n$ represents the predicted values, and n denotes the indices of observed samples, the metrics are defined as follows.

The mean absolute error (MAE) and the root mean squared error (RMSE):

$$MAE = (\frac{1}{n}) \sum_{i=1}^{n} |\hat{y}_i - y_i| \tag{1}$$

$$RMSE = \sqrt{(\frac{1}{n}) \sum_{i=1}^{n} (\hat{y}_i - y_i)^2} \tag{2}$$

4.4   Result and Discussion

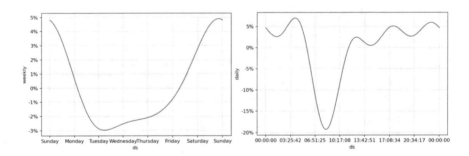

Fig. 3. Weekly seasonality (left) and daily seasonality (right)

Weekly and daily seasonality are first extracted from the model to analyze congestion time in a week and in a day, as shown in Fig. 3. Throughout a week, heavy traffic happens on Tuesdays, Wednesdays and Thursdays, and on a daily basis, from 7 AM to 10 AM.

Next, we evaluate MAE and RMSE when testing on the first week after training data, specifically, training data is taken from January 1st to March 31st, so the test week data is the first week of April, we obtained an MAE of 3.64 (see Table 1), which is better compared to 5.02 when testing on the whole 3 months. The same performance difference applies for RMSE. The main reason for this MAE increment must be the unreliability usage of the old model.

Following on this catch, we actually retrained data every week on its previous 3 months data (for example, if the current time is on week n, its model would be trained on data from week n-13 to week n-1). Then, we calculate weekly MAE on both methods which are using old dataset and retraining every week to compare on Fig. 6. It turns out that MAE is kept below 4.5 when updating our model every week, unlike using an old model, which its MAE could go up high forever since the model becomes less and less accurate.

Table 1. Model evaluation on PEMS-BAY dataset

Model	MAE	RMSE	Year
Prophet (test on 1 week)	3.64	5.31	2021
Prophet (test on 3 months)	5.02	7.09	2021

Look closely at Fig. 4, since the model uses data from the first 3 months and is not updated or retrained on new data, it becomes inaccurate on predicting data in a far future.

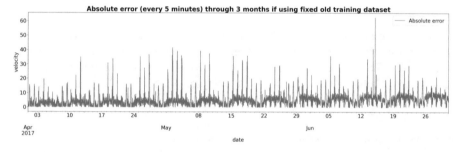

Fig. 4. Absolute error tends to increase over time if the model is keep outdated

As seen in the result section, or at Fig. 7, high congestion happen from 7 AM to 10 AM. Surprisingly, high errors happen at the same time with congestion, which is shown in Fig. 5. This leads to two explanations:

– During congestion time, velocity values may vary from even 0 (stopped vehicle when waiting for the vehicle in front of it to move) to high velocity (the vehicles have some space to move forward), thus sensor may as well captured the velocity that is way much different from the average velocity around that specific time. Therefore during congestion time, after every time window of 5 min, the current captured velocity can fluctuate up and down far from the previous velocity value making the whole ground truth velocity flow unexpectedly inconsistent. And surely the predicted velocity flow is much more smoother making the errors at congestion time to be higher than normal.

– Another explanation is that both approaches mentioned are using models that are based on at least 1 week old data, and basically does not consider new data in a range of previous hours or previous minutes, for which would causes high errors because the model does not adapt to current congestion situation. We can as well consider this as well-fitting overall but under-fitting at congestion time, this is understandable since Prophet was initially made for sales forecasting and was not specifically made to predict velocity.

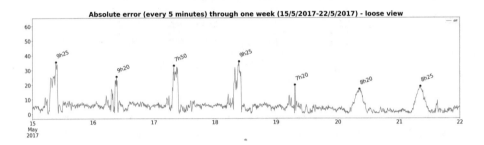

Fig. 5. High errors happens at a specific time in a day

The "high errors happen at the same time with congestion" statement once again appear to be true on a week long perspective. Figure 8 clearly shows how congestion affects errors by raising the errors high on mid-week and lowering the errors on weekend. Interesting how the left plot of Fig. 8 is the condensed view of Fig. 5.

4.5   How Our Framework Works in Action

This is where Spark Streaming comes in to play. Since our work focuses solely on analyzing data, we decided to stick with old PEMS-BAY 2017 dataset to demonstrate how our framework helps users in practice, which also means we are using simulated streaming and not actual streaming. But do not get us wrong, our streaming framework just needs a little tweak to work with real-time data receive from California Department of Transportation. Every 5 min, the framework receive new velocities across all 325 sensors in Bay Area, it will write velocity data and produces congestion predicts on the whole map for the next 5, 30 and 90 min.

– Data written will be stored to be ready for retraining model every Monday.
– Congestion are presented in 5 difference levels corresponding to lowest to highest congestion, smallest red points as lowest congestion and largest red points as highest congestion. Note that these 5 levels are divided by 4 quantiles which are 0.2, 0.4, 0.6, 0.8 and not based on a specific "congestion threshold" to determine which places have congestion. This way users will be able to choose best routes to avoid congestion, or simply the routes that has the "least congestion" even if the current time is not congestion time.

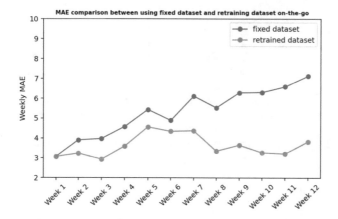

Fig. 6. Weekly MAE comparison between a model trained on old dataset and a model updated to include new dataset

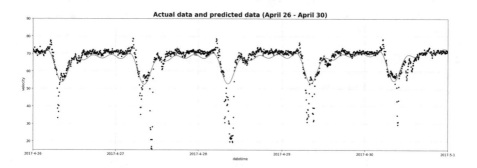

Fig. 7. Ground truth velocity and predicted velocity within 5 days

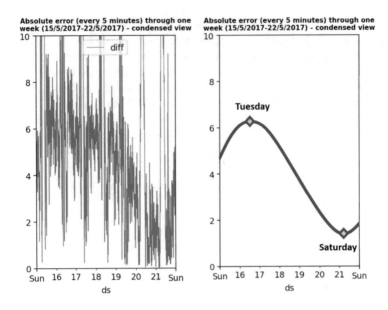

Fig. 8. High errors also happens at a specific time in a week

5   Conclusion

In this study, we propose a real-time traffic congestion prediction system based on Prophet and Spark Streaming. The system captures long term with high accuracy that we have experimented on real datasets with the results presented in Sect. 4. In addition, integrating Spark framework in the system helps the scheme to scale with an enormous amount of data in the future according to actual demands if there is an increase in traffic in the city. If the amount of data increases rapidly in the future, structured streaming can process large amounts of data in real time.

Acknowledgement. Code and dataset are available at https://github.com/thongvh oang/Real-Time-Traffic-Congestion-Forecasting and video how our system works at https://youtu.be/6gE3cyNJYLM.

References

1. Amini, S., Gerostathopoulos, I., Prehofer, C.: Big data analytics architecture for real-time traffic control. In: 2017 5th IEEE International Conference on Models and Technologies for Intelligent Transportation Systems (MT-ITS), pp. 710–715. IEEE (2017)
2. Anveshrithaa, S., Lavanya, K.: Real-time vehicle traffic analysis using long short term memory networks in apache spark. In: 2020 International Conference on Emerging Trends in Information Technology and Engineering (IC-ETITE), pp. 1–5. IEEE (2020)

3. Armbrust, M., et al.: Spark SQL: relational data processing in spark. In: Proceedings of the 2015 ACM SIGMOD International Conference on Management of Data, SIGMOD 2015, pp. 1383–1394. Association for Computing Machinery, New York, NY (2015). https://doi.org/10.1145/2723372.2742797

4. Fouladgar, M., Parchami, M., Elmasri, R., Ghaderi, A.: Scalable deep traffic flow neural networks for urban traffic congestion prediction. In: 2017 International Joint Conference on Neural Networks (IJCNN), pp. 2251–2258. IEEE (2017)

5. Jiang, W., Luo, J.: Big data for traffic estimation and prediction: a survey of data and tools. arXiv preprint arXiv:2103.11824 (2021)

6. Li, Y., Yu, R., Shahabi, C., Liu, Y.: Diffusion convolutional recurrent neural network: data-driven traffic forecasting. arXiv preprint arXiv:1707.01926 (2017)

7. Makino, H., Tamada, K., Sakai, K., Kamijo, S.: Solutions for urban traffic issues by its technologies. IATSS Res. **42**(2), 49–60 (2018)

8. Meng, X., et al.: MLlib: machine learning in apache spark (2015)

9. Saraswathi, A., Mummoorthy, A., GR, A.R., Porkodi, K.: Real-time traffic monitoring system using spark. In: 2019 International Conference on Emerging Trends in Science and Engineering (ICESE), vol. 1, pp. 1–6. IEEE (2019)

10. Taylor, S.J., Letham, B.: Forecasting at scale. Am. Stat. **72**(1), 37–45 (2018)

11. Yin, X., Wu, G., Wei, J., Shen, Y., Qi, H., Yin, B.: Deep learning on traffic prediction: methods, analysis and future directions. IEEE Trans. Intell. Transp. Syst. **23**(6), 4927–4943 (2021)

12. Zaharia, M., et al.: Resilient distributed datasets: a fault-tolerant abstraction for in-memory cluster computing. In: Proceedings of the 9th USENIX Conference on Networked Systems Design and Implementation, NSDI 2012, p. 2. USENIX Association, USA (2012)

13. Zaharia, M., Chowdhury, M., Franklin, M.J., Shenker, S., Stoica, I.: Spark: Cluster computing with working sets. In: Proceedings of the 2nd USENIX Conference on Hot Topics in Cloud Computing, HotCloud 2010, p. 10. USENIX Association, USA (2010)

Real-Time Traffic Flow Prediction Using Big Data Analytics

Dang-Khoa Tran[1,2], Dinh-Quang Hoang[1,2], Viet-Thang Le[1,2],
Minh-Duc Nguyen Thai[1,2], and Trong-Hop Do[1,2(✉)] (iD)

[1] University of Information Technology, Ho Chi Minh City, Vietnam
{18520936,18521294,18520356,18520267}@gm.uit.edu.vn
[2] Vietnam National University, Ho Chi Minh City, Vietnam
hopdt@uit.edu.vn

Abstract. Traffic congestion is always a big problem to be solved in the world because of its negative effects. There are many ways to solve traffic congestion based on its causes. And one of the important factors needed to reduce traffic congestion is the control of traffic on the road at a reasonable level. There have been many algorithms proposed to deal with this problem. However, any algorithm has its own limitations such as slow processing time, no timeliness, complicated implementation methods, etc. In recent years, with the explosion of data, many tools and methods are also developed to handle big data. This has opened up many more solutions for traffic-related problems. In this paper, a real-time traffic flow prediction system is proposed with high accuracy, simple method and vivid visualization. The performance of the proposed system is verified through experiments results.

Keywords: Real-time traffic flow · Spark Streaming · Big data analytics

1 Introduction

Our world is developing more and more, along with the rapid development of large cities, which has increased the number of people and vehicles in these urban areas. This has caused a very serious traffic congestion. The negative effects of traffic congestion on our lives are immense. It is estimated that Ho Chi Minh City loses about 1.2 million working hours each year, 1.3 billion USD/year due to traffic congestion and 2.3 billion USD due to environmental pollution from motor vehicles. According to an assessment from the Institute of Transport Strategy and Development (Ministry of Transport) also said that congestion causes damage to Hanoi every year about 1-1.2 billion USD. That's just statistical data in 2 cities in Vietnam. In addition, traffic congestion was also negative impacts on the environment and human health because it is one of the causes of air and noise pollution. Therefore, improving traffic conditions is an important issue in Vietnam and around the world. It is not enough to plan urban areas properly, expand the road system or reduce the traffic volume, a intelligent traffic regulation system will be a useful solution and optimization.

Recent years have witnessed a rapid development of information technology and especially artificial intelligent (AI). AI has been shown to be tremendously beneficial to many aspects of life, including traffic. More specifically, deep learning, an outstanding technique in artificial intelligent has been developed and solved problems in traffic such as vehicle detection, vehicle tracking, traffic flow prediction and many orther problems. Among them, traffic flow prediction is an important prerequisite for traffic regulation. Traffic flow prediction is a combination of time series prediction and Big Data analysis. There are many approaches to time series prediction problem based on deep learning, machine learning algorithms, etc. For example, using spatial temporal graph neural network [1], which can comprehensively capture spatial and temporal patterns and effectively aggregate information from adjacent roads. Another approach using model combining the attention Conv-LSTM and Bi-LSTM [2], this model extracts daily and weekly periodic features so as to capture variance tendency of the traffic flow from both previous and posterior directions based on the short-term as well as long-term spatial and temporal features. On the other hand, currently, almost everything is now interconnected due to the growth of the Internet of Things (IoT) which has led to an explosion of data. And traffic data is also a kind of big data, collected from numerous sensors on a lot of different roads. Not only that, they are also streaming data sources. Towards solving the traffic flow prediction problem over a wide range (predicted simultaneously on multiple paths) requires the tools and powerful technique can handle aforementioned big data [3]. Apache Hadoop is an example, it is one of the first open source frameworks for storing and processing Big Data, enables distributed processing of large datasets on clusters of computers. Hadoop works on an algorithm called MapReduce. This algorithm will break the jobs into small parts and divide them among the machines in the distributed system. It then aggregates to the final result. Apache Spark is another framework for big data processing, it is a data processing engine for batch and streaming modes featuring SQL queries, Graph Processing and Machine Learning. Spark can process real-time data which from real-time event streams at a rate of millions of events per second. Compared to Hadoop, Spark's processing speed is many times faster. This makes Spark more suitable for many problems, especially in traffic.

In this paper, traffic flow is predicted in real time on Big Data platform, applied simultaneously to many roads, the prediction results are visualized on the heat map. Firstly, a Prophet model is built to predict traffic flow at next time steps in the future. Training data is collected from sensors on the roads, including information about the average speed, time, location of the sensors and so on. And Prophet is a procedure for forecasting time series data based on an additive model where non-linear trends are fit with yearly, weekly, and daily seasonality, plus holiday effects. It works best with time series that have strong seasonal effects and several seasons of historical data. Prophet is robust to missing data and shifts in the trend, and typically handles outliers well [8]. After that, the results predicted from the model are displayed on a real-time heat map. Experimental entire process are used PySpark, an interface for Apache Spark in Python. Details of the methods and experimental results are presented in the following sections of this paper.

2   Related Works

Hitherto, there have been many research works on traffic flow prediction problem. In 2014, Yisheng Lv et al. proposed a deep-learning-based traffic flow prediction method [4]. They used a stacked auto encoder (SAE) model is trained in a layerwise greedy fashion to extract generic traffic flow features. The spatial and temporal correlations are also considered in the model. Following that, a method using deep learning approach proposed by Rui Fu et al. (2016) [5], Long Short Term Memory Neural Network (LSTM NN) and Gated Recurrent Neural Network (GRU NN) models are applied to predict traffic flow. They used Adam optimizer with adaptive learning rates, the results show that GRU NN model perform a little than LSTM NN model. Nicholas G. Polson et al. proposed a deep learning architecture that combines a linear model that is fitted using l_1 regularization and a sequence of tanh layers [6]. Another method used machine learning-based models for real-time traffic prediction was proposed by Sun et al. in 2020. They used many models such as Artificial Neural Network (ANN), Support Vector Regression (SVR), Long Short Term Memory Neural Network and compare their results [7].

3   Proposed Real-Time Traffic Flow Prediction Architecture

3.1   Proposed System Architecture

The proposed system architecture consists of four parts: Data source, Batch Processing, Stream Processing and Stream Visualizing as described in Fig. 1.

The proposed system architecture consists of four parts: Data source, Batch Processing, Stream Processing and Stream Visualizing as described in Fig. 1. The entire data of the system are collected from a reliable data source, which is published on IEEE DataPort. Data is collected in two ways. The first way is to download a large amount of data enough to training the prediction model in Batch Processing part. The other way, data is received by Spark Structure Streaming in Streaming Processing part as input to the real-time traffic flow prediction model trained in Batch Processing part. In Batch Processing, the data collected from the Data Source is saved in csv format and goes through preprocessing steps before put in to training model. The preprocessing steps are also performed similarly for the data in Stream Processing part. The model is trained in Batch Processing part will get these data as input to make predictions about traffic flow in the future using Spark Streaming. Final results are shown on the heatmap.

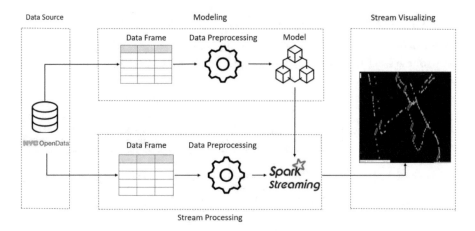

Fig. 1. System architecture

3.2   Traffic Flow Prediction Model

As the core of the problem, the model determines most of the system's performance. In this paper, Prophet is the model chosen to train for traffic forecasting. Prophet is a procedure for forecasting time series data and works well with time series with strong seasonal effects and several seasons of historical data, even in the absence of data, outliers, and the trend of the data has changed. Prophet is based on an additive model where non-linear trends are consistent with annual, weekly, and daily seasonality, plus holiday effects. [8]. The procedure makes use of a decomposable time series model with three main model components: trend, seasonality, and holidays, which can be represented as in Eq. (1).

$$y(t) = g(t) + h(t) + s(t) + \epsilon t \tag{1}$$

g(t) is trend models non-periodic changes (linear or logistic), s(t) is seasonality represents periodic changes (weekly, monthly, yearly) and h(t) is ties in effects of holidays. The error term e(t) represents any idiosyncratic changes which are not accommodated by the model. Using time as a regressor, Prophet is trying to fit some linear and non-linear functions of time as components. Modeling seasonality as an additive component is a similar approach taken by exponential smoothing in the Holt-Winters technique. Prophet is framed forecasting problem than an exercise adjusted curve is more obvious considering the dependence on time of each observation in a time series. According to research by Taylor and Letham [8], Prophet is used in many applications on Facebook to provide reliable forecasts and works better than any other approach in the majority of cases. In this research, 132 Prophet models were trained corresponding to 132 roads in New York City. These models are combined into one main model for inclusion in Stream Processing part (Fig. 2).

3.3   Online Real-Time Traffic Flow Prediction Pipeline

After the model is trained, it will be combined with real-time data get from the data source's API to perform traffic flow predictions according to the pipeline in Fig. 2

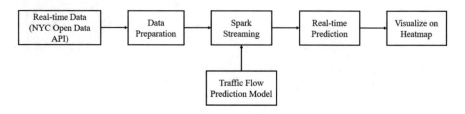

Fig. 2. Online real-time traffic flow prediction pipeline

Spark Streaming is used to crawl real-time data from the NYC Open Data API, then the data will be preprocessed to transform into the input format of the prediction model. The model will make predictions about the traffic flow of the roads for the next time period based on these current data. Folium framework is used to derive predictions from the model, and combine with the coordinate information of each road to visualize a heat map.

4   Experiment

4.1   Experiment Procedure

The experimental steps of this research are shown according to the procedure described in Fig. 3. According to the procedure, the data collection and preprocessing are divided into two separate parts. The first part is to collect historical data for training and testing the predictive model. The rest will collect real-time data for the online real-time traffic flow prediction phase. The model used for training is Prophet. The trained Prophet model will be evaluated by two measures, root mean square error (RMSE) and mean absolute error (MAE), if the performance of the model is not satisfactory, the model will be updated. The process of training and evaluating and updating the model is repeated until the model achieves the desired results. The final model selected will be used to predict real-time online traffic with clean real-time data during the Aggregation and Prepare phase as input. Final results is the traffic flow in consecutive time intervals of the roads in the dataset. These results are combined with available coordinates of the roads to visualize a real-time heat map. The entire experimental process is processed on Colab Pro.

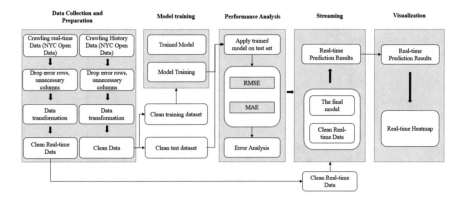

Fig. 3. Experiment procedure

4.2 Experiment Result and Discussion

The experimental results show that the trained model has relatively good performance. From the heat map in Fig. 4, the high-traffic flow of the road are shown as orange and red areas and low-traffic areas in green.

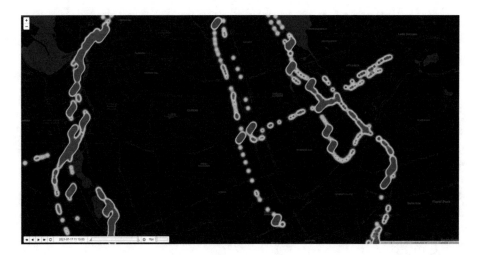

Fig. 4. Traffic flow heatmap

Model performance is also shown in Fig. 5 and Fig. 6. Figure 5 is the result of model evaluation based on RMSE and MAE and Fig. 6 is the results of the model's speed prediction compared to the actual speed in a day. Accordingly, the predicted speed results are not really close to reality, there are still many differences and the model performance is lower when predicting for further time points. Therefore, in order for the model to work with a stable performance, it is

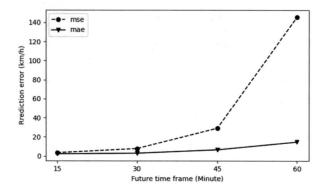

Fig. 5. Model performance over time

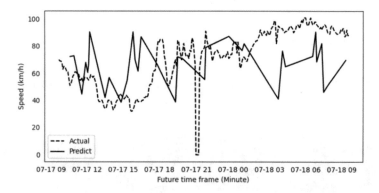

Fig. 6. The model's speed prediction results compared to the actual speed for a day

necessary to update the model after a certain period of time. This helps the model capture more recent road information and make more accurate predictions. In addition, the model performance is not really high, partly because the Prophet model is suitable for trend forecasting problems based on long-term information. Meanwhile, the traffic flow is highly dependent on short-term information in the past.

5   Conclusion

Traffic flow prediction problem is an important part of intelligent traffic system. Many published researches use traditional machine learning methods, others use more modern methods such as deep learning. However, the methods are still limited when they cannot be optimized for real-time application. In this paper, a traffic prediction method based on a combination of big data analytics and modern machine learning algorithms is capable of making real-time predictions and can be visualized on a real-time heat map. Experimental results show that the proposed system has relatively good performance.

References

1. Wang, X., et al.: Traffic flow prediction via spatial temporal graph neural network. In: Proceedings of the Web Conference 2020 (2020)
2. Zheng, H., Lin, F., Feng, X., Chen, Y.: A hybrid deep learning model with attention-based conv-LSTM networks for short-term traffic flow prediction. In: IEEE Transactions on Intelligent Transportation Systems, pp. 1-11 (2020)
3. Taylor, S.J., Letham, B.: Forecasting at scale (2017)
4. Lv, Y., Duan, Y., Kang, W., Li, Z., Wang, F.-Y.: Traffic flow prediction with big data: a deep learning approach. In: IEEE Transactions on Intelligent Transportation Systems, pp. 1-9 (2014)
5. Fu, R., Zhang, Z., Li, L.: Using LSTM and GRU neural network methods for traffic flow prediction. In: 2016 31st Youth Academic Annual Conference of Chinese Association of Automation (YAC) (2016)
6. Polson, N.G., Sokolov, V.O.: Deep learning for short-term traffic flow prediction. Transp. Res. Part C Emerg. Technol **79**, 1–17 (2017)
7. Sun, P., Aljeri, N., Boukerche, A.: Machine learning-based models for real-time traffic flow prediction in vehicular networks. IEEE Netw. **34**(3), 1–8 (2020)
8. Taylor, S.J., Letham, B.: Forecasting at scale. Am. Stat. **72**(1), 37–45 (2017)

SSL-MedImNet: Self-Supervised Pre-training of Deep Neural Network for COVID-19 Diagnosis

Tran Nhat Minh Hoang[1,3] , Tran The Son[2(✉)] , Nguyen Duy Nghiem[1] ,
and Le Minh Tuan[4]

[1] FPT University, FPT Greenwich Center (University of Greenwich - Alliance
with FPT Education), Danang, Vietnam
hoangtnmgcd17037@fpt.edu.vn, nghiemnd@fe.edu.vn
[2] Vietnam-Korea University of Information and Communications Technology,
Danang, Vietnam
ttson@vku.udn.vn
[3] THOR Lab, D-Soft JSC, Danang, Vietnam
hoangtnm@d-soft.com.vn
[4] Paradox, Inc., Danang, Vietnam
tuan.le@paradox.ai

Abstract. This paper applies self-supervised learning to diagnose coronavirus disease (COVID-19) among other pneumonia and normal cases based on chest Computed Tomography (CT) images. Being aware that medical imaging in real-world scenarios lacks well-verified and explicitly labeled datasets, which is known as a big challenge for supervised learning, we utilize Momentum Contrast v2 (MoCo v2) algorithm to pre-train our proposed Self-Supervised Medical Imaging Network (SSL-MedImNet) with remarkable generalization from substantial unlabeled data. The proposed model achieves competitive and promising performance in COVIDx CT-2, which is a well-known and high-quality dataset for COVID-19 assessment. Besides, its pre-trained representations can be transferred well for the diagnosis task. Moreover, SSL-MedImNet approximately matches its supervised candidates COVID-Net CT-1 and COVID-Net CT-2 by small distinctions. In particular, with only some additional dense layers, the proposed model achieves COVID-19 accuracy of 88.3% and specificity of 98.4% approximately, and competitive results for normal and pneumonia cases. The results advocate the potential of self-supervised learning to accomplish highly generalized understanding from unlabeled medical images and then transfer it for relevant supervised tasks in real scenarios.

Keywords: COVID-19 · Medical imaging · Self-supervised learning · Artificial intelligence · Computed Tomography Scan

N.-T. Nguyen et al. (Eds.): ICIT 2022, LNDECT 148, pp. 406–415, 2022.
https://doi.org/10.1007/978-3-031-15063-0_39

1   Introduction

It has been more than two years since the first COVID-19 outbreak occurred in the world, Too many people still have severe lung damage and even death from the COVID-19 pandemic [1]. It rapidly spreads out with multiple waves and various dangerous variants and put hospital systems in an overloaded circumstance [2] resulting in more and more deaths recorded. As reported by Johns Hopkins University [3], at least 258,994,000 cases of infection and more than 5,170,000 deaths around the world have been recorded. Motivated by the challenges mentioned above, worldwide research and collaboration have been made to apply cutting-edge techniques such as Reverse transcription polymerase chain reaction (RT-PCR) to identify the virus's RNA, which is its unique genetic material. Besides, with the tremendous advancement of Artificial Intelligence (AI), AI-based approaches were proposed to diagnose whether a patient is infected with COVID-19 among other lung conditions given a CT image [4–6], or used for COVID deterioration prognosis [7–9] with a high reliability.

Whereas, as a result of ethical and privacy concerns, most clinically curated datasets employed by those research papers such as X-Rays, CT images have not been made accessible to the public, which prevents other scientists from reproducing such aforementioned approaches and advancing novel improvements to develop more reliable models for the task of COVID-19 diagnosis. Moreover, in real-world medical imaging scenarios, the amounts of data samples of distinct diseases come from different forms and distributions, which causes an imbalance in datasets [10] and bias in supervised models. Furthermore, curating and assigning labels for a tremendous dataset in the field of medical imaging requires years of clinical experience is costly [11]. Regarding AI-based methods for COVID-19 diagnosis, the conventional approach is transfer learning, which leverages knowledge from ImageNet and then transfer it for supervised tasks afterward such as classification, detection, which is also called downstream tasks. Although this approach has been applied extensively as mentioned above, it depends on the quality and balance of the applied datasets to accomplish reliable results. Besides, the problem in domain shift, in which the task or the distributions of pre-trained and fine-tuned data are different from each other, can cause unsatisfactory outcomes. In order to deal with these challenges, we apply self-supervised learning (SSL) to take advantage of massive unlabeled radiographic images to train a deep neural network with highly generalized representations in the medical imaging domain, and later fine-tune it with less annotated data for COVID-19 determination via supervised learning. In this paper, we adopt the previous work on COVID-19 prognosis on chest X-rays via Momentum Contrast (MoCo) [12] and apply the model of *MoCo v2* [13] (i.e. an improved version of the original MoCo) to our proposed model for diagnosing based on chest CT images.

2   Related Work

Since the first SARS-CoV-2 outbreak occurred in 2019, RT-PCR has been known to be the best method of diagnosing COVID-19, as it provides very high precision and reliability for diagnosis. Besides, imaging diagnosis is also considered as a good alternative method with a good accuracy and reliability in a very quick time based on radiographic images. Various models have been proposed for imaging diagnosis such as COVNet [4] and COVIDNet CT-2 [6]. The COVNet is a deep neural network based on ResNet-50 architecture [14] for COVID-19 detection and developed in partnership with six healthcare facilities in China. It can reach Area Under The Curve (AUC) of 0.96, and the sensitivity and specificity of COVID-19 are 90% and 96%, respectively, which was a breakthrough in the diagnosis of COVID-19 based on image processing. However, medical data is a sensitive topic and related to personal information, which cannot be shared without patients' permission; therefore, other scholars cannot validate the results and make novel improvements building upon them. Therefore, the insufficiency of well-verified open-access datasets [10]. became the first difficulty in advancing AI-based approaches for the task of COVID-19 diagnosis.

In response to the pandemic, the COVID-Net Open Initiative was launched to accelerate machine learning-based advances worldwide and provide open access datasets for the challenge. Inspired by the initiative, Wang, Lin, and Wong [15] applied a machine-driven design exploration strategy to develop COVID-Net specialized for COVID-19 detection from chest X-Rays. Just less than a year later, COVIDNet CT-2 [6] was proposed as the successor to COVIDNet-CT [5] with huge improvements in COVID-19 sensitivity, specificity at 98,1% and 98,8%, respectively. These open up new capabilities and novel AI-based approaches to the pandemic. Regardless of these optimistic outcomes, most proposed solutions still rely on supervised learning that has some limitations as mentioned above, which becomes a bottleneck in designing superior models reaching human-level performance in the future without immense human-annotated data.

As opposed to supervised candidates, self-supervised learning aims to discover general representations with self-defined pseudo labels that are generated spontaneously from existing unlabeled data during training. In practice, contrastive self-supervised learning achieves this objective by learning representations that decrease the gap between positive pairs and vice versa [16]. Moreover, its ability to derive generalized representations in a high-dimensional space that is independent of downstream tasks makes it optimal for this objective. MoCo [17] can be considered as a successful method with a contrastive loss for this purpose. During training, encoders are trained to execute dictionary look-up in which a query q should match its positive key k_+ and be far from negative ones k_-. Its objective can be formulated by the InfoNCE [18] function:

$$\mathcal{L}_q = -\log \frac{\exp(q \cdot k_+/\tau)}{\sum_{i=0}^{K} \exp(q \cdot k_i/\tau)} \tag{1}$$

Closest to our work is research about COVID-19 prognosis [12] which leverages a transformer-based model trained by MoCo to predict patient deterioration given multiple chest XRays. We adapt this idea to COVID-19 diagnosis given chest CT images with some improvements. To begin with, we replace the CXR dataset with the COVID-Net CT-2 dataset so that the model can learn clinically visual symptoms from chest CT images. Secondly, MoCo v2 [13], which is a stronger baseline over the original MoCo, is used instead.

3   Dataset

3.1   Pre-training Unlabeled Data

Regarding the self-supervised pre-training phase, it aims to extract and learn generalized representations from medical images without pre-defined interest. Therefore, we assume that the quality, diversity of pre-training data plays a crucial role in the pre-trained model's generalization. For this reason, COVIDx CT-2A dataset [6] is chosen. Compared to other public datasets for COVID-19 diagnosis, it is recognized as the most enormous, diverse dataset comprising more than 194,000 chest CT images from various sources and from at least 15 countries around the world. Furthermore, human-provided annotations of its samples are verified via clinical methods such as RT-PCR and radiologist-confirmed, which ensures the reliability of the images and labels during pre-training and the fine-tuning stage afterward. Especially during the pre-training stage, only raw chest CT images are needed. The dataset is splitted into training, validation, and test subsets.

3.2   Fine-tuning Labeled Data

In order to fine-tune the pre-trained backbone, the same training data with its three corresponding diagnosis labels is used. Each label corresponds to one of three findings or chest conditions: normal, common pneumonia, and COVID-19. Especially, to evaluate the robustness of our approach in real-world scenarios, a ablation study with different ratios of data sampling from 50% to 100% are used to investigate the relationship between model performance and how data is needed to fine-tune the model to achieve an acceptable level of accuracy and specificity.

4   The Proposed Model: SSL-MedImNet

In this section, we depict our proposed *SSL-MedImNet* architecture and its training strategies such as self-supervised pre-training and supervised fine-tuning for COVID-19 diagnosis. The model builds upon the DenseNet-121 [19], which is a deep learning model that has been applied extensively in medical imaging [8, 20] and has implicit deep supervision.

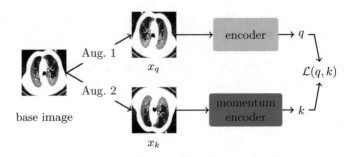

Fig. 1. Simplified MoCo training strategy.

4.1   Self-Supervised Pre-training Using Momentum Contrast

Figure 1 illustrates the simplified pre-training approach at each iteration step, in which the loss is calculated from a given chest image. At each step, two random augmentations transform each CT image x from the mini-batch separately to generate transformed images x_q and x_k. After that, two different encoder networks (encoder and momentum encoder) compute its corresponding representations q and k. Then, the contrastive loss as (see Eq. 1) is computed given these representations. Furthermore, to improve the generalization of representations, the output latent vector of the model is replaced with a MLP projection head and the training paradigms from MoCo v2 [13] were used instead of the original MoCo.

4.2   Supervised Fine-tuning

Figure 2 depicts our proposed SSL-MedImNet model's architecture. It combines the pre-trained encoder and three fully connected (FC) layers for the classification task. Since the backbone model (encoder) was pre-trained to learn generally good representations without concerning a specific chest disease or condition, only two FC layers followed by Swish activation and Dropout, and one final FC layer followed by Softmax is required to fine-tune it. The reason why Swish is chosen is that its smooth, non-monotonic characteristics and similarity to ReLU were verified to lead to better accuracy in a variety of benchmarks.

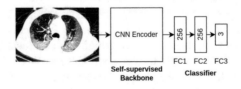

Fig. 2. Architecture design of SSL-MedImNet.

It is clear that by training the encoder (backbone) in the self-supervised manner, the pre-trained backbone can be fine-tuned for the downstream COVID-19 diagnosis task without considerable modifications. This illustrates the flexibility of our proposed approach, in which a more generalized model can be pre-trained via self-supervised learning while being able to be fine-tuned for downstream tasks in a short time rather than training in an end-to-end supervised manner.

5   Experimental Results

5.1   Self-Supervised Pre-training

This section describes results in the pre-training stage. Figure 3 describes the pre-training accuracy of SSL-MedImNet on the training dataset. For the purpose of the experiment, well-known CNN architectures such as ResNet variants and our proposed model are used to examine our hypothesis of contrastive self-supervised learning. For this purpose, accuracy and contrastive loss (see Eq. 1) are used.

Accuracy: is the quantity of queries q matching its positive k_+ on the the set of K keys. A key is considered as a positive one for a query if they are generated from the same image.

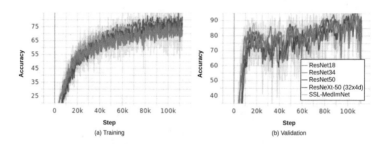

Fig. 3. Accuracy on (a) training and (b) validation set during pre-training.

Figure 3 demonstrates the training and validation progress for 200 epochs for the pre-training stage. It can be seen that SSL-MedImNet can improve accuracy in a more stable way while achieving higher accuracy compared to its candidates. It can be seen that with the MoCo training strategy, the overall pre-training performance witness an upward trend through training and reach the training and validation accuracy of approximately 82% and 94%, respectively. These results demonstrate the high generalization of data representation for the pretext task without telling the model what the interesting features are required for the fine-tuning step afterward. Furthermore, even though the pre-training data are just raw and relatively uncurated data samples to simulate the real-world scenario, the ability to achieve high pre-training accuracy indicates the feasibility to apply this training strategy to actual applications.

Therefore, our proposed model could achieve remarkable generalization for the pre-training task of matching q and k_+ with the more steady improvement compared to its competitive architectures. Moreover, regarding the improvement with respect to current supervised learning approaches, via self-supervise MoCo, SSL-MedImNet can capture more general representations while only needing a large number of unlabeled data, which cannot be done with the supervised-learning methods.

5.2   Supervised Fine-tuning

Table 1 describes the fine-tuning results of our proposed model in terms of accuracy and specificity. Furthermore, to compare our results with the supervised learning approach, we choose COVID-Net CT variants [5,6], which are the most cutting-edge models trained on this dataset, as the candidate:

$$Accuracy = \frac{TP}{P} = \frac{TP}{TP + FN} \tag{2}$$

$$Specificity = \frac{TN}{N} = \frac{TN}{TN + FP} \tag{3}$$

where TP is True Positive, FP is False Positive, TN is True Negative and FN is False Negative.

Table 1. Accuracy and specificity on the test dataset.

Architecture	Accuracy (%)			Specificity (%)		
	Normal	Pneumonia	COVID-19	Normal	Pneumonia	COVID-19
SSL-MedImNet (Ours)	98.7	96.8	88.3	98.9	96.3	98.4
COVID-Net CT-1 [5]	98.8	**99.0**	80.2	96.3	95.7	**99.4**
COVID-Net CT-2 L	**99.0**	98.2	**96.2**	**99.5**	**98.8**	99.0
COVID-Net CT-2 S	98.9	98.1	95.7	99.3	98.8	98.9

It is obvious that our SSL-MedImNet is 8.1% higher than COVID-Net CT-1 in terms of COVID-19 accuracy, while the specificity is just lower with 1%. Moreover, even though COVID-Net CT-2's COVID-19 accuracy surpasses SSL-MedImNet with 7.9% higher, our proposed model can nearly match the candidates' performance in both accuracy and specificity for normal and pneumonia cases with the differences are just about 1%. Though MoCo [17] and MoCo v2 [13] are originally proposed to deal with the challenges of real-life images such as RGB ones in ImageNet and COCO datasets rather than healthcare purposes, with our customization and adaptions, it seems that it can converge well with medical imaging to learn more general representations without pre-defined interest and be fine-tunning task as expected. These results indicate the potential of self-supervised learning in general and SSL-MedImNet can almost close the gap with supervised learning approaches in the field of medical imaging in some days, respectively.

Ablation Study: Impact of Fine-tuning Data. In order to measure the relative performance of our approach in the actual scenario, in which the amount of fine-tuning data may vary a lot and be extremely fewer than the pre-training data, we conducted several ablation experiments on SSL-MedImNet trained for 200 epochs on the full training data and a proportion of it from 90% to 50% and evaluated it on the full validation data.

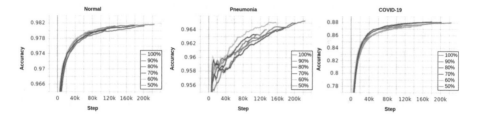

Fig. 4. The significance of data size to accuracy during fine-tuning.

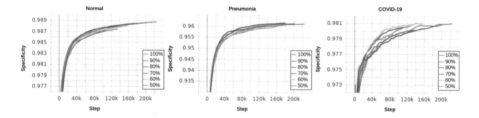

Fig. 5. The significance of data size to specificity during fine-tuning.

The ablation experiments (see Fig. 4 and 5) reveal the superior capability of SSL-MedImNet with respect to different training data ratios. It can be seen that even though the model experiences a dramatic decrease of training data from 100% to 50%, the values of accuracy and specificity just drops approximately by 1% in maximum. This is extremely competitive with supervised learning methods which demand on many labeled samples and can cause difficulties as mentioned above [21]. Take COVID-Net CT-2 [6] for example, which is also trained on this data set. It is trained with complex augmentations in a supervised learning manner to achieve precision and specificity of the three diseases of approximately between 96% and 99%, which is difficult to achieve in the real-world scenario with limited public and labeled datasets. Although our approach in terms of normal and pneumonia nearly matches with COVID-Net CT-2 and only COVID-19 accuracy is still far from it with about 10%, it is still a big improvement overall when our model can adapt training data of different ratios and is able to approximately match the results of COVID-Net CT-2.

6   Conclusion

In this paper, we proposed Self-Supervised Medical Imaging Network (SSL-MedImNet) for the task of COVID-19 diagnosis among other pneumonia and normal cases given single CT images. First, we pre-trained the proposed model given a large amount of medical images available without any explicit label, so that it can attain exceptional generalization for the downstream task afterward. Then SSL-MedImNet is fine-tuned for the diagnosis task via supervised learning. By doing so, our approach can reach COVID-19 accuracy and specificity of 88,3% and 98,4%, respectively. Furthermore, its competitive generalization is also proven through the contraction of data with different ratios for the fine-tuning stage, in which its accuracy and specificity for the three lung conditions witness with the degradation of no more than 1%, which is hard to reach even by cutting-edge end-to-end supervised models. Regarding future work, we plan to adapt the approach to downstream tasks of medical imaging such as detecting lung damaged regions, using multi-images over time to improve the diagnosis results or prognosticate deterioration in the injured regions.

Acknowledgment. We would like to acknowledge the authors who built the COVIDx CT-2 dataset to be publicly available.

References

1. World Health Organization. WHO characterizes COVID-19 as a pandemic (2020). https://www.who.int/emergencies/diseases/novel-coronavirus-2019/events-as-they-happen
2. Nguyen, L.H., et al.: Risk of COVID-19 among front-line health-care workers and the general community: a prospective cohort study. Lancet Publ. Health **2667**(20) (2020). https://doi.org/10.1016/s2468-2667(20)30164-x. ISSN 24682667
3. Johns Hopkins University. COVID-19 Dashboard by Johns Hopkins University (2020). https://coronavirus.jhu.edu/map.html
4. Li, L., et al.: Using artificial intelligence to detect COVID-19 and community-acquired pneumonia based on pulmonary CT: evaluation of the diagnostic accuracy. Radiology **296**(2), E65–E71 (2020)
5. Gunraj, H., Wang, L., Wong, A.: COVIDNet-CT: a tailored deep convolutional neural network design for detection of COVID-19 cases from chest CT images (2020). arXiv: 2009.05383 [eess.IV]
6. Gunraj, H., Sabri, A., Koff, D., Wong, A.: COVID-Net CT-2: Enhanced deep neural networks for detection of COVID-19 from chest CT images through bigger, more diverse learning (2021)
7. Shamout, F.E., et al.: An artificial intelligence system for predicting the deterioration of COVID-19 patients in the emergency department. arXiv preprint. arXiv:2008.01774 (2020)
8. Kwon, Y.J., et al.: Combining initial radiographs and clinical variables improves deep learning prognostication of patients with COVID-19 from the emergency department. Radiol. Artif. Intell. **3**(2), e200098 (2020)

9. Zhang, K., et al.: Clinically applicable AI system for accurate diagnosis, quantitative measurements, and prognosis of COVID-19 pneumonia using computed tomography. Cell **181**(6), 1423–1433 (2020)

10. Islam, M., Karray, F., Alhajj, R., Zeng, J., et al.: A review on deep learning techniques for the diagnosis of novel coronavirus (COVID-19). arXiv preprint. arXiv:2008.04815 (2020)

11. Chartrand, G., et al.: Deep learning: a primer for radiologists. Radiographics **37**(7), 2113–2131 (2017)

12. Sriram, A., et al.: COVID-19 prognosis via self-supervised representation learning and multi-image prediction (2021). arXiv: 2101.04909 [cs.CV]

13. Chen, X., Fan, H., Girshick, R., He, K.: Improved baselines with momentum contrastive learning. CoRR, vol. abs/2003.04297. arXiv:2003.04297 (2020)

14. He, K., Zhang, X., Ren, S., Sun, J.: Deep residual learning for image recognition. In Proceedings of the IEEE conference on computer vision and pattern recognition, pp. 770–778 (2016)

15. Wang, L., Lin, Z.Q., Wong, A.: COVID-Net: a tailored deep convolutional neural network design for detection of COVID-19 cases from chest x-ray images. Sci. Rep. **10**(1), 1–12 (2020)

16. Falcon, W., Cho, K.: A framework for contrastive self-supervised learning and designing a new approach (2020). arXiv: 2009.00104 [cs.CV]

17. He, K., Fan, H., Wu, Y., Xie, S., Girshick, R.: Momentum contrast for unsupervised visual representation learning. In: Proceedings of the IEEE/CVF Conference on Computer Vision and Pattern Recognition, pp. 9729–9738 (2020)

18. Oord, A.V.D., Li, Y., Vinyals, O.: Representation learning with contrastive predictive coding. arXiv preprint. arXiv:1807.03748 (2018)

19. Huang, G., Liu, Z., Van Der Maaten, L., Weinberger, K.Q.: Densely connected convolutional networks. In: Proceedings of the IEEE conference on computer vision and pattern recognition, pp. 4700–4708 (2017)

20. Rajpurkar, P., et al.: CheXNet: radiologist-level pneumonia detection on chest x-rays with deep learning (2017)

21. Irvin, J., et al.: CheXpert: a large chest radiograph dataset with uncertainty labels and expert comparison. In Proceedings of the AAAI Conference on Artificial Intelligence, vol. 33, pp. 590–597 (2019)

Vietnamese Sentence Paraphrase Identification Using Sentence-BERT and PhoBERT

Quoc Long Phan[1,2](✉) ⓘ, Tran Huu Phuoc Doan[1,2] ⓘ, Ngoc Hieu Le[1,2] ⓘ, Ngoc Bao Duy Tran[1,2] ⓘ, and Tuong Nguyen Huynh[1,2] ⓘ

[1] Ho Chi Minh City University of Technology (HCMUT), 268 Ly Thuong Kiet Street, District 10, Ho Chi Minh City, Vietnam
{long.phan2810,phuoc.doan0412,hieu.le6102,duytnb, htnguyen}@hcmut.edu.vn
[2] Vietnam National University Ho Chi Minh City, Linh Trung Ward, Thu Duc District, Ho Chi Minh City, Vietnam

Abstract. In 2019, Reimers et al. proposed SBERT to derive sentence embedding for many purposes. It highly reduced the time complexity of finding the most similar pair of sentences from traditional BERT/RoBERTa to SBERT, while the accuracy is maintained. There are many English SBERT models, but lacking the other languages ones. In this publication, we develop our Vietnamese SBERT model for Vietnamese sentence embeddings, using PhoBERT as our main transformer for Vietnamese token embeddings. For the training processes, we use the Vietnamese NLI and STSb datasets, and for the evaluation of sentence paraphrase identification task to compare with other models, we use the VnPara dataset in. Our model has achieved an accuracy of 95.33% and F1 of 95.42%, slightly outperforming many recent methods in Vietnamese.

Keywords: Sentence-BERT · SBERT · PhoBERT · Sentence embeddings · Vietnamese · Pre-trained model · Sentence transformer · Paraphrase

1 Introduction

Measuring the semantic similarity between two texts is a common task in NLP. There are many techniques to achieve the results, and can be divided into 4 categories: *string based techniques, vector space model (VSM) techniques, syntax and semantic-based techniques and structured based techniques* [4].

Word embedding represented texts in a vector space, that can be used in many NLP tasks. *One hot encoding, Word2Vec, GloVe, TF-IDF, Fast-Text, BERT, etc.* are some common word embedding techniques. Bidirectional Encoder Representations from Transformers (BERT) is introduced by Google [5], which is a pre-trained model that represents a text in a vector space based on the contexts. Quickly after being presented, BERT becomes one of the best models according to SQuAD Leaderboard [10]. With BERT, the NLP tasks can be done to obtain more efficient results.

© The Author(s), under exclusive license to Springer Nature Switzerland AG 2022
N.-T. Nguyen et al. (Eds.): ICIT 2022, LNDECT 148, pp. 416–423, 2022.
https://doi.org/10.1007/978-3-031-15063-0_40

In 2019, Nils Reimers and Iryna Gurevych introduced *Sentence-BERT (SBERT)*, which is a modification of the BERT network using siamese and triplet networks to derive semantically meaningful sentence embeddings [1].

SBERT models can be used to measure the semantic similarity of two texts after embedding calculations. SBERT models greatly reduce the time complexity to find the similarity sentences than using the traditional BERT models, while the accuracy is maintained [1]. Moreover, SBERT is also flexible, which is easy for researchers to create their models with customized components.

Recently, more and more Monolingual and Multilingual Sentence BERT models have been developed, but lacking models that support non-English languages, including Vietnamese, except Multilingual models. However, Multilingual models support many languages, not only Vietnamese, which leads to inaccuracies and imbalances in the results. In this research, we will develop and conduct experiments on a Monolingual SBERT model combined with PhoBERT using Vietnamese corpus.

We present previous studies that relate to our works about Sentence-BERT and PhoBERT in Sect. 2. Then, we describe our model and conduct our experiments in Sect. 3. In Sect. 4, we evaluate our models with many tasks and use VnPara Corpus in paraphrase identification task to compare with other recent models in Vietnamese. Finally, Sect. 6 concludes the work and discusses future directions.

2   Related Works

2.1   SBERT Model Architecture

The architecture of an SBERT model has been introduced in [1]. There are three main components: a single BERT model as the transformer, the pooling operation, and the siamese and triplet networks. Figure 1 shows the architecture with classification objective function, i.e. for Natural Language Inference (NLI) task. Figure 2 shows the architecture with regression objective function, i.e. for Semantic Textual Similarity (STS) task.

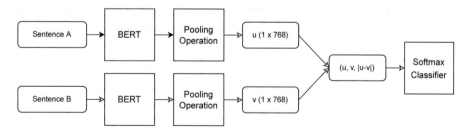

Fig. 1. SBERT architecture with classification objective function, for example, to predict label in NLI dataset, Reimers et al. 2019 [1]

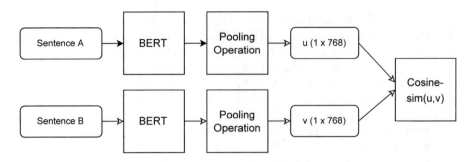

Fig. 2. SBERT architecture at inference, for example, to compute similarity scores, Reimers et al. 2019 [1]

The pair of input sentences must be tokenized before going through the BERT model. Note that each tokenized sentence must be embedded with the same BERT model. After the embeddings, the model performs a pooling task, which reduces the presence of features with a specific pooling strategy. The output of the above sequence is two sentence embeddings u, v. Finally, the model uses the siamese and triplet networks for updating the weight of the sentence embeddings and can be used to calculate cosine-similarity.

2.2   PhoBERT - A BERT Model for Vietnamese

PhoBERT is a pre-train language models for Vietnamese [2] is based on architecture BERT [6] use *RoBERTa* method [8]. They train the first large-scale monolingual BERT-based using 20 GB word-level Vietnamese corpus. Experimental results show that PhoBERT consistently outperforms the recent best pre-trained multilingual model XLM-R [7] and obtain state-of-the-art (SOTA) in multiple Vietnamese-specific NLP tasks.

3   Model

Our model, i.e. Vietnamese SBERT model[1], is based on the original SBERT model, which includes two layers: transformer and pooling. The model architecture is described in Fig. 3.

---

[1] Our Vietnamese SBERT model: https://huggingface.co/keepitreal/vietnamese-sbert.

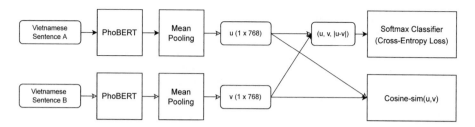

Fig. 3. Our Vietnamese SBERT model architecture

In training process for our models, we use labeled Vietnamese sentence pairs which the label value is annotated with respectively type of corpus (e.g. NLI and STS datasets).

We have configured the transformer layer using a pre-trained PhoBert-base model for extracted sentence representation vectors. This layer has a max sequence length of *256* and embedded *768* features. Since PhoBERT by Dat et al. 2019 [2] has proved to be the state-of-the-art model for Vietnamese word embedding, we use PhoBERT as a component in our Vietnamese SBERT model.

The embedding vector for sentence representation will go to the pooling layer, which we choose the *mean method*, due to the best performance in Reimers et al., 2019 [1]. This will reduce the presence of features of each sentence to one fixed length (*768* features) embedding vector.

The sentence transformer receives 2 above layer and it adds a siamese and triplet networks for updating and extracted semantic meaningful features for sentence representation. For fine-tuning model with NLI dataset, we use *Classification Objective Function* i.e. 3-way *Softmax function*, which Nils et al. used in [1], for 3 labels. For fine-tuning model with STS dataset, we use *Regression Objective Function* which used *cosine-similarity function* between sentence pair and *mean-squared-error* loss (MSE) [1] is objective function.

4   Experiments

4.1   Training Dataset

Many state-of-the-art deep learning models based on corpora for common semantic textual similarity tasks often use NLI and STS datasets for the training process. It has been proved in Reimers et al., 2019 [1] that multitask learning or continuous training on both NLI (i.e. Natural Language Inference) and STS (i.e. Semantic Textual Similarity) datasets is significantly better. What works better is to first train on NLI dataset, then take that model and train on STS dataset.

Since the lack of Vietnamese corpora on both quantity and quality, we decided to use NLI and STS datasets in English, then translated to Vietnamese. This is probably the most plausible way in training many Vietnamese NLP tasks, thanks to the strength of Google translate API and its performance and accuracy are getting better and better.

In our experiments, we used 2 corpora for fine-tuning our Vietnamese SBERT model, which is NLI and STSb. We used NLI dataset by author Dat Quoc Nguyen [11], which is the collection of 392.703 pair of sentences translated from English NLI dataset, labeled to 3 types: *entailment, neutral* and *contradiction.*

We also used STS datasets, which contain 5.749 sentence pairs and its label ranging from 0 to 5, annotated for the semantic similarity score between the pair of sentences. This STS dataset is also translated from English STS Benchmark dataset collected between 2012 and 2017 [12].

4.2   Training Process

In the first training, we fine-tuned our model with the NLI dataset and optimized its weight with a 3-way Softmax objective function corresponding to 3 labels. We trained our model with 1 epoch and a batch size of 32. It took about 15 h for the training and validating process.

After training the model with the NLI dataset, we fine-tuned the model with the STSb dataset as a continuous training task. In this process, we use the Cosine Similarity as the loss function with l2 distance function. We trained the model for about 2 h with 4 epochs and a batch size of 32.

5   Evaluation

5.1   Evaluation on NLI and STS Tasks

We evaluate our *Vietnamese SBERT* model on the training process with *Natural language inference (NLI)* tasks and *Semantic Textual Similarity (STS)* tasks. For NLI, we used the trained softmax function to evaluate the classification of labels as *entailment, neutral, contradiction* in the NLI test dataset. For the STS task, we use the method proposed by SBERT, which is the cosine similarity function, to calculate the similarity between two sentence embeddings [1]. Then, *Spearman's rank* was used to compute a correlation between the *cosine-similarity* of the sentence embeddings and the gold labels.

We experimented with two setups: Only training on NLI data (392.703 pairs of sentence), and first train on NLI data, then continue training on STS data (5.749 pairs of sentence). In addition, we also use the *pre-trained Multi-Lingual model of SBERT* to evaluate the effectiveness of our Vietnamese monolingual model to the available pre-trained Multi-Lingual model, and the latest model up to this point is *paraphrase-multilingual-mpnet-base-v2*. The results are depicted in Table 1.

We observed that combining training on both STS and NLI datasets gave a slight improvement of *2–3 points* for the NLI task and *14–15 points* for the STS task. On NLI tasks, our model evaluation on the NLI dataset has outperformed the pre-trained Multi-Lingual model of SBERT (*paraphrase-multilingual-mpnet-base-v2*).

However, on STS tasks, the pre-trained Multi-Lingual model provided by SBERT still has better performance. In addition to model architecture, aspects

Table 1. Evaluation Vietnamese-SBERT on the NLI test set and STS test set.

Model	NLI tasks (accuracy)	STS tasks (spearman)
Trained on NLI data	31.98	51.44
Trained on STS data	**34.19**	76.30
Trained on NLI data + STS data	33.11	76.32
Pre-trained Multi-Lingual model (*paraphrase-multilingual-mpnet-base-v2*)	32.14	**82.48**

that affect the evaluation of a model include the number of epochs for training and dataset. The language models to get good results often need to train using large corpora, while the model that we experimented with only training on a translated dataset, not the original Vietnamese dataset.

5.2 Evaluation on Paraphrase Identification Task

Corpus: To easily compare with other methods and models in the Vietnamese Paraphrase Identification task, we have evaluated our model on VnPara corpus by Bach et al. [9], the first original Vietnamese corpus for Paraphrase Identification Task. This corpus contains 3.083 sentence pairs, manually labeled by two people with around half labeled paraphrased and another labeled non-paraphrased.

Evaluation Method: To evaluate our model with VnPara corpus, we evaluate Accuracy and F1-score on this corpus. Let call TP is true positive values (number of correct paraphrased predictions), TN is true negative values (number of correct non-paraphrased predictions), FP is false positive values (number of incorrect paraphrased predictions) and FN is false negative values (number of incorrect non-paraphrased predictions). The formulas for calculating Accuracy and F1-score are as follow:

$$Accuracy = \frac{TP + TN}{TP + TN + FP + FN} \tag{1}$$

$$Precision = \frac{TP}{TP + FP} \tag{2}$$

$$Recall = \frac{TP}{TP + FN} \tag{3}$$

$$F1 = \frac{2 * Precision * Recall}{Precision + Recall} \tag{4}$$

To predict for a paraphrased or non-paraphrased label of sentence pairs, we proposed a method that calculating a threshold for cosine similarity score of sentence pair to determine whether paraphrased or not. This method will initialize a default threshold (about 0.5), then it will increase the threshold by a rate (about 0.01) for maximizing accuracy and F1-score.

Result and Discussion: We have tested VnPara corpus on our models, which are trained and fine-tuned with NLI and STSb dataset, compared to other models that have been evaluated in Dien et al. 2021 [3], Bach et al. 2015 [9] and Multi-Lingual SBERT model (supports 50+ languages including Vietnamese). The results are depicted in Table 2.

Table 2. Evaluation our Vietnamese SBERT model on VnPara corpus.

Model	Accuracy	F1-score
vnPara [9]	89.10	86.70
Feature vector (BERT) + Semantic vector + POS vector [3]	94.27	94.38
Feature vector (XLM-R) + Semantic vector + POS vector [3]	93.67	93.85
Feature vector (PhoBERT) + Semantic vector + POS vector [3]	94.86	94.97
Pre-trained Multi-Lingual model (*paraphrase-multilingual-mpnet-base-v2*)	95.04	95.15
Our model: Vietnamese SBERT (based on PhoBERT)		
Training on NLI dataset	92.77	92.78
Training on STSb dataset	95.29	95.36
Continuous training on NLI + STSb dataset	**95.33**	**95.42**

From the result table, our model continuously training on NLI + STSb dataset has achieved the best accuracy (95.33%) and F1-score (95.42%) with threshold calculating by our method is equal 0.715. This result has outperformed many recent models for paraphrase identification tasks on VnPara Corpus. This proved our model with continuous training on NLI and STSb dataset has worked well with strongly supported by PhoBERT, a pre-trained language model for Vietnamese.

However, since VnPara corpus, which is collected from many news web pages, is not many nor diverse in topics. Also, this corpus is small and does not have non-trivial cases (cases that are deeply paraphrased or non-paraphrase but contain many similar words). This is also the common defect that most Vietnamese Corpus encounters.

Moreover, since our models are implemented corpus-based, we need to improve our models in terms of quality and quantity of training corpus and enhance our architecture model for better learning from a not large corpus.

6   Conclusion

In this paper, we have introduced our method which implemented SBERT model with the strong support of PhoBERT in sentence representation for many downstream tasks in Vietnamese such as Paraphrase Identification, Semantic Searching,... In our experiments and evaluations on VnPara Corpus, a popular Vietnamese paraphrase corpus, our models have proved to slightly outperform many recent methods apply in Vietnamese text in terms of accuracy and f1-score. In the future, we will continue to enhance our model architecture and improve many shortcomings of the model that are encountered such as the lack of accurate corpus, small corpus or the lack of non-trivial cases. We would also consider many methods which enrich semantic meaningful representation for Vietnamese text or sentence.

Acknowledgement. We are very grateful to our instructors who helped us review our works and our friends for their valuable support. Also, we acknowledge the support of facilities from Ho Chi Minh City University of Technology (HCMUT), VNU-HCM for this study.

References

1. Reimers, N., Gurevych, I.: Sentence-BERT: sentence embeddings using siamese BERT-networks. In: EMNLP 2019 (2019). arXiv:1908.10084
2. Nguyen, D.Q., Nguyen, A.T.: PhoBERT: pre-trained language models for Vietnamese, pp. 1037–1042. Association for Computational Linguistics (2020). https://doi.org/10.18653/v1/2020.findings-emnlp.92
3. Dinh, D., Le Thanh, N.: Vietnamese sentence paraphrase identification using pretrained model and linguistic knowledge. Int. J. Adv. Comput. Sci. Appl. (IJACSA) **12**(8), 2021 (2021). https://doi.org/10.14569/IJACSA.2021.0120891
4. Vani, K., Gupta, D.: Study on extrinsic text plagiarism detection techniques and tools. J. Eng. Sci. Technol. Rev. (2016). https://doi.org/10.25103/jestr.095.02
5. Devlin, J., Chang, M.-W., Lee, K., Toutanova, K.: BERT: pre-training of deep bidirectional transformers for language understanding. arXiv:1810.04805v2 [cs.CL], 24 May 2019
6. Vaswani, A., et al.: Attention is all you need. arXiv:1706.03762v5 [cs.CL], 6 December 2017
7. Conneau, A., et al.: Unsupervised cross-lingual representation learning at scale. arXiv:1911.02116 [cs.CL], 5 November 2019
8. Liu, Y., et al: RoBERTa: a robustly optimized BERT pretraining approach. arXiv:2003.00744 [cs.CL], 26 July 2019
9. Bach, N.X., Oanh, T., Hai, N., Phuong, T.: Paraphrase identification in Vietnamese documents. In: 2015 IEEE International Conference on Knowledge and Systems Engineering, KSE 2015, pp. 174–179 (2015). ISBN: 9781467380133. https://doi.org/10.1109/KSE.2015.37
10. SQuAD2.0 The Stanford Question Answering Dataset. https://rajpurkar.github.io/SQuAD-explorer/
11. Vietnamese NLI Dataset, Dat Quoc Nguyen. https://github.com/DatCanCode/sentence-transformers/tree/master/DataNLI
12. Semantic Textual Similarity Wiki. http://ixa2.si.ehu.eus/stswiki

Author Index

Printed in the United States
by Baker & Taylor Publisher Services